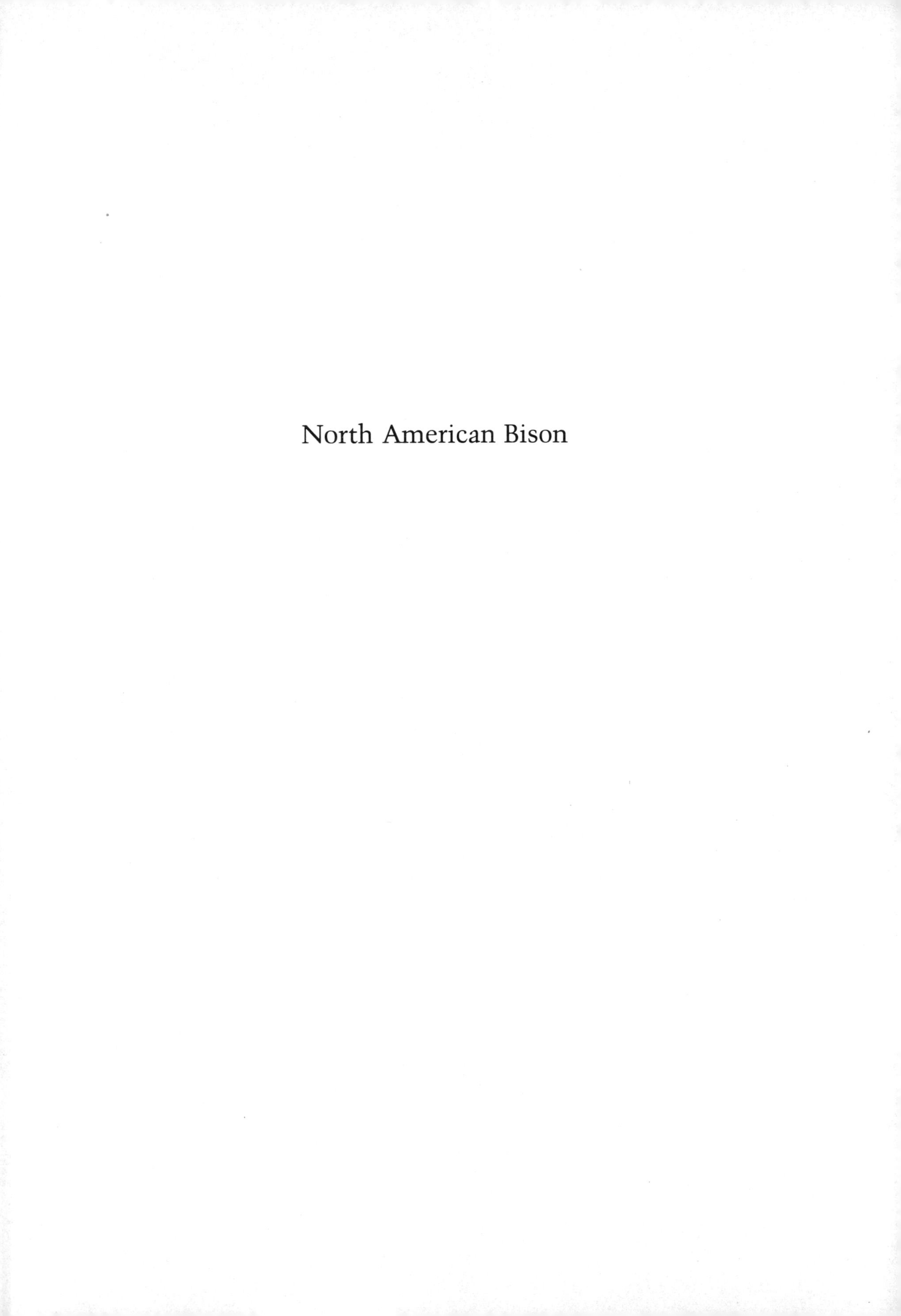

North American Bison

NORTH AMERICAN BISON

THEIR CLASSIFICATION AND EVOLUTION

Jerry N. McDonald

UNIVERSITY OF CALIFORNIA PRESS
Berkeley Los Angeles London

University of California Press
Berkeley and Los Angeles, California

University of California Press, Ltd.
London, England

Library of Congress Cataloging in Publication Data

McDonald, Jerry N.
North American bison.

Bibliography
1. Bison, American—Classification. 2. Bison,
American—Evolution. 3. Bison, Fossil.
4. Mammals—North America. I. Title.
QL737.U53M27 569′.73 80-36831
ISBN 0-520-04002-3

Printed in the United States of America

123456789

To
Chris and Jay,
who, more than anybody else,
kept me going

CONTENTS

PREFACE

This is a study of bison taxonomy and evolution—an attempt to bring together in one place a comprehensive and logical conceptualization of the classification, variation, adaptation, general ecology, distribution, and chronology of bison, especially in North America. This is admittedly an ambitious undertaking, but it is one which I believe is necessary if an understanding of bison evolution is to advance along the broad front now made possible by modern information, methods, and concepts. This study is based on the complementary beliefs that a comprehensive factual foundation is necessary to support a realistic understanding of bison evolution, and that an interdisciplinary analysis of the facts is necessary in order to approximate a realistic interpretation of the data. I have attempted to assemble here information on bison morphology and distribution in time and space for all of North America, to organize this information into biologically meaningful groups, and to relate the observed patterns of morphology and distribution to contemporaneous environments. Thereafter, drawing from the fields of paleobotany, paleoecology, sociobiology, geography, and archeology, I have attempted a Darwinistic interpretation of bison evolution.

I have several specific objectives for this study; each will, I hope, either fill a current need or introduce new possibilities in the study of North American bison evolution. These objectives are as follows.

To revise the classification of North American bison, recognizing evolutionary species, as opposed to morphological species.—The most recent systematic classification of North American bison (Skinner and Kaisen, 1947) has served as the standard reference for matters of North American bison taxonomy for thirty years. As a result of new information, however, many of the conclusions reached by Skinner and Kaisen have become suspect, and a new modernized classification system is needed. An evolutionary species is here considered to be a morphologically variable but canalized population adapted to a relatively stable environment. This population continues to be variable yet canalized through space and time as long as the correlated selection regime prevails. Each species must first be recognized as a biological reality, adapted to a relatively stable environment, and then assigned its valid systematic name. This differs from the morphological species concept in which the size and shape of skeletal characters have been the basis for species differentiation.

To provide measures of morphological variation for both sexes of each taxon.—Measurements of variation have been presented infrequently for North American bison (e.g., Bedord, 1974; Hillerud, 1970; Shackleton, Hills, and Hutton, 1975) and never systematically and comprehensively for all North American representatives of the genus. Variation among female groups has been especially neglected. Measures of variation provide insight into the absolute, relative, and statistical nature of variation among selected characters for each group.

To identify morphological characters that differ significantly among taxa, and to consider the adaptive significance of these differences.—Certain characters differ among taxa more than others. Presumedly, significant morphological differentiation indicates a relatively different adaptive value and/or use of each character. The identification of differing characters and a consideration of their respective adaptive significance are necessary preludes to an analysis of the adaptive value of observed character differences.

To develop models of adaptation to environments for each taxon based on observed character differences. —Characters presumedly differentiate during adaptation to new environments and become canalized once morphological-environmental equilibrium has been reached by a viable species population. Consideration of the adaptive value of character differences, distribution of species, and reconstruction of species' environments assist in modeling species' adaptations to environment. Historic records of recent bison distribution patterns and morphological trends permit testing some aspects of these models, and current management practices offer the opportunity to test others.

To examine the nature of morphological abnormalities found in individual specimens.—Certain characteristics appear infrequently among bison yet are sufficiently common to deserve analysis. Some specimens show typical characters of two taxa and probably represent hybrid individuals. Other infrequent characters tend to appear near the edge of species' ranges or in what could conceivably have been small populations and probably represent the consequences of genetic drift.

To develop a dynamic model of North American bison evolution.—Evolution is a complex process; it takes place as a species' gene pool potential progresses to equilibrium with the composite selective forces in the species' environment. The model presented here integrates considerations of morphological variation, population dynamics, adaptation, ecology, dispersals, and extraneous environmental patterns and processes, and attempts to present a logical conceptualization of the process of North American bison evolution.

I fully realize that the holistic approach I am using here cannot deal with every aspect of the classification and evolution of bison; nor can all facets of the problem be studied with the same depth and technical refinement that would be possible in a specialized study employing fewer methods and seeking more restricted but detailed answers. My extremely superficial treatment of vegetation changes might appear inadequate to some students, as might my limited use of quantitative analyses to others. The same could be said for other aspects of this study. My primary purpose is to provide a logical framework for the holistic study of North American bison evolution based on a systematic and comprehensive study of available bison materials compatible with modern biological concepts.

The study of bison taxonomy and evolution has traditionally been controversial, but perhaps never more so than at present. There currently exists a blend of several exciting, provocative, dynamic theories, several enduring dogmatic assumptions, and, occasionally, more speculation than documentation at the megatheory level. As a corollary to the stated objectives of this study, I will also attempt to test the following major hypotheses that are currently jousting to dominate the theoretical sphere of bison evolution research. (These hypotheses are only briefly outlined here; except for the first one presented, they are much more complex than stated and, to properly understand them, the reader should consult their stated source.)

The Horn Core Attrition Theory (Allen, 1876).—This theory maintains that the horn size (and, by implication, the body size) of North American bison decreased with each successive species. The largest horned species, *B. latifrons*, is therefore the oldest, and the shortest horned species, *B. bison*, is the youngest.

The Orthogenetic Theory (Schultz and Frankforter, 1946).—This theory, given the name used here by Guthrie (1970), originally maintained that a single immigration of bison into North America occurred before the middle Pleistocene. The immigrating bison were either *B. latifrons*, or they gave rise to *B. latifrons*. This event was followed by a progressive stepwise attrition of horn (and body?) size synchronous with the appearance of successive glacial or interglacial phases. Thus, each glacial and interglacial phase, beginning at the latest with the Illinoian, had a typical species of bison, each younger species shorter horned than the preceding species. A recent revision of this hypothesis (Hillerud, 1978; Schultz and Hillerud, 1977) states that the single lineage descent of all North America bison is no longer necessarily accepted, but an alternate explanation has not been provided.

This theory was developed with special reference to the central Great Plains, but it included most taxa then and since recognized from all of midlatitude North America, and it has, in practice, been considered applicable to most of the continent.

The Wave Theory (Skinner and Kaisen, 1947).—The Wave Theory (also named by Guthrie, 1970) maintains that two waves of bison immigrated into North America during the Pleistocene. The first wave entered Alaska from Siberia during the Aftonian interglacial and was forced south into midlatitude North America during either the Kansan or Illinoian glacial phase. This wave either consisted of several different species or one or more of the species evolved into other species. No fossil evidence of this wave was left in Alaska. The second wave reached Alaska during the interglacial following the southward displacement of the first wave (i.e., during the Yarmouth or Sangamon interglacial). This wave also consisted of either more than one species or of some immigrating species that evolved into new species. The two lineages (i.e., bison of the first and second wave) were allopatric for most of the Pleistocene. Only during the Holocene did they come together, in central North America. The first lineage became extinct soon after sympatry was established, whereas the second lineage is still represented by *B. bison*.

The Synthetic Theory (Guthrie, 1970).—This theory constitutes a synthesis of several theoretical statements on bison phylogenesis and dispersal which appeared before the late 1960s. It generally maintains that all midlatitude North America bison originated from a common Palearctic ancestor, *B. priscus*, but at different times. Three (possibly four) species of midlatitude bison are recognized in this theory in addition to *B. priscus*. *B. latifrons* evolved during the Illinoian directly from *B. priscus*. *B. alleni*, a species of questionable validity, occurred (if at all) during the Sangamon as either a dwarfed phyletic descendant of *B. latifrons* or a larger phyletic descendant of *B. priscus*. *Bison antiquus* appeared during the Wisconsin as either a dwarfed descendant of *B. latifrons* (by way of *B. alleni*) or *B. priscus*. *Bison bison* evolved from *B. priscus* during the late Wisconsin in the Beringian region and dispersed southward during the early Holocene, at about which time *B. antiquus* became extinct. Among the evolutionary trends this theory recognizes are northern sources for midlatitude bison and the typical enlargement of horns among southerly species. The enlargement of horns is attributed to a higher quality diet in the south more conducive to horn growth.

The Dispersal Theory (Geist, 1971*a*).—The Dispersal Theory maintains that pioneering populations that disperse into a previously unoccupied habitat, especially recently deglaciated territory where there exists an abundance of highly nutritious forage, are exposed to different selective pressures than are established populations in a stable environment. Consequently, these populations became morphologically differentiated from their parent populations. The longer a species is exposed to the frontier environment the more it will differ from its parent species. The pioneering population will be characterized by rapid development of individuals, early sexual maturation, large body size, rigorous social interactions (especially frequent and intense combat), and a short life expectancy. When the pioneering role is completed (i.e., when the habitat carrying capacity is reached), body size is reduced, individual development is slow, life expectancy increases, and, presumedly, social interactions are less rigorous. As generally applied to North American bison, the Dispersal Theory predicts that the first bison species to disperse into and across the continent furthest should have the most elaborately developed social organs (i.e., horns and/or body hair). Later species, or those which dispersed less far, should be smaller and have less elaborate social organs.

The Clinal Theory (Fuller and Bayrock, 1965; Wilson, 1974*a*).—This theory maintains that northerly bison were characterized by narrow frontals, high orbital protrusion, and posteriorly deflected and distally twisted horn cores, all of which graded into broader frontals, low orbital protrusions, and more laterally directed horn cores with little or no distal twist in southerly bison. In addition to this north-south morphological chorocline, the Clinal Theory recognizes a progressive chronoclinal diminution in characters over time among both northerly and southerly bison. Thus, whereas all North American bison were supposedly becoming smaller, the north–south chorocline was preserved. Wilson, who is actually responsible for developing this model into a working hypothesis, has suggested that the chorocline shifted southward during the early Holocene and was preserved in the historic distribution of *B. bison.*

The results of the tests of the aforementioned hypotheses will be combined with other ideas on bison evolution and a new hypothesis of North America bison evolution will be presented. This, too, no doubt will be controversial, but if I can either disprove or fail to disprove elements of the preceding megatheories and I stimulate additional research and endeavors to falsify my hypothesis, then some measure of success will be realized. I do hope, however, that this study makes clear the fact that bison evolution is a complex process, much more so than the essence of the above-mentioned theories recognizes, and that it consists of much more than simply eating nutritious food, having well-traveled ancestors, fighting frequently, living in warmer climates, or becoming spontaneously smaller with every change of climate.

I begin this study by presenting background information on the chronology and changing environments of the late Cenozoic and a brief review of the evolution and dispersal of the Bovini, with emphasis on the bison. Thereafter, I consider specific aspects of bison systematics, progressing from the simplest to increasingly complex treatments of each taxon. This organization requires a certain amount of tolerance; there is simply no unequivocally *best* organization to use in a study of this type. Although I recognize the inherent circularity in treating diverse but mutually dependent aspects of bison systematics separately, I believe that the organization used here is logical in that it represents a progressive expansion of the concept of each taxon throughout the book. This requires that the reader accept my taxonomic conclusions from the start. Although I apologize for this necessity, I hope that I show satisfactorily in later chapters that my classification is rational and acceptable.

This work is a revision of my doctoral dissertation, which was completed during the summer of 1978. Most of the data was collected during the summers of 1975, 1976, and 1977 from museums and universities throughout Canada, Mexico, and the United States. I was able to study additional collections, previously inaccessible to me, during the early summer of 1979, including the important collections at Idaho State University and the University of Nebraska State Museum, Division of Vertebrate Paleontology. I have revised the conclusion presented in my dissertation very little, but subsequent thought and discussions with colleagues have, I hope, resulted in a more adequate treatment of critical aspects of my thesis.

A copy of the data used in this study has been deposited with the George C. Page Museum of Rancho La Brea Discoveries in Los Angeles, California, and the U.S. National Museum Department of Paleobiology in Washington, D.C. These data are available for use by other researchers.

ACKNOWLEDGMENTS

Don Brand and Tom Campbell independently sparked my interest in bison while I was completing my Master's degree at the University of Texas in 1972. Two years later, Ernest Lundelius, Jr., called my attention to the need for a comparative reexamination of all available bison skulls, suggesting that this was the most promising method by which to reduce the existing confusion in bison systematics. That same year Larry Agenbroad indicated the need among archeologists for a modern systematic study of bison. Stimulated by the comments of these men, I undertook this study of bison evolution early in 1975 as my doctoral research problem. The research problem and methods were developed through discussions with, especially, George Jefferson, Everett Olson, Jonathan Sauer, and Dennis Stanford. George Jefferson, Everett Olson, Jonathan Sauer, Hartmut Walter, and Susan Woodward provided extensive advice, insight, and criticism during subsequent years of discussions. Many other people contributed to the refinement of my ideas. I especially appreciate thoughts shared by Rainer Berger, Bob Bright, Dick Estes, George Frison, Val Geist, Stephen Gould, Russell Graham, Harvey Gunderson, Dale Guthrie, Dick Harington, Art Harris, Bob Hoffmann, Jack Hughes, Mike Kaczor, Jose Luis

Lorenzo, Dale Lott, Ernest Lundelius, Jr., Paul Martin, Chuck Reher, Dale Russell, Orrin Shane, Dick Tedford, Waldo Wedel, and John White. Bob Hoffmann, Everett Olson, Jonathan Sauer, Susan Woodward, and an anonymous reviewer provided detailed criticism of earlier drafts of this manuscript. I, of course, assume full responsibility for the information and ideas as they are presented herein.

Access to faunal collections in numerous museums and universities was in most cases freely provided (see Appendix 1 for institutional names and abbreviations including those appearing below in parentheses). I am grateful to the following for allowing me time and space to study specimens, and for assistance with those specimens: Sydney Anderson, Phil Goldstein, and Dick Tedford (AMNH); Earl Shapiro and L. Gay Vostreys (ANSP); James Cleek (BBL); Bob Bright (BeMNH); John Guilday (CM); Charles Oehler (CMNH); Bob Akerley, K. Don Lindsey and Richard Stucky (CoMNH); Larry Agenbroad (CSC); Jessie Robertson (DM); Gary Coovert (DMNH); Bill Turnbull and Gayle Ziegler (FMNH); Jose Luis Lorenzo and Demetrio Porras Dias (INAH); Jeff Saunders (ISM); David Fortsch, John White, and David Wright (IMNH); Dan Adams, Orville Bonner and Larry Martin (KU); Farish Jenkins, Jr., Barbara Lawrence, and Charles Schaff (MCZ); George Lammers (MMMN); Alexander J. Lindsay, Jr. (MNA); Robert West (MPM); Federico A. Solorzano (MRG); C. van-Zyll de Jong, Dick Harington, and Dale Russell (NMC); Bill Akersten, Larry Barnes, George Jefferson, and David Whistler (LACM); Donn Davids and Suzanne Lime (OHS); Stig Bergstrom (OSU); Ron Mussieux and Hugh Smith (PMA); Billy Harrison and Rolla Schaller (PPHM); Bob Slaughter (SM); John Storer and Gil Watson (SMNH); Karl de Rochefort-Reynolds, Robert Hoge and Lennis Moore (SMP); Tom O'Brien (SPSM); William Evett and Jeff Stein (SU); Richard Shutler, Jr. (SUI); Wann Langston, Jr., Ernest Lundelius, Jr., and Bob Rainey (TMM); Elayne Brown, Eileen Johnson, and Mike Kaczor (TTU); David Krause (UA1); Holly Chaffee (UASM); Paul Martin and Kevin Moodie (UALP); Dick Forbis and Len Hills (UCy); Joe Ben Wheat (UCM); Howard Hutchinson (UCMP); Mike Frazier and S. David Webb (UF); Bill Melton, Bart O'Gara, and Sewall Young (UM); Bob Habetler (UMMP); James Stitt (UMo); George Corner, Harvey Gunderson, James Gunnerson, Tom Myers, and Mike Voorhies (UNSM); Henry Fahl (UP); Robert O'Connell and G. Edward Lewis (USGS); Bob Purdy, Clayton Ray, Henry Setzer, and Dennis Stanford (USNM); Art Harris (UTEP); George Frison, Chuck Reher, and Danny Walker (UW); David Hughes, Jack Hughes, Gerald Schultz, and Roberta Speer (WTSU); Mary Ann Turner (YPM).

Unpublished data were generously provided by Don Brand, Bob Bright, Barbara Eikum, Dick Forbis, T. Weber Greiser, Frank Hibben, John Hillerud, David Hughes, Jack Hughes, Jim Keyser, Leslie Marcus, Jeff Saunders, Hugh Smith, Dennis Stanford, Alan Stevens, John Storer, Joe Ben Wheat, and John White.

Special thanks are due Jim Burt for advice and assistance in the use of a computer, and Mary Ann Turner for going above and beyond the call of duty in locating misplaced specimens and elusive information at the Yale Peabody Museum. Much of the value of this study lies in the illustrations, almost all of which were drawn by Susan Woodward. Irving and Alicia Frank of Guadalajara deserve a special thank you for coming to my aid after local hustlers had rendered my car inoperable. The hospitality of Danielle and Maurice Boisvert, Gloria and David Bradbury, Linda and Bob Bradford, Mike Donohue, Linda and Robin Doughty, Sue and Tom Eller, Benita and Roy Halliday, Frank Hanes, Carolyn and Paul Leskinen, Donna and Scott Lockwood, Mike Montanaro, Linda and Denny Pierce, Mary Ellen and Dick Schubel, Dorothy Woodward, and Susan Woodward is immensely appreciated. Marsha Bedwell typed the manuscript. Beverly contributed much to this effort, and her assistance is truly appreciated.

Financial support for parts of this research was provided by the Smithsonian Institution, the University of California Graduate Student Patent Fund, and the UCLA Department of Geography.

Back Pond J. N. McD.
North Waterford, Maine
August 21, 1979

CHAPTER 1

THE WORLD OF BISON

Bison is among the most recently evolved genera of large mammals. It is first known from the late Pliocene of Asia, where it represents the northernmost genus of the oxlike bovids—the Bovini. *Bison* occurred throughout most of Eurasia and North America, the only Bovini genus to achieve widespread distribution across more than one continent without human assistance. The appearance of bison and the expansion and contraction of the bison range correspond with the environmentally dynamic Ice Age of the late Cenozoic. During this time oscillation of climate and its physical consequences drastically modified the area, shape, and biotic characteristics of terrestrial habitats. Vegetation associations were periodically displaced, shuffled, and resorted, resulting in significant changes in the character of vegetation through time. Faunal evolution was typified by increased body size in many genera, especially among ungulates and their predators, but it also included the appearance of advanced human forms with their unprecedented level of intelligence and complexity of behavior. The fossil record indicates that bison were quite flexible in responding to the evolutionary stimuli presented by the late Cenozoic environments in the Northern Hemisphere. Exposure to differing environments in both North America and Eurasia resulted in differential selection for certain adaptive characters, hence the appearance of different species or subspecies in separate parts of the overall bison range. A necessary prelude to a Darwinistically meaningful understanding of bison evolution, then, is a review of the important environmental characteristics of the late Cenozoic, especially those affecting the distribution and composition of habitat and dispersal opportunities.

CENOZOIC CHRONOLOGY AND THE MOST RECENT ICE AGE

The Cenozoic—the Age of Mammals—comprises the past sixty-five million years or so (table 1). The physical environment of the earth changed

Table 1
Cenozoic Chronology and Bovini Evolution

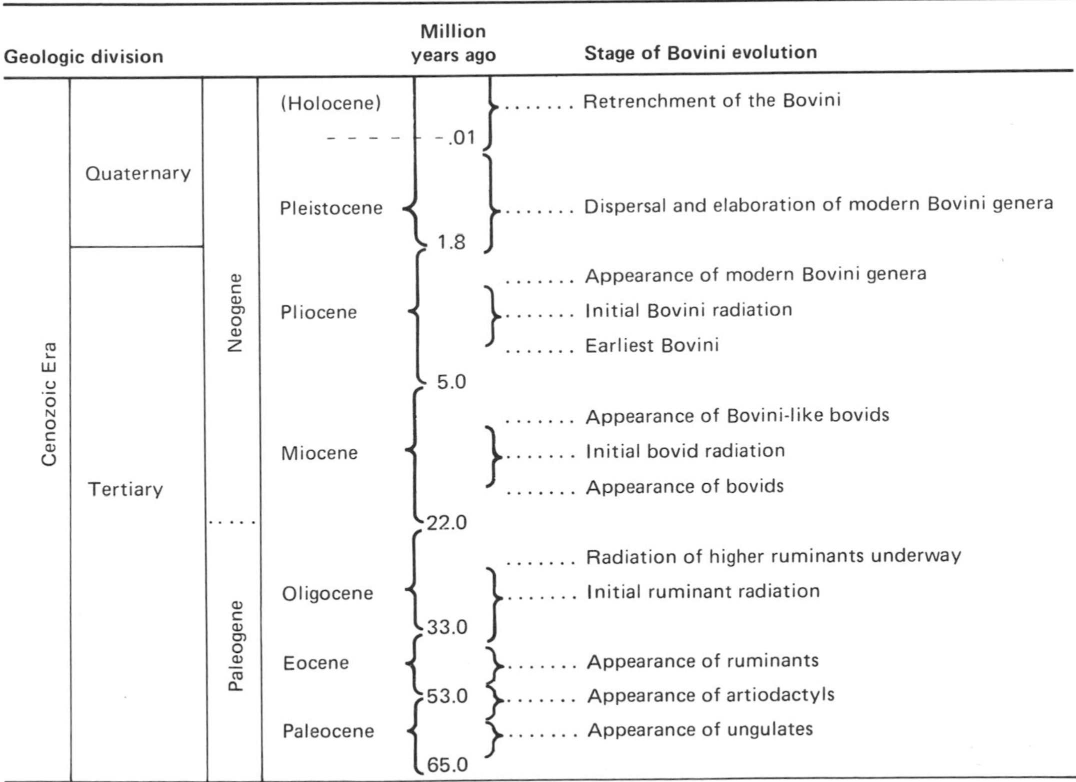

Geologic division				Million years ago	Stage of Bovini evolution
Cenozoic Era	Quaternary	Neogene	(Holocene)		Retrenchment of the Bovini
				.01	
			Pleistocene		Dispersal and elaboration of modern Bovini genera
				1.8	
	Tertiary		Pliocene		Appearance of modern Bovini genera
					Initial Bovini radiation
					Earliest Bovini
				5.0	
			Miocene		Appearance of Bovini-like bovids
					Initial bovid radiation
					Appearance of bovids
				22.0	
		Paleogene	Oligocene		Radiation of higher ruminants underway
					Initial ruminant radiation
				33.0	
			Eocene		Appearance of ruminants
				53.0	Appearance of artiodactyls
			Paleocene		Appearance of ungulates
				65.0	

Sources: The geochronology follows Berggren and Van Couvering (1974) and Wolfe (1978). Bovini evolution follows Kurten (1972), Pilgrim (1947), Romer (1966), Sinclair (1977), and Van Valen (1971).

Table 2
Late Neogene Geochronology for Northern Hemisphere Regions

Epoch	Million years ago	North America (continental)	North American land mammal age	Siberia	Europe (alpine/continental)	European land mammal age
Pleistocene		(Holocene)	Rancholabrean	(Holocene)	(Holocene)	Oldenburgian *et al.*
	.010					
		Wisconsin		Sartan (Karginsky) Zyrianka	Wurm/Warthe-Weichsel	
	.075					
		(Sangamon)	Irvingtonian	(Kazantsevo)	(Eem)	
	.125					
		Illinoian		Bakhtian/Samarov-Taz	Riss/Saal (Holstein) Mindel/Elster	
	.500					Biharian *et. al.*
		(Yarmouth)		(Tobol)	(Cromer)	
	.700					
		Kansan		Demyanka (?)	Gunz II (Waal) Gunz I	
	.900					Villafranchian — Late
		(Aftonian)			(Tiglian)	
	1.350					
		Nebraskan	Blancan		Donau	Villafranchian — Middle
	1.600					
	1.800					
Pliocene						Villafranchian — Early
	5.000					

Sources: Berggren and Van Couvering, 1974; Kind, 1967; Richmond, 1970.
Note: Interglacials are shown in parentheses; glacials without parentheses. Land mammal ages for North America and Europe are shown vertically.

substantially during this era as, generally, continental land areas became larger and increasingly interconnected, tectonic and volcanic processes built high mountain chains, and climate became cooler, less equable, and regionally more arid.

The late Neogene chronology used here follows Berggren and Van Couvering (1974). The traditional glacial and interglacial geologic divisions of the Pleistocene—the last 1.8 million years—will be the primary terms of reference used for specific phases of that epoch. The Holocene is here considered an interglacial phase of the Pleistocene. Glacial-interglacial phases for North America, Siberia, and Europe are named and correlated in table 2. Although Pleistocene chronology and intercontinental correlations are far more complex and far less definite than the classical generalized three- or four-glacial phase, direct intercontinental correlation concept used here, the vagueness of the bison record for pre-Wisconsin time makes greater detail of chronology or correlation superfluous.

The climatic alternations that eventually resulted in the pronounced continental glaciations of the Pleistocene began as early as the late Eocene–early Oligocene, when the world climate became cooler and the Antarctic ice cap formed. The formation of the Arctic ice cap about 3.0 to 2.5 million years ago (MYA) initiated a distinctly colder climate for the earth. Progressive cooling followed; montane and high latitude glacial activity increased, signaling the dawn of the most recent Ice Age. Then, about 1.6 MYA, continental glaciation began in North America, initiating the sequence of major glaciations, separated by interglacial phases, which have continued to the present in the Northern Hemisphere (Berggren and Van Couvering, 1974; Savage and Curtis, 1970; Wolfe, 1978).

The first North American continental glaciation, the Nebraskan, lasted from about 1.6 to 1.35 MYA. This was accompanied by alpine glaciation in the Alps (the Donau glaciation), but no correlated continental glaciation is recorded in Europe or Asia. The duration of the subsequent Aftonian interglacial is indefinite, but it extended from approximately 1.35 MYA to about 0.9 MYA. The Aftonian is correlated with the Tiglian warm climate of Europe. The Kansan continental glaciation occurred between about 0.9 and 0.7 MYA and correlates with the Gunz alpine glaciation of Europe. The first lowland glaciation in Siberia might have occurred at this time as the Demyanka. The Yarmouth/Cromer/Tobol interglacial occurred between about 0.7 and 0.5 MYA. The Illinoian and its correlates, the Mindel/Elster and Riss/Saal of Europe and the Bakhtian of Siberia, represent a long period of glaciation lasting from 0.5 to 0.125 MYA. This is not only the longest glacial phase recognized, but it was the first to glaciate lowland Europe and it was the most extensive and harsh glaciation of the entire Pleistocene for Europe, Asia, and Alaska. It has also been one of the most difficult to correlate (cf., e.g., Berggren and Van Couvering, 1974; Richmond, 1970). The subsequent Sangamon/Eem/Kazantsevo interglacial, possibly the warmest of the Pleistocene, lasted from about 0.125 to 0.075 MYA. The last glacial period—the Wisconsin/Wurm/Zyrianka-Sartan—extended from about

0.075 to 0.01 MYA. Each of these continental glaciations was punctuated by one or more interglacials; one of significance extended from about 35,000 to 22,000 years before the present time (BP). The late Wisconsin, which reached its maximum about 20,000–18,000 BP, represented the most extremely cold period of the North American Pleistocene. The beginning of the Holocene has been established by international agreement at 10,000 BP (Berggren and Van Couvering, 1974; Kind, 1967; Richmond, 1970).

THE BERING LAND BRIDGE

The waxing and waning of glaciers and the associated lockup and release of substantial volumes of the earth's water resulted in significant fluctuations of sea level. One consequence of these fluctuations was the appearance and disappearance of a land connection between eastern Siberia and Alaska (fig. 1). These two regions are now separated by the Bering Sea, but a lowering of sea level by some 50 meters would expose two narrow land connections between Siberia and Alaska, and a reduction by 100 meters would expose most of the 800–1,000 mile wide Bering-Chukchi Platform (Creager and McManus, 1967; Hopkins, 1967*a*).

Uncertainty exists as to exactly when, to what extent, and for how long a land bridge existed in this region during the Pliocene and the first million years of the Pleistocene, but the periodic dispersal of land mammals between the Palearctic and Nearctic confirms the existence of such a land bridge during at least part of his time (Hoffmann, 1978; Repenning, 1967, 1978). The Pliocene history of the bridge is very poorly known. Presumedly, during the Pleistocene, a bridge existed only during a glacial stade, but the frequency, intensity, and duration of these stades within the glacial phases is also poorly known (cf., e.g., Einarsson, Hopkins, and Doell, 1967; Fink and Kukla, 1977). The probability also exists that regional uplift and accumulating glacial deposits elevated the surface of the western Bering-Chukchi Platform as time passed and created a situation where less reduction of sea level was necessary to expose the platform surface (Creager and McManus, 1967; Petrov, 1967; Sainsbury, 1967). The maximum lowering of sea level during the Pleistocene probably occurred during the Illinoian glaciation, when sea level was lowered 135 to 160 meters and the land bridge attained its maximum size (Donn, Farrand, and Ewing, 1962). Considering the long duration of the Illinoian and the likelihood of periodic interstadial transgressions, however, there is no reason to infer that the land bridge existed without interruption during Illinoian time. In fact, the Einahnuhtan and Kotzebuan marine transgressions, absolutely dated at between 320,000 to 175,000 BP and 175,000 to 120,000 BP, respectively (Hopkins, 1967*b*), may both fall within the Illinoian as conceived herein. The Bering Strait was reestablished during the Sangamon interglacial. Sea level was lowered again about 115 to 135 meters during the early Wisconsin, which reestablished the land bridge, but this was again inundated from about 35,000 to 25,000 BP. Sea level was lowered from 100 to 120 meters during the late Wisconsin maximum, about 20,000–18,000 BP (Blackwelder, Pilkey, and Howard, 1979; Curray, 1965; Donn, Farrand, and

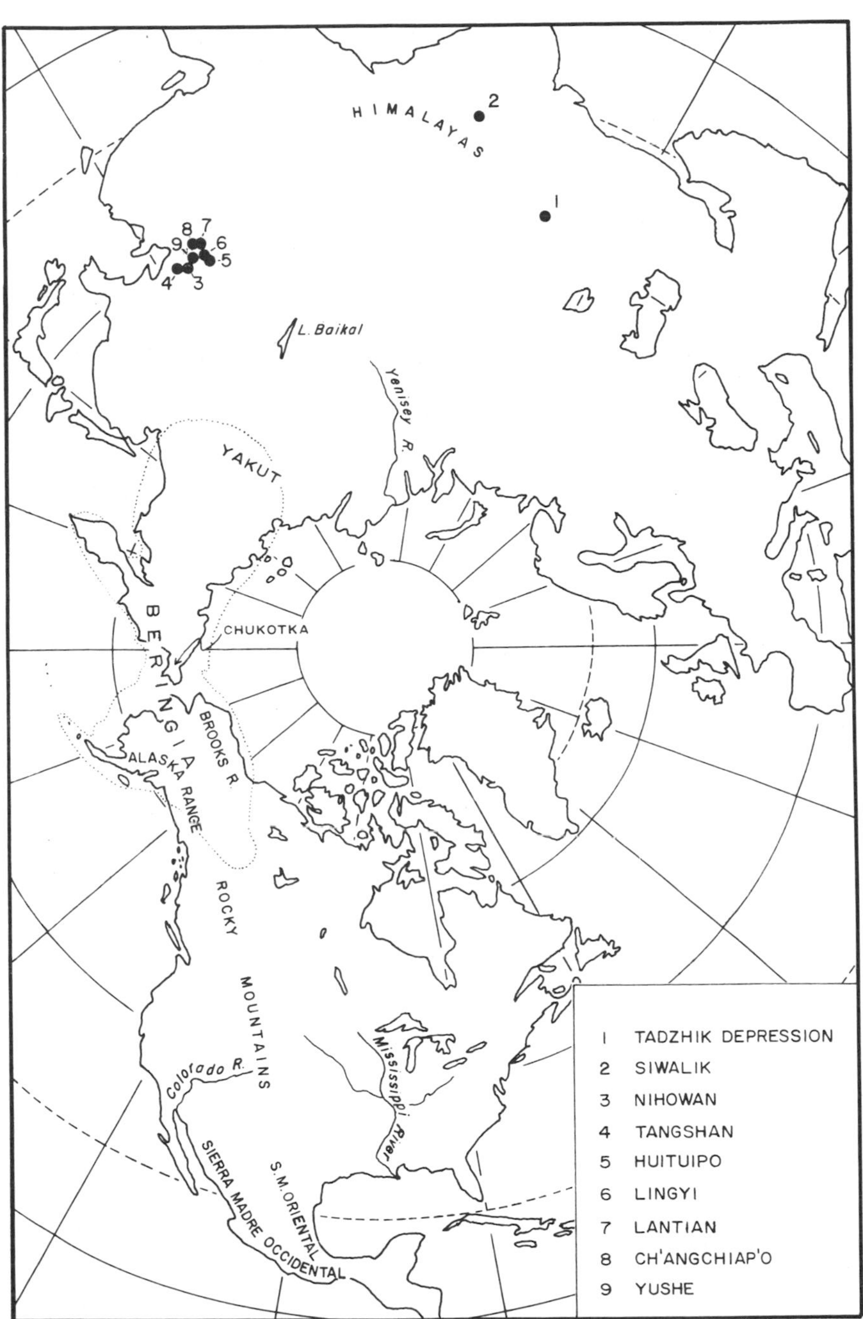

FIGURE 1 Major Holarctic place names and early Palearctic bison localities. Glacial Beringia is enclosed with a dotted line.

Ewing, 1962; Hopkins, 1967*b*; Stearns, 1976).

The late Wisconsin history of the Bering land bridge is not precisely known, but is important in evaluating certain aspects of late Wisconsin bison dispersal. Hopkins (1967*a*) concludes that the late Wisconsin land bridge could have existed until about 14,000 BP, but that it was flooded at that time. Limited information suggests that regressions of about fifty meters occurred at about 13,000 BP and 11,000 BP, and land connections between the continents could have been reestablished at those times. A rapid rise in sea level drowned the bridge for the last time about 10,000 BP. According to Mercer (1972), however, sea level probably continued to rise after about 14,000 BP, so once the land bridge was inundated at or after that time, it probably remained closed as a land corridor (although winter passage over a relatively narrow ice bridge might have been feasible).

LATE NEOGENE VEGETATION PATTERNS

An overall simplification of floras, especially at higher latitudes, higher elevations, and in desiccating continental interiors was one trend characteristic of the Cenozoic. Formerly widespread tropical, subtropical, and warm temperate floras were increasingly relegated to lower latitudes or more moist regions with relatively equable climates and were replaced by deciduous, xeric, coniferous, and/or herbaceous floras. The trend was toward a general vegetation mosaic with less broadleaf forest, more coniferous forest, more savanna, and more open vegetation (table 3).

Pliocene Vegetation Patterns

The Pliocene vegetation of much of Eurasia is not well known, but a general pattern is suggested by evidence from scattered localities. Subtropical or warm deciduous broadleaf

Table 3
Major Vegetation Types Mentioned in Text

Forest	Trees are dominant and closely spaced; canopy is normally complete; exists in association with relatively humid climate.
Woodland	Trees or large shrubs are dominant; canopy is normally not complete.
Savanna	Trees and/or shrubs and herbs are codominant over extensive area; often, but not always, associated with semiarid or seasonally dry climates.
Brushland	Smaller shrubs are dominant; canopy is normally not complete.
Parkland	Relatively large openings in forests or woodlands; herbs are dominant with trees and/or shrubs present; ground cover normally complete.
Grassland	Grasses are dominant; forbs are present but trees or shrubs are rare or absent. In steppes, short grasses adapted to semiarid to arid conditions dominate; ground cover can be complete or incomplete. In prairies, tall grasses adapted to subhumid to humid conditions dominate; ground cover is normally complete.
Tundra	Herbs, mosses and lichens are dominant; shrubs or small trees are occasionally present and codominant; exists under continuing cold conditions with short growing season.
Desert	Shrubs and herbs are dominant; exists under arid conditions; ground cover normally incomplete.

forest probably dominated China south of about 37°N in the early Pliocene, but the northern boundary of this vegetation apparently shifted some 5° farther south by the late Pliocene as deciduous and/or xeric vegetation, with mesic conifers at higher elevations, became more widespread (Aigner, 1972; Movius, 1944). Kurten (1950) has inferred from paleontological evidence that a southeast–northwest transition from forest to steppe occurred from lowland to highland north-central China, with a continuation of steppe and desert across great areas of the continental interior. Coastal east-central Asia probably retained a mesic broadleaf forest. A savanna zone presumedly separated the interior steppe and desert from surrounding forests (Frenzel, 1968; Kurten, 1972). A sequence of conifer-dominated forest zones occurred between the interior desert/steppe/forest steppe and the Arctic coast of Asia. Alpine tundra possibly existed at this time in limited areas on higher latitude mountains of Asia. Most of Europe was vegetated by a mixed conifer-broadleaf forest (Frenzel, 1968; Yurtsev, 1972).

In North America continental drying displaced the interior border of the mesic broadleaf forest zone to the south and east. Xeric broadleaf and herbaceous plant associations became increasingly dominant at lower elevations in the west, whereas mesic conifer-dominated forests developed and spread at higher elevations and latitudes and along the northwest Pacific coast (Axelrod, 1975; Graham, 1972; Waring and Franklin, 1979). Webb (1977) has inferred from paleontological and geological evidence that the vegetation of central North America changed from a woodland savanna to an open steppe during the Pliocene. Alaska, at least that area south of the Arctic Circle, was dominated by a rich conifer forest early in the Pliocene. This forest later withdrew eastward from the Bering Sea area and was replaced by low-growing shrubs and herbs. Tundra is not known from Pliocene Alaska, although it might have existed at this time on the arctic slope (Wolfe, 1972).

Pleistocene Vegetation Patterns: Eurasia

The Pleistocene vegetation history of the Palearctic was characterized by an alternating dominance of forests during interglacial phases and tundras and steppes during glacial phases (Fink and Kukla, 1977; Frenzel, 1968; Kurten, 1968; Woillard, 1978). Vegetation in Europe during the Donau was dominated by subarctic parklands and forest steppes. Subsequent glacial phases produced sufficiently severe climates to force the withdrawal of forests as regionally significant plant associations from north of the Alps, leaving a cold, dry, and barren landscape between the alpine and Fennoscandian glacial systems. Tundra might have persisted in moderately significant areas along the coast, and along the periglacial zone during glacial phases, but elsewhere it was probably patchy in distribution. Steppe communities dominated by *Artemisia*, chenopods, grasses, and halophytes extended east from northwestern Europe. During at least the last two major glaciations (Mindel-Riss and Wurm) permafrost was widely distributed. South of the Alps steppe communities dominated the lowland vegetation from Iberia to Caucasia, although uplands might have supported subalpine forests. Significant broadleaf forests might have existed in the Black and Caspian basins, but elsewhere broadleaf trees probably survived only as local relict populations (fig. 2).

The Tiglian forests that covered most of Europe after the Donau glaciation retained many subtropical genera that had been part of the region's Pliocene forest, but the Cromerian interglacial forests were definitely temperate in character, dominated by conifers in the highlands and high latitudes and deciduous broadleaf species, especially oak and elm, in the lowlands. The

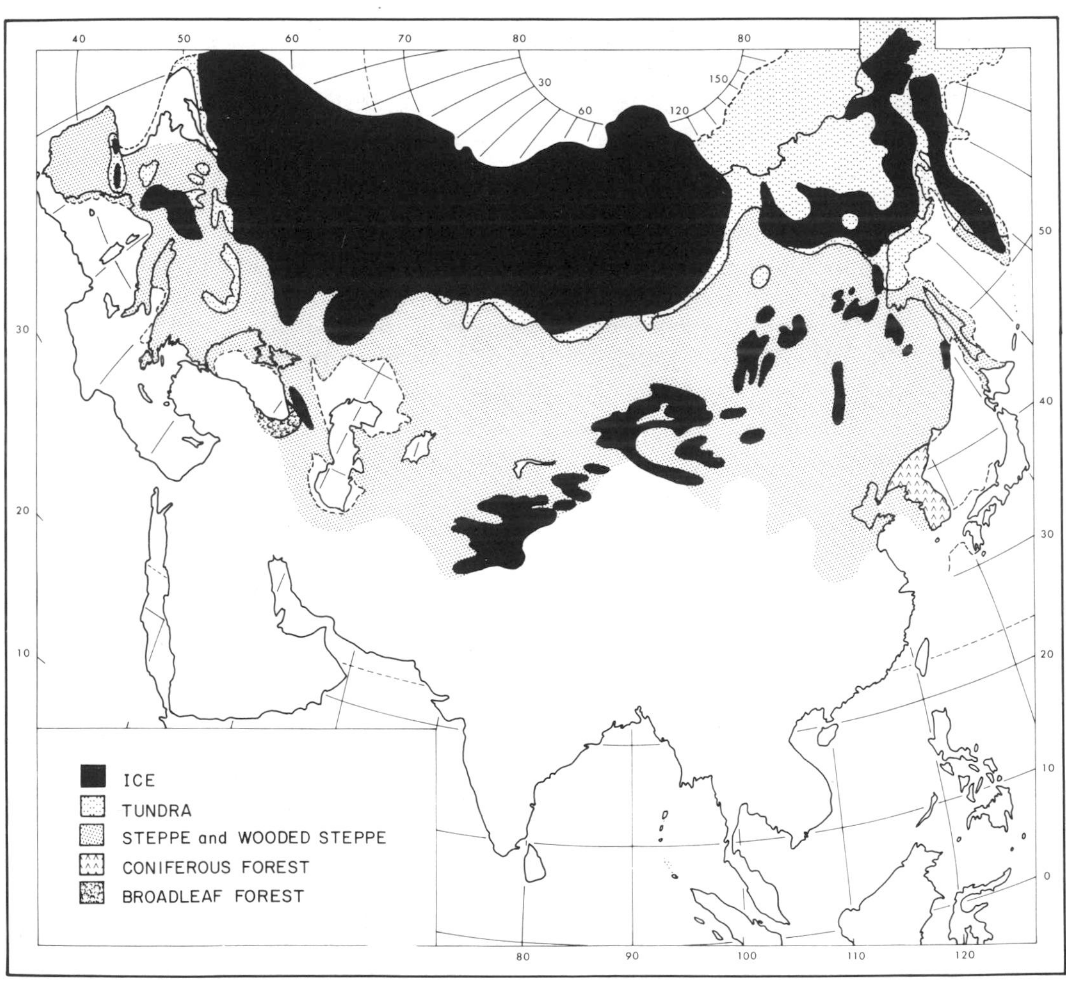

FIGURE 2 Generalized vegetation patterns of mid- and high-latitude Eurasia during the Riss/Saal/Taz full glacial, the most extensive Eurasian glaciation of the Pleistocene (after Frenzel, 1968).

Holsteinian phase appears to have been dominated by conifers, especially spruce, across Eurasia, even in the lowlands. The Eemian forests were again primarily broadleaf deciduous, although the dominance of oak and elm was more restricted to the west and linden was dominant in central and western Europe. The natural Holocene vegetation was primarily broadleaf deciduous forest, but this has been greatly modified or eliminated by human activity.

The glacial vegetation of northern Asia was broadly similar to that of Europe. Forest steppe vegetation apparently dominated the continent west of the Yenisey River during the Donau, whereas a conifer forest of pine, spruce, and larch dominated between the Yenisey River and the Sea of Japan. The regional climate of subsequent glacial phases was more severe, and dry open steppe, dominated by *Artemisia,* was widespread. The extent in which tundra existed during glacial maxima is equivocal, but there is general agreement that open xeric vegetation dominated the region. Forest steppe was locally restricted to more moist mountain ranges and protected river valleys. The interglacial forests of northern Asia effectively disintegrated in the extremely dry, cold continental climate of glacial phases.

Pollen evidence from Yakutia, Baikalia, and the south-central Siberian Plateau led Giterman and Golubeva (1967) to suggest a three-phase sequence of glacial vegetation in eastern Siberia. The onset of glaciation with a cold moist climate produced tundra and forest tundra as birch, grass, and herbs became widespread and *Sphagnum* and sedge bogs formed. During the second (cold, dry) phase, xerophytes such as *Artemisia, Ephedra, Selaginella sibirica,* and chenopods became dominant. The third phase, a period more moist than the second but drier than the first, corresponding to the melting of glaciers, is characterized by an increase of birch, *Sphagnum,* and green mosses, and a reconstitution of forests. The Sartan glacial phase was, according to Giterman and Golubeva, too localized to produce the sequence of regional vegetation change characteristic of the Bakhtian and Zyrianka, although local change did occur. Kind (1967) presents evidence that there was a marked deterioration in the forest and spread of birch and *Artemisia* (i.e., cold tolerant deciduous broadleafs and steppe vegetation) in the northern Siberian lowland during the Sartan. Aigner (1972) indicates that some of north central China was cool or cold and dry during at least some part of the glacial phases.

Giterman and Golubeva also recognized three phases in the development of interglacial vegetation. The Kazantsevo interglacial began with the development of swamp tundra and forest tundra in the north and open parklike pine-birch and spruce-birch forests with some larch to the south. Steppes predominated even further south on the interfluves and pine-spruce-birch forests occurred in the valleys. During the second (optimal) phase of the interglacial, taiga occupied central and western Siberia, whereas eastern Siberia was dominated by larch and pine. The vegetation to the south of these forests varied with elevation and aspect, but generally had taigalike forest on north slopes and larch forests and steppes on southern slopes. The Far East coastal region supported conifer forests with broadleaf elements intermixed. Vegetation trends of the third phase are not described by Giterman and Golubeva, but the vegetation apparently graded into moist tundra, birch, and bog vegetation as the Zyrianka glacial phase waxed. A pollen profile from the Lantian area (southern Shensi) of northeast China, tentatively assigned to the Kazantsevo interglacial, revealed five phases grading from a cool moist environment with spruce, fir, and birch to a warm dry environment with high Compositae and Labiatae, increasing *Artemisia*, and some oak and pine (Aigner, 1972). This evidence suggests a different trend in vegetation than occurred further north, but the extent of the southerly warm, dry interglacial vegetation is not known. The Lantian vegetation cycle does seem reasonable for middle latitudes (i.e., interglacial warming and drying in the continental interior).

The early Holocene vegetation of Siberia was characterized by reconstituted and expanded forests of larch, birch, and pine, with pine-spruce forests in the valleys. During the climatic optimum, the forest zone extended further north than now, fingers of forest penetrated into the Arctic tundra, and the upper timberline was higher in the mountains than it now is. Oak and elm were present in Baikalia at this time. Deteriorating climate has brought about a southward shift of the northern tree line, a lowering of the montane forests, and general reduction of the broadleaf elements. A narrow zone of Arctic tundra occurs between the taiga and the Arctic Ocean, alpine tundra is found in higher mountains, and a subalpine shrub formation (stlaniks) separates the taiga and the Bering Sea south of Chukotka (Yurtsev, 1972) (fig. 3).

Pleistocene Vegetation Patterns: Alaska

The Pleistocene climate of interior Alaska alternated between colder glacial climates with shorter summers and warmer interglacial climates with longer summers. Western Alaska probably had a more continental climate during glacials than during interglacials. Glacial systems developed in the Brooks, Alaska, and southeastern ranges. Pre-Illinoian glacial activity in the region is not sufficiently well documented to allow a regional reconstruction of glacial activity for any given glacial/stadial period, but evidence does indicate that pre-Illinoian glaciation occurred and that it might have equaled that of Illinoian time in some places. The Illinoian glaciation was the most extensive of the Pleistocene in Alaska; both eastern and western Alaska (about 50 percent of the state) were glaciated at this time. Tree line was lowered by 400 to 600 meters. Wisconsin glaciers were less extensive in western Alaska than were Illinoian glaciers, but glaciation was extensive in central and eastern Alaska. During the late Wisconsin, which peaked here about 20,000 to 17,000 BP, the tree line in central Alaska was lowered by some 500 to 600 meters. Permafrost was more widely distributed throughout Alaska during glacial periods than at present. A significant late Wisconsin glacial readvance occurred about 14,000 BP, when, during a period of ca. 1,500–2,000 years, glacial activity was expanded. Maximum withdrawal of glaciers occurred during the thermal maximum (ca. 5,000–3,000 BP). Substantial glacier withdrawal also occurred during earlier interglacials (Hamilton and Porter, 1975; Péwé, 1975).

The vegetation of unglaciated Alaska during glacial periods was dominated by a mosaic of different tundras and, perhaps, steppes, ranging from the most xeric tundras north of the Brooks Range to more mesic herbaceous and shrub tundras in the western and central regions. Birch and alder were common constituents of mesic shrub tundras, with *Artemisia* and grass elements ranging from present to perhaps dominant elements in the vegetation of more xeric or better-drained environments. *Artemisia* and grass were probably most widespread on vegetated outwash plains, loess fields, dunes, or south-facing slopes. Forest vegetation disintegrated into relict tree populations in sheltered lowlands as the tree line was lowered or was displaced to a southern refugium.

Interglacial vegetation was characterized by the reconstruction of

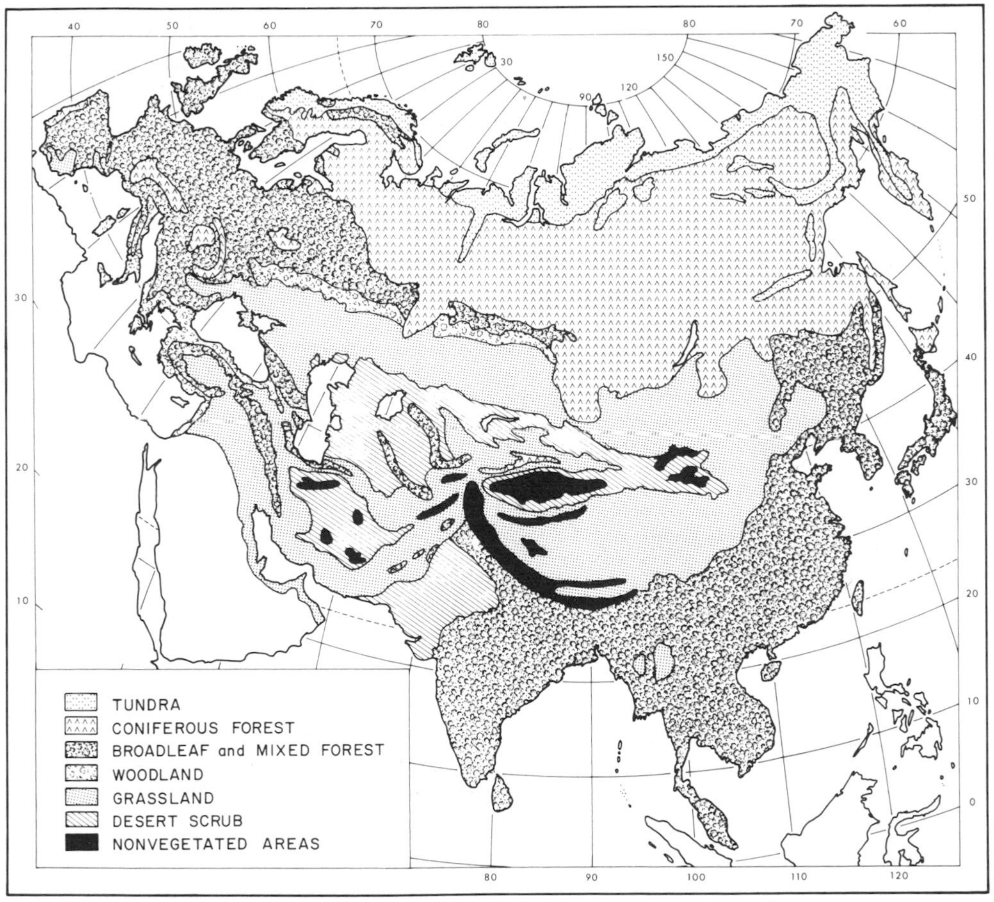

FIGURE 3 Generalized natural vegetation patterns of Eurasia during the Holocene interglacial (after Espenshade, 1960).

forest and reduction of tundra. The spruce forest in interior arid western Alaska was more extensive in the Yarmouth (?) interglacial when it reached the Bering Coast on the Seward Peninsula than during the Sangamon or Holocene. Alder and other shrubs, such as willow and aspen, have extended their range beyond that of the spruce forest in the Holocene, reaching fingerlike into the tundra. Fossil evidence indicates that similar extensions probably occurred during previous interglacials. Coastal southern and southeastern Alaska are now dominated by spruce (*Picea sitchensis*) and western hemlock forests, northern extensions of the coastal evergreen forest which, during glacial times, served as a refugium for their floras. Interglacial tundra/steppe covered substantially less area than did glacial tundras, and the steppe elements, *Artemisia* and grass, were substantially reduced during interglacials (Colinvaux, 1967; R. Guthrie, 1968; Guthrie and Matthews, 1971; Hopkins, 1970; Matthews, 1976; Waring and Franklin, 1979; Young, 1976).

Pleistocene Vegetation Patterns: The Bering Land Bridge

Reconstructions of the Bering land bridge vegetation must be inferred primarily from fossil vegetation evidence from Chukotka and Alaska. Beringia would probably have been under the influence of the Siberian high pressure cell during glacial phases and its climate would have been relatively continental, except possibly for the coastlines that would have been more oceanic. Portions of the Bering land bridge were glaciated from Chukotka, at least during the Illinoian. Biotically, the Bering bridge represented an extension of the arctic floristic region, as a majority of the flora, after dispersal equilibrium was attained, was of Asiatic rather than Alaskan origin (Yurtsev, 1972).

The Bering land bridge probably supported open vegetation, certainly during the Illinoian/Bakhtian and Wisconsin/Zyrianka-Sartan and perhaps also during earlier marine regressions when the bridge existed. Colinvaux (1967) states that a continuous steppe probably occupied the land bridge plains from somewhere east of Seward Peninsula to somewhere in central Siberia to the west. Flerow and Zablotski (1961; Flerow, 1967) hold a similar view. Johnson and Packer (1967:263) "envision a flat-to-slightly undulating tundra plain whose primary plant constituents were lowland species—cotton grass, sedges, grasses, and moisture tolerant herbs and low shrubs." A three-zone pattern of vegetation distribution has been proposed by Yurtsev (1972) in which the northern coast was characterized by arctic tundra; the broad interior was dominated by steppe, tundra steppe, and other tundras locally; and the southern coast contained some boreal conifers, willow, alder, bog plants, and various tundras. Forest vegetation was probably nonexistent on the bridge, particularly during the Illinoian and Wisconsin glacial phases, although pollen and macrofossil evidence from the Seward Peninsula indicate that spruce was near the Bering Sea coast during the Sangamon. Hopkins (1970) even considers the southeastern land bridge area a possible refugium for the spruce forest. The concept emerges, then, of a land bridge supporting open herbaceous and low-growing woody vegetation, probably mosaiclike, superimposed on a steppe-tundra dominated Yurtsevian zonation under a cold climatic regime, drier inland and more moist along the coast.

Pleistocene Vegetation Patterns: North America South of Alaska

The vegetation of North America did not fluctuate in the same way as did the vegetation of Eurasia and Alaska. Contrary to the Eurasian situation, forests and woodland vegetation were widespread in North America during glacial periods, at which times open vegetation was relatively insignificant. Open vegetation increased considerably during interglacial phases, but extensive forests and woodlands also existed during this period. Several factors contributed to this difference of vegetation response to the glacial-interglacial oscillations. North America is smaller than Eurasia and, receiving an adequate supply of moist marine air from the Pacific Ocean and the Gulf of Mexico, did not develop an intensely cold continental climate over the unglaciated portion of the continent during glacial phases. Also, the north-south trending North American mountain ranges acted as corridors for the latitudinal migration of plants consonant with climatic oscillations.

Much more information is available for the late Wisconsin and Holocene vegetation patterns than for earlier glacial or interglacial periods, so this late Pleistocene period will be used here to illustrate the glacial and interglacial vegetation patterns of North America. Vegetation information available for earlier glacial and interglacial phases suggests that the late Wisconsin and Holocene vegetation patterns are generally representative of North American glacial and interglacial vegetation patterns, respectively (Berti, 1975; E. Grüger, 1972*a*, 1972*b*; Kapp, 1970; Kapp and Gooding, 1964; Watts, 1970, 1973; Wright, 1971).

The Late Wisconsin

The late Wisconsin glacial stade comprises the period from 24,000 to 10,000 BP. The maximum glacial extent and most extreme physical environment occurred about 20,000–18,000 BP. A later and significant readvance of glaciers occurred between 15,000–14,000 BP, after which time general ablation in both the Cordilleran and Laurentide glacial systems began (Mercer, 1972). The existence of an ice blockade somewhere along the northern Rocky Mountain front during the late Wisconsin is still equivocal (Rutter, 1978; Stalker, 1978), but some investigators have concluded that such a barrier probably did exist (Mathews, 1978). In unglaciated regions of North America where internal drainage basins existed, numerous and sometimes extensive lakes formed and constituted important components of the physical environment. These lakes generally reached their maximum extent during some part or parts of the glacial phase, although the degree to which maximum water levels are tied to glacial maxima is not yet resolved (Benson, 1978; Van Horn, 1979).

Glacial ice covered about one-half of North America during the late Wisconsin maximum. Forest and woodland vegetation dominated at this time. Spruce and pine forests containing scattered deciduous elements, primarily oak, dominated much of the midwest and most of the east. The spruce forest extended south of the glaciers from the

Atlantic Coast to northeast Kansas, the Dakotas, and across the prairie provinces (at least after glacial withdrawal had begun, if not before) (Bernabo and Webb, 1977; Sirkin, 1977; Wright, 1970, 1971). It is uncertain whether the spruce forest continued westward across Nebraska and the Dakotas to merge with the coniferous montane forest during the glacial maximum, but relict populations of boreal species and fossil evidence suggest that at least part of the northern plains contained boreal trees, even if it did not support forest. The spruce dominance extended at least as far south as the Ozark Plateau (King, 1973; Mehringer et al., 1968; VanDevender and King, 1975), and spruce was possibly present in the lower Mississippi Valley (Delcourt and Delcourt, 1977; Otvos, 1978). Spruce and pine were dominant in the Appalachian Mountain forests, as well as along the Atlantic Coastal Plain as far south as North Carolina (Whitehead, 1973). Pine was dominant on the northern Georgia Piedmont, although spruce and other boreal species (as well as oak) were also present (Watts, 1970). Much of unglaciated eastern North America, then, was dominated by a boreal coniferous forest. The actual extent of this forest is not known because of limited or nonexisting data for the southeastern and southern prairie-southern plains regions of the United States. Wright (1971) has suggested that spruce could have extended westward from northern Georgia, possibly until it met with the montane forest east of the Rockies, but data to confirm or refute this are lacking. Deciduous elements, particularly oak, were distributed through much of the eastern and midwestern conifer forest, although apparently in small numbers, especially to the north (E. Grüger, 1972*a*, 1972*b*; J. Grüger, 1973; Sirkin, 1977; Whitehead, 1973). Whitehead (1973) thinks that southern coastal plain vegetation might have contained primarily deciduous species but, again, supporting data are lacking. Watts (1975) found a xeric scrub vegetation in south central Florida, and Delcourt and Delcourt (1977) found traces of deciduous broadleaf species in a small pollen sample from the lower Mississippi Valley (fig. 4).

Western montane forests were lowered zonally by some 300 to 1,000 meters along mountain flanks; they extended onto the adjacent plains or plateaus where today steppes, woodlands, or brushlands exist. Evidence of this shift comes from the western Great Plains (Hafsten, 1961; Wendorf, 1970), the southwestern United States (Martin and Mehringer, 1965; VanDevender and Spaulding, 1979), and the margins of the great Basin (Bright, 1966; Leskinen, 1975; Mehringer and Ferguson, 1969; Roosma, 1958; VanDevender and Spaulding, 1979). Woodlands of juniper or pinyon-juniper occurred at middle elevations, below montane forests, and throughout the southwestern United States and northern Mexico to elevations as low as 300 meters above sea level. These woodlands contained significant broadleaf associates (including oak), many of which (e.g., big-leaf sage: *Artemisia tridentata*; Joshua Tree: *Yucca brevifolia*; blackbrush: *Coleogyne ramosissima*) now occur only further north or at higher elevations (VanDevender and Spaulding, 1979). True desert brush communities appear to have been restricted to elevations below 300 meters, especially in the lower

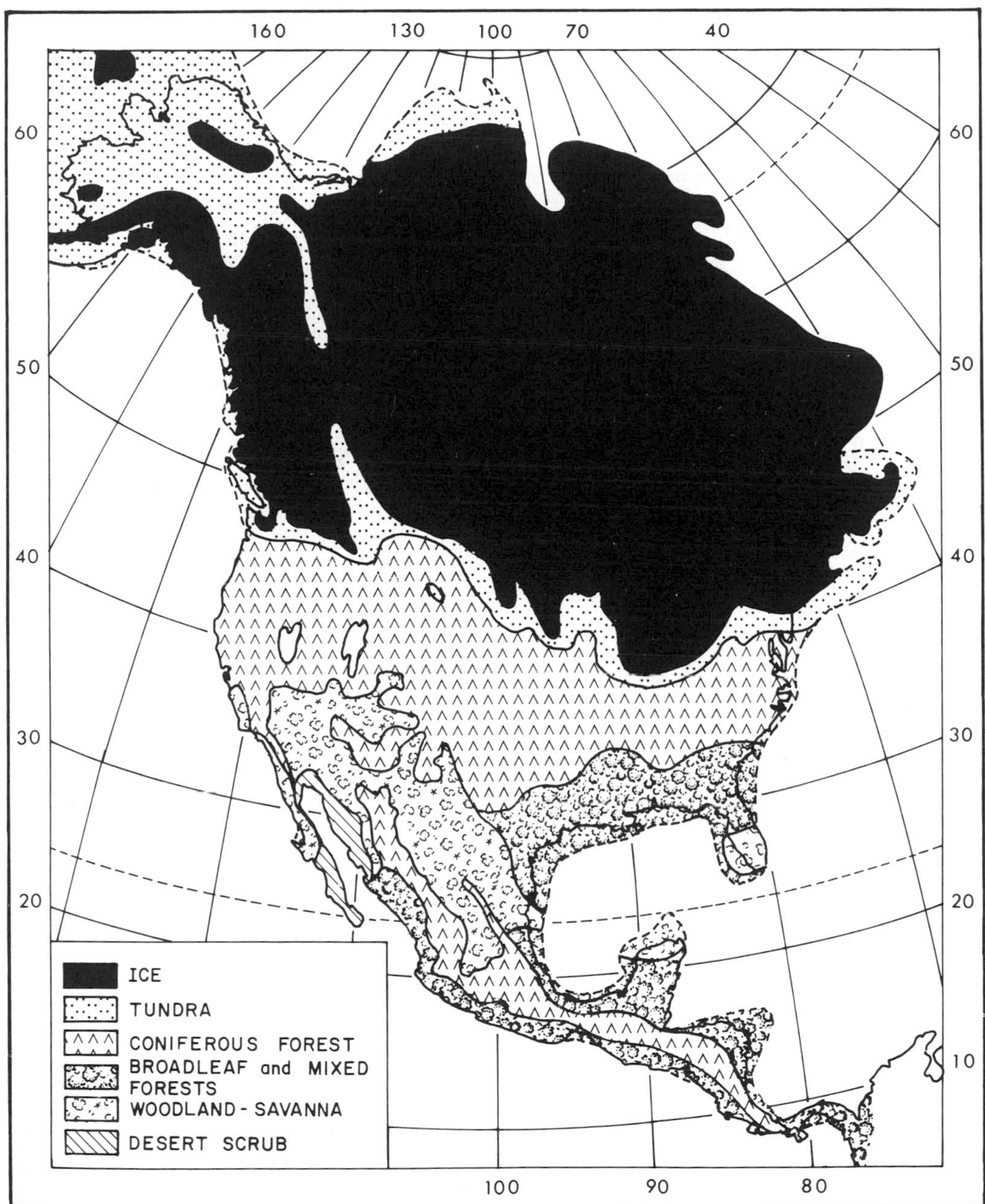

FIGURE 4 Generalized vegetation patterns of North America during the late Wisconsin glacial stade, the most extensive North American glaciation of the Pleistocene (after Canby, 1979; Wright, 1971).

Colorado River drainage and near the head of the Gulf of California (VanDevender and Spaulding, 1979).

Forest and woodland species were also widely distributed through the Great Basin, but the nature of the vegetation (whether forest, woodland, or savanna) is not known for most areas. *Artemisia* was more abundant in the southern and central Great Basin than during the Holocene (Martin and Mehringer, 1965; Mehringer, 1977). Forests of the Pacific Northwest were also lowered and shifted southward during glacial periods (Heusser, 1966). Coastal southern California probably had a summer-dry sclerophyllous vegetation similar to that of today, although summers were possibly more moist and cooler then than now (Johnson, 1977).

The nature of glacial vegetation in the central Great Plains is not well understood; conjecture ranges from boreal forest to open steppe (Webb, 1977; Whitehead, 1973; Wright, 1970). The presence of spruce across much of the northern plains during the very late Wisconsin, after ablation of the Laurentide glacial system had begun; evidence of spruce and pine in the central southern plains; relict populations of pine and spruce across the northern and central plains; and relict populations of thermophillous broadleaf species across much of the plains in east-west trending stream courses and north-south trending scarps suggest that much of the plains was probably wooded if not actually forested (J. Grüger, 1973; Hafsten, 1961; McAndrews, 1967; Moir, 1958; Ritchie, 1964, 1978; Ritchie and de Vries, 1964; Shay, 1967; Watts and Bright, 1968; Watts and Wright, 1966; Wells, 1965; Wendorf, 1970; Wright, 1968, 1970). The argument that the Great Plains supported open vegetation is based primarily on the presence of the Nebraska Sand Hills and adjacent, intergrading loess deposits lying to the southeast. It is argued that these eolian deposits could not have formed if their source area had been heavily vegetated, as with forest. However, these deposits probably accumulated over a relatively short time, perhaps only a couple thousand years, and their formation cannot, therefore, be used to argue against the presence of forests for the entire late Wisconsin.

Open steppe vegetation (grass and *Artemisia*) might have existed in the central plains by 12,600 BP (Watts and Wright, 1966). The evidence for this, however, comes from the northern sandhills and could represent a provincial xeric community that formed on well-drained, stabilized sand dunes (Wright, 1970) rather than an advance of the regional steppe, which probably started forming and actively expanding sometime later.

Little information on vegetation patterns is available for Mexico. Lowering of vegetation zones in southwestern Texas (Wells, 1966) and the Cuatras Cienegas region of Coahuila (Meyer, 1973) suggests that vegetation zones were probably lowered along both ranges of the Sierra Madre. Dillon (1956) felt that pluvial environmental alterations extended as far south as the Durango area. Canby's (1979) representation of Mexican and Central American vegetation patterns for 18,000 BP indicates that all but minor areas were forested or wooded. There is a possibility, however, that tropical Mexico and Central America experienced environmental conditions negatively correlated with the mid- and high latitude glaciations

(Heine, 1973; J. L. Lorenzo, oral comm., July 15, 1976; Martin, 1958), a circumstance that would have resulted in vegetation shifts to higher elevations during the late Wisconsin. If glacial activity in these two regions was, in fact, out of phase, northern Mexico was probably a transitional area between the environmental alterations of middle latitude glacial phases and those of tropical America.

Tundra probably existed in several places along the front of the Laurentide glacier system and at higher elevations in mountains during the late Wisconsin (Hoffmann and Taber, 1967; Moran, 1976; Wright, 1971). Evidence for tundra exists for New England (Davis, 1967; 1969) (probably) New Brunswick and southern Quebec (Bernabo and Webb, 1977), the middle Atlantic region (Martin, 1958; Maxwell and Davis, 1972; Sirkin, 1977), the midwest (Frye and Willman, 1958; Wayne, 1967), the western Great Lakes region (Black, 1969; Fries, 1962; Watts, 1967), southern Alberta ("treeless assemblages": Ritchie, 1978), and the Yukon (Ritchie, 1978; Ritchie and Hare, 1971). Generally, then, some form of tundra occurred along much of the glacial front and probably in the mountains as well, but the tundra zone was relatively narrow, especially in the eastern and central United States.

The Holocene

The Holocene vegetation pattern developed in response to the general continental warming and drying trend that occurred between about 11,000 and 7,000 BP. Thereafter, a period of relatively stable climate existed until about 4,000 BP, after which time, at least in central North America, a cooling trend developed. Other regions of North America apparently experienced a somewhat different thermal and moisture regime; for example, the southeast might have experienced a dry interval from 10,000 to 6,000 BP, and the northeast could have had a similar dry period from 4,000 to 1,500 BP (Wright, 1971). The early Holocene climatic changes and vegetation movements accompanied the withdrawal of glaciers. This general withdrawal began after about 14,000 BP (Mercer, 1972), and probably accelerated about 12,000 to 11,000 BP. Most or all of the Laurentide ice had disappeared by about 6,500 BP (Ogden, 1977), whereas mountain glaciers still exist at higher elevations and latitudes. A period of renewed glacial advance, the Little Ice Age, occurred from about 400 to 100 BP. Bryson and colleagues (Bryson, Baerreis and Wendland, 1970; Bryson and Wendland, 1967) have classified the lower-order climatic episodes of the Holocene; their classification (table 4) will be used when reference is made to specific periods within the Holocene.

The increase in open vegetation is perhaps the most distinguishing difference between late Wisconsin and Holocene vegetation (fig. 5). The greatest expanse of open vegetation is the central grassland, which began to form during the very late Wisconsin and reached its maximum extent about 7,000 BP. The expansion of this grassland and its ecotone has been well documented in the western Great Lakes region. A steppelike vegetation was recognized for Rosebud, South Dakota, as early as 12,600 BP (Watts and Wright, 1966), but this could represent only the provincial open xeric vegetation of the Nebraska Sand Hills (Wright, 1970) and, perhaps, eastern Wyoming. By 11,000–10,000 BP prairie openings were appearing at Pickerel Lake, South Dakota; a prairie was established there by 9,500 BP (Watts and Bright, 1968; Wright, 1971). Oak-dominated savanna moved steadily eastward into southeastern and eastern Minnesota in advance of the prairie, which reached southern Minnesota about 7,200 BP (Wright, 1971). At this time, prairie openings probably existed in the forest as far as north-eastern Minnesota (Fries, 1962; Wright, 1971). The prairie-deciduous forest ecotone remained relatively fixed at this position until about 5,000–4,000 BP, at which time the effects of a climate change toward cooler and wetter conditions resulted in a southwestward shift of the ecotone. Generally, a mixed forest expanded into what were formerly the ecotone and the prairie edge (Bernabo and Webb, 1977; Davis, 1977; Wright, 1971).

Grasslands expanded in a similar way in other parts of central North America during the late Wisconsin and the early Holocene. Between 11,000 and 10,000 BP, herbaceous vegetation (not tundra) developed in the plains of southern Alberta (Ritchie, 1978) and southern Manitoba (Ritchie and Lichti-Federovich, 1968), in northeastern Kansas (J. Grüger, 1973) and on the high plains of the Texas panhandle (Wendorf, 1970). Between 10,000 and 9,000 BP, prairie advanced outward rapidly, especially eastward, and then slowed until reaching its maximum expansion about 7,000 BP (Bernabo and Webb, 1977; Davis, 1977; Wright, 1971). The location and composition of the southeastern prairie-forest ecotone remains undetermined, but it is possible that much of this area, especially the eastern half of Texas and southeastern Oklahoma, retained a forest or woodland character throughout the Holocene. The prairie-deciduous forest ecotone would

Table 4
Major Climatic Episodes of the Holocene

Episode	Duration
Late glacial	Before 10,030 BP
Pre-Boreal	10,030 – 9,300 BP
Boreal	9,300 – 8,490 BP
Atlantic	8,490 – 5,060 BP
Sub-Boreal	5,060 – 2,760 BP
Sub-Atlantic	2,760 – 1,680 BP
Scandic	1,680 – 1,260 BP
Neo-Atlantic	1,260 – 850 BP
Pacific	850 – ca. 400 BP
Neo-Boreal	ca. 400 – ca. 100 BP
"Recent"	ca. 100 BP – present

Source: Wendland, 1978.

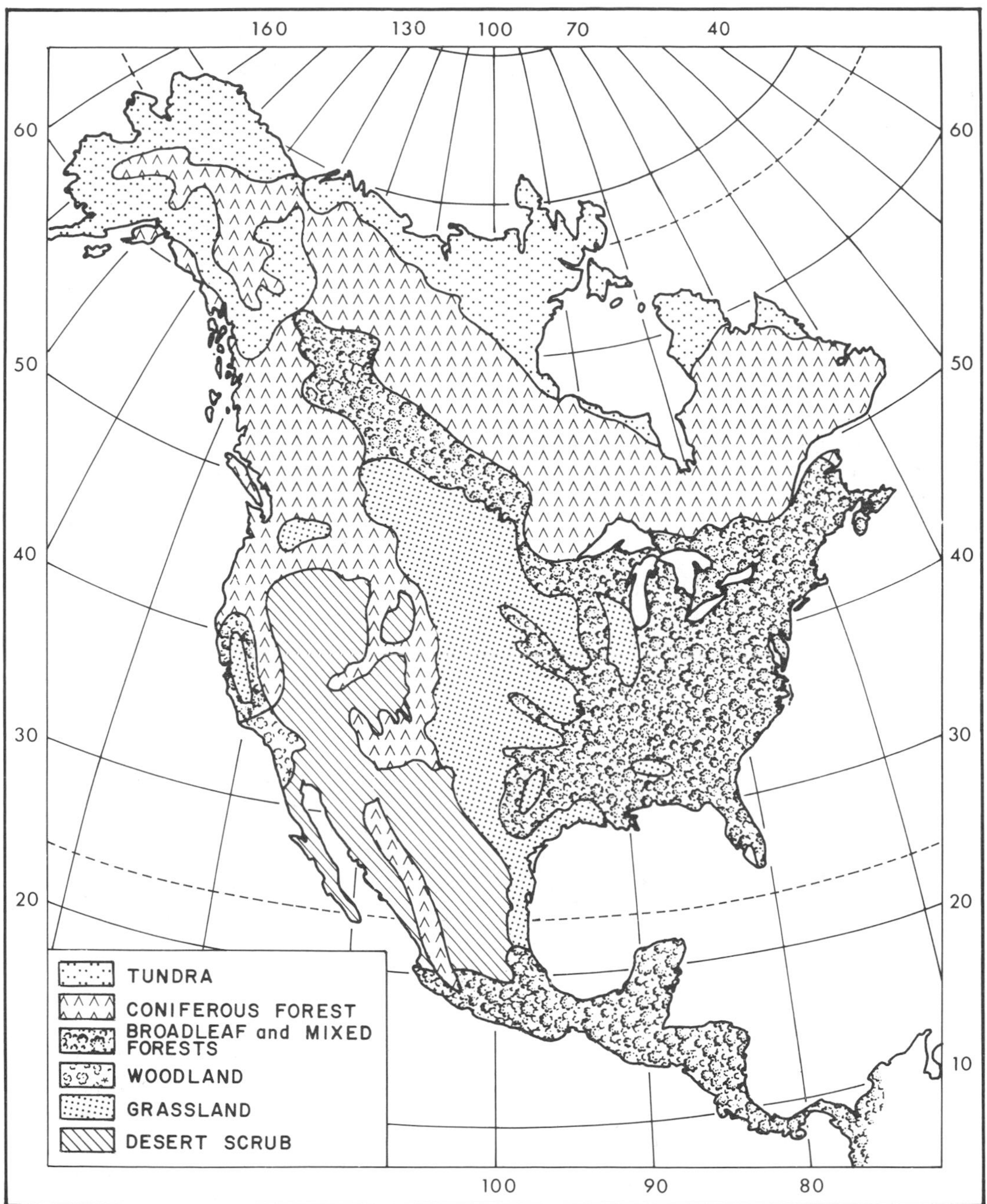

FIGURE 5 Generalized natural vegetation patterns of North America during the Holocene interglacial (after Espenshade, 1960).

then be far to the west with open steppe possibly restricted to the Edwards Plateau and the high plains. The western boundary of the central grassland during the Atlantic climatic episode was essentially the base of the Rocky Mountains, although, if the Rockies were influenced to an appreciable extent by the Atlantic episode climate, as they probably were, then xeric parklike openings in the montane forests probably became more numerous and extensive. To the southwest the central grasslands probably articulated with those of the Mexican Plateau and the southwestern basins. This would have been facilitated by the existence of a northern Chihuahuan Desert grassland which VanDevender and Spaulding (1979) postulate existed until the late Holocene, and which would conform to the increased precipitation predicted for the southwest at this time by Bryson, Baerreis, and Wendland (1970). After about 4,000 BP climatic changes brought about a slight constriction of the central grassland. Deciduous woodlands and forests expanded slightly westward onto former prairie or prairie-forest ecotone all along the eastern prairie border from Alberta to Missouri, and probably southwestward into Texas as well. Increasing dryness in the southwest resulted in the replacement of the Chihuahuan Desert grassland by desert scrub (Bernabo and Webb, 1977; Ritchie, 1978; VanDevender and Spaulding, 1979).

The northeastward expansion of grassland during the Holocene occurred simultaneously with the dramatic northward shift of the Pleistocene conifer forests onto formerly glaciated areas. The most conspicuous and well-documented shift was that of the spruce forests, which began to deteriorate in southern Minnesota as early as 12,000 BP. The rate of spruce withdrawal across eastern and central North America accelerated about 11,000 BP, and by 8,000 BP the southern border of the forest had reached its most northerly position. Spruce continued to spread northward, however, until it reached its maximum latitude about 5,500–4,500 BP. Between 4,000 and 3,500 BP the northern boundary of the forest withdrew southward because of colder climatic conditions (Bernabo and Webb, 1977; Moran, 1973; Nichols, 1975; Sorenson, 1977; Wright, 1971). Eastern pine, too, spread northward during the very late Wisconsin and early Holocene from its primary refugium in and near the Appalachian Mountains to north of the Great Lakes, then west across Canada. As the conifers withdrew from the east, the temperate deciduous forest reconstituted, largely *in situ* from elements already present in the conifer forest. Some eastern deciduous species also migrated northward, at different rates, and contributed to the forests of formerly glaciated areas. Florida and the Atlantic Coast region as far north as Virginia might have supported a xeric woodland with prairie openings (oak with *Artemisia*, *Ambrosia*, and grass) until about 6,000 BP, after which time pine became dominant (Bernabo and Webb, 1977; Wright, 1971).

In the mountain areas of the west, Holocene climatic amelioration was accompanied by zonal migration up the slopes, although different species' responses resulted in somewhat different species mixes at new elevations. The area of alpine tundra and montane forest was reduced as a result of these zonal shifts. The Great Basin developed a more xeric vegetation dominated by such brush species as *Artemisia*, blackbrush, *Atriplex*, and *Yucca brevifolia* at lower elevations, with a probable increase in grass locally, and juniper or pinyon-juniper woodland at higher elevations where forests formerly stood. In both the mountains and Great Basin there is evidence of a warm, dry mid-Holocene from about 8,000 BP to 4,000–3,000 BP (Bright, 1966; Butler, 1972*a*, 1972*b*, 1976; LaMarche, 1973; Mack et al., 1978; Mehringer, 1977; VanDevender and Spaulding, 1979; Waddington and Wright, 1974; Wells, 1966; Wells and Berger, 1967; Wells and Jorgensen, 1964).

Late Wisconsin and Holocene vegetation change in the southwestern mountains was under way by 11,000 BP when mesic conifers had essentially completed a shift from middle to higher elevations, and some brush species had shifted northward. By 8,000 BP the modern vegetation patterns had been established, but the increased summer precipitation postulated for the middle Holocene probably expanded the region's grasslands (Bryson, Baerreis, and Wendland, 1970; Martin, 1963; VanDevender and Spaulding, 1979). Increased aridity during the late Holocene has probably reduced the grasslands and favored the spread of desert brush and succulents.

Arctic tundra remained in northern Canada following the disappearance of glaciers. It did expand onto formerly glaciated areas as glaciers withdrew, especially at higher elevations and latitudes. Arctic tundra probably occupied its minimum area during the Atlantic episode (Nichols, 1975; Ritchie and Hare, 1971; Sorenson, 1977).

Although changing climate was the primary cause of vegetation changes during the late Wisconsin and Holocene, the Holocene vegetation was probably modified to an extent by prehistoric human activities, especially those involving the use of fire. Burning would have the immediate effect of increasing the openness of habitat. Frequent burning, especially in arid to subhumid climates, could have altered the composition and physiognomy of vegetation considerably by generally eliminating or reducing larger, slower-maturing woody plants and favoring smaller, more rapidly maturing herbs. This would have had the general effect of confining woody plants to more humid climates or protected localities, and, if burning was frequent enough, could have created and maintained openings in mesic forests (Lewis and Schweger, 1973; Sauer, 1956; Stewart, 1951, 1956). The human use of fire in North America was not restricted to the Holocene, but it probably was more intense during the Holocene than ever before and it, along with natural fires, probably contributed to the floral simplification of vegetation in many arid to subhumid areas.

THE EVOLUTIONARY PATHWAY LEADING TO BISON

The Paleocene radiation of mammals produced an abundance of generally small, structurally primitive ungulates, some of which possessed hoofed feet and squared-up molars. Artiodactyls—even-toed ungulates with a double-trochleared tarsus and selenodont cheek teeth—appeared in the late Paleocene or early Eocene. Both groups of ruminant artiodactyls—Tylopoda and Pecora—appeared in the late Eocene, and a major ruminant radiation early in the Oligocene resulted in the ruminants becoming the dominant medium-sized herbivores for the remainder of the Cenozoic. This successful ruminant radiation contrasts sharply with the general reduction of mammalian diversity at the family level which characterized the Oligocene, and speaks strongly for ruminant adaptability when presented with the marked environmental changes then occurring. Bovids first appeared in the early Miocene of Europe where they are represented by *Eotragus*, a primitive woodland antelope with upright horns situated directly over the orbits, selenodont molars, and molariform premolars. *Pachyportax*, from the late Miocene of India, possessed more advanced bovid characters; a broadened occipital, a more triangular horn, more hypsodont molars with compressed selenodont cusps and cement covering the enamel, and more robust feet. There is no evidence that *Eotragus* or *Pachyportax* females had horns (Gazin, 1955; Janis, 1976; Kurten, 1972; Pilgrim, 1947; Romer, 1966; Van Valen, 1971; Wolfe, 1978). Ungulate evolution through the early bovids represents a manifestation of, and improvement on, cursorial habits among numerous taxa that were primarily herbivores. The artiodactyl tarsus presumedly facilitates rapid acceleration and thus more efficient escape from predators (D. Guthrie, 1968). Selenodont teeth provide more effective grinding surfaces for chewing fibrous vegetation, whereas hypsodonty provides longer tooth-life. Ruminant digestion permits an efficient use of available vegetation, requiring less overall intake and variety of nutrient sources (Janis, 1976). Increased body size indicates an ability, and probably an adaptive strategy, to obtain increasing supplies of nutrients and energy and to deter predation. It probably also indicates greater longevity. Horns—social and defensive organs—provided greater flexibility in methods of social organization and effective weapons against predators and competitors. The Miocene bovid radiation corresponded with the radiation of other higher ruminants—early cervids, giraffids, and antilocaprids—and represents an adaptive response of medium-sized herbivores to the expansion and increasing availability of herbaceous and low-growing woody vegetation resulting from the cooling and drying of increasingly large areas of terrestrial habitat. The mosaic of habitats which formed along latitudinal, longitudinal, and altitudinal gradients created ample opportunity for local speciation to occur among the phyletically plastic ruminants, as epitomized by the Bovidae. By the late Miocene, then, the major evolutionary antecedents of the Bovini had appeared. Only elaboration was necessary.

The early Bovini are represented by *Parabos*, from Africa and Europe, and *Proamphibos*, from south Asia. Both are from the late Miocene-early Pliocene. *Hemibos*, abundant in the late Pliocene and early Pleistocene of south Asia, probably gave rise to *Bubalus*, the extant Asian water buffalo, during the early Pleistocene. The Pleistocene range of *Bubalus* extended from southern Europe through southeast Asia and into northern China.

Two closely related lineages apparently diverged from the early Bovini stock and gave rise to several African genera. One branch, now extinct, led to the bizarre-horned African buffalo. *Simatherium*, from the late Pliocene and about the size of a modern buffalo, but with flattened and posteriorly placed horns, is considered the earliest known member of this line. The longhorned *Pelorovis* of the middle Pleistocene and *Homoioceras* of the late Pleistocene are more recent genera in this line. The other branch of African buffalo evolution has culminated in *Syncerus*, the living African or Cape buffalo. *Syncerus* is known from Africa, perhaps as early as the very late Pliocene, and might have derived from *Ugandax*, a possible bovine recently classified as a hippotragine (A. W. Gentry, pers. comm., cited in Sinclair, 1977). The evolutionary history of *Syncerus* is, however, poorly known (Gentry, 1967; Sinclair, 1977).

A fourth major lineage of Bovini evolution led to the *Bos-Bison* group. *Proleptobos*, from the early Pliocene, is the recognized ancestor of this group. *Leptobos* appeared later in the Pliocene and became widely distributed throughout southern and central Eurasia during the Pliocene and Pleistocene. *Leptobos* might be the common ancestor and contemporary of both *Bos* and *Bison*, or it might only be a closely related contemporary taxon. *Leptobos* did survive into the late Pleistocene, long after *Bos* and *Bison* appeared in the latest Pliocene—earliest Pleistocene of Asia. *Bos* is known from Pleistocene Asia, Europe, and (rarely) North America. *Bison* is known from Pleistocene Asia, Europe, and North America. It is the only Bovini genus to have become widely established in the Western Hemisphere. The earliest absolutely dated *Bison* remains are from the late Pliocene of Central Asia (Ranov and Davis, 1979), with other possible middle Villafranchian occurrences known from northern India and northern China (Aigner, 1972; Berggren and Van Couvering, 1974; Kowalski, 1967; Skinner and Kaisen, 1947). In this book I am tentatively accepting a late Pliocene/middle Villafranchian age for the appearance of *Bison*.

Sinclair (1977) has integrated data on Bovini amino acid differences from Mross and Doolittle (1967), differences in agonistic behavior, and fossil evidence in deriving a model of phylogenetic relationships among the Bovini genera (fig. 6). The basic radiation among the Bovini was a Pliocene event. By the early Pleistocene the now-extant lineages, and some now extinct, were in existence. The general phenotypic trend of this radiation was toward larger, heavier body size. At least once in each lineage, except possibly *Ugandax-Syncerus* (perhaps the most tentative and least well-known lineage), a giant-horned species existed. Geographically, the radiation produced a generally predictable pattern of diversity. More genera occurred in Eurasia (especially south and east Asia), the probable center of Bovini radiation and the largest land area, with Africa second in diversity. Only one genus (*Bison*) became well established in North America—the result of a relatively narrow dispersal corridor located in an area of rather strong ecological filtering on a distant corner of the Eurasian Bovini range.

Ecologically, extant Bovini show considerable latitude in diet, habitat associations, and environmental tolerance. All graze and all browse to some extent; none feeds exclusively by

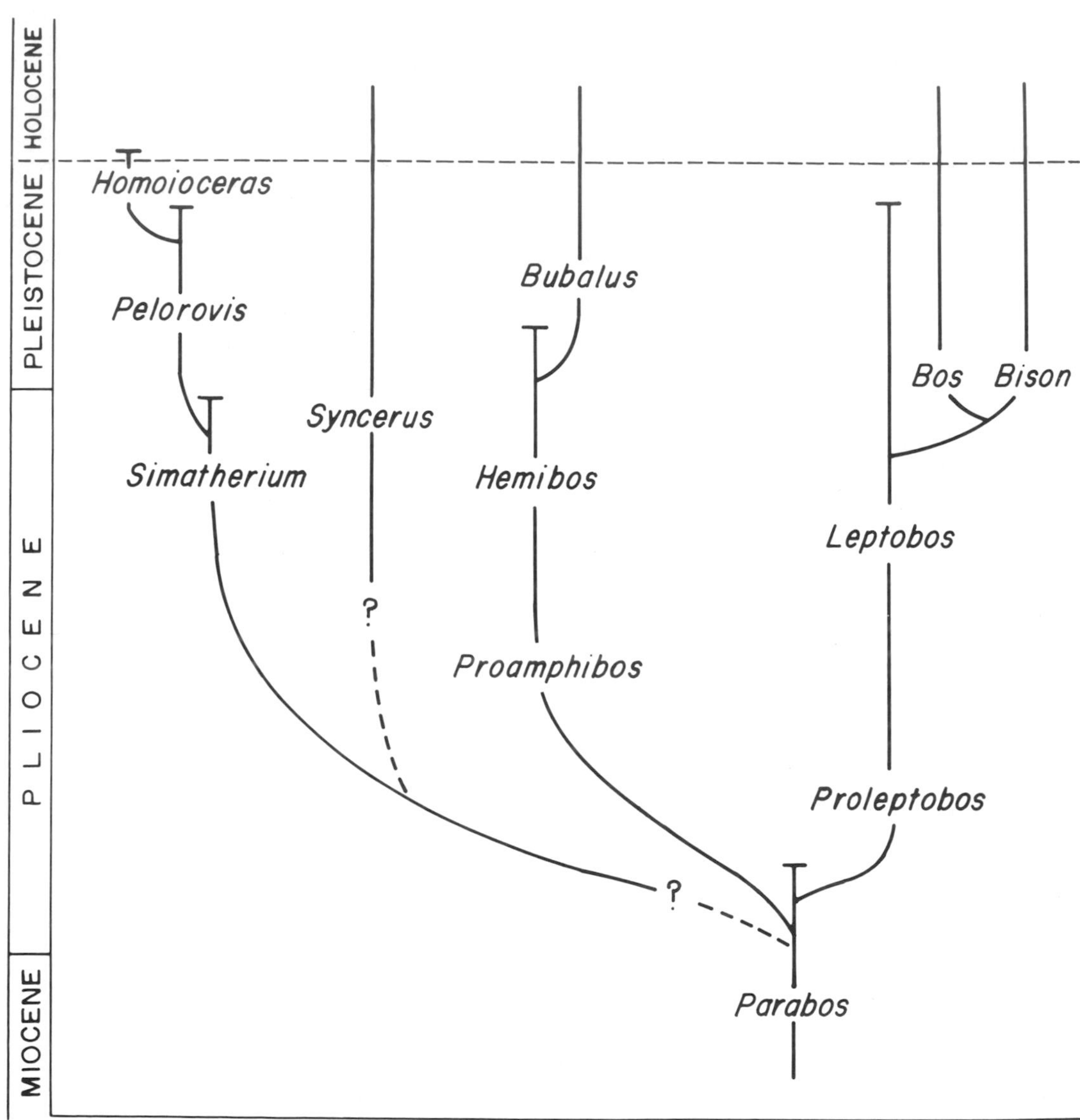

FIGURE 6 A model of Bovini phylogenesis (modified from Sinclair, 1977). Four genera, representing three of the Pliocene lineages, are extant, but the tribe attained its greatest diversity during the pre-Holocene portion of the Pleistocene.

one manner or the other. *Bubalus*, the most aquatic and tropical genus, and perhaps the most primitive in terms of distribution and evolution, normally occupies humid forests and swamps where it feeds mainly on herbaceous matter and secondarily on woody material (McKay and Eisenberg, 1974; Sinclair, 1977). *Syncerus* is probably the most euryecious (ecologically flexible) of living Bovini, occupying as it does much of sub-Saharan Africa from tropical lowland rainforests to the arid Sahel and east African grasslands to high mountains (Sinclair, 1977). *Bos* and *Bison* are intermediate to *Bubalus* and *Syncerus* in terms of habitats now occupied. The several species of wild cattle are currently confined to southeast and central Asia where populations occupy open areas in rain forests, uplands, and high mountains, and feed by both grazing and browsing (Schaller, 1977; Simoons, 1968; Sinclair, 1977). *Bison* now inhabit limited forested and mountain areas in eastern Europe and grasslands and parklands in North America. *Bison* is the only living Bovini that does not have, at least in part, a tropical distribution. *Bison* feed by both grazing and browsing, but the North American grassland form—*B. bison bison*—is probably nearest the grazing end of the feeding spectrum of all living Bovini, although grassland populations of *Syncerus caffer* would be equally dependent grazers. In assessing the modern ecologies of living Bovini, the fact must be kept in mind that modern distribution patterns are, for all genera, much reduced from former ranges. In the not-so-distant past all genera (with the possible exception of *Syncerus*, which is still obviously euryecious) occupied more diverse habitats than they do today, and it seems reasonable to infer that they probably had somewhat different ecologies and social characteristics in those habitats than they do today.

The end of the Pliocene and early Pleistocene, then, had witnessed the radiation of the Bovini and the appearance of all four modern genera now representing the tribe, as well as other genera now extinct. This radiation occurred at a time when megaherbivore forage quality was generally deteriorating (e.g., evergreen broadleaf vegetation diminished as coniferous vegetation expanded in high latitudes and high elevations), but as accessible forage supplies, although more fibrous and abrasive, were increasing (e.g., in the form of enlarged savanna and steppe areas). Competition among large herbivores was also increasing; new, presumedly more specialized and efficient grazers and browsers appeared and more taxa were becoming larger bodied. Predators, too, were keeping pace and increasing in body size. The adaptive response of the Bovini was to increase their body size and to accentuate this with flexible, often allometrically enlarged social organs—horns in both sexes and/or body hair. These developments increased the Bovini ability to compete for resources, to prevent or defend against predation, and to experiment with new schemes of social organization. The ruminant digestive system facilitated the enlargement of body size by permitting the intake of large volumes of fibrous forage, and allowed great latitude in the variety of diet elements, hence of habitats successfully occupied. *Bison* appear to have evolved further north than any other Bovini genus and to have been well adapted to the temperate environment of Asia the genus occupied during the late Pliocene and early Pleistocene.

DISPERSAL AND DIVERSIFICATION OF BISON

Bison were apparently confined to temperate and upland east Asia until after the Donau glaciation. They appear in southeastern and central Europe during the Gunz glacial phase, and are relatively common as fossils and, later, cave paintings, throughout much of Europe until the late Wurm, when they either disappeared for a brief period or were considerably reduced in numbers. Relict populations of now-extinct bison might have survived into the Mesolithic of southwest Asia. Pleistocene European bison have generally been referred to two groups—the smaller, rare, shorter horned woodland wisent, *B. schoetensacki*, and the common, larger steppe wisent, *B. priscus*. Some authorities (e.g., Geist, 1971*a*) consider the latter group ancestral to *B. bonasus*, the modern wisent, which is smaller than any Pleistocene form and associated with a woodland habitat in Europe and, formerly, with upland habitat in the Caucasus and Carpathian mountains. Other authorities (e.g., Kurten, 1968) feel that *B. bonasus* might have evolved from a late immigrant from North America.

Bison reached northern Eurasia during the Mindel glacial phase and spread quickly across the Palearctic, into Alaska, and probably into central North America. The North Eurasian bison were very similar to those that occupied Europe, but taxonomists in the Soviet Union have recognized numerous subspecies (Flerow, 1971; Gromova, 1935). These bison tended to be larger horned than the European bison, and were probably larger bodied as well, although information on postcranial characteristics is generally not available. Pleistocene bison survived in northern Eurasia until the very late Wurm/Sartan glacial phase, when smaller bodied bison appeared in both northeastern Siberia and along an arc from Caucasia north into what is now eastern Poland (Flerow, 1971).

The bison that entered mid-latitude North America (as distinct from Alaska, which were more nearly like those of northern Eurasia) apparently differentiated into a smaller horned form, *B. antiquus*, and a larger horned, but only slightly larger bodied form, *B. latifrons*. *Bison antiquus* is most common in the southwestern United States and Mexico, and probably inhabited the relatively open woodland or savanna habitats of this region. *Bison latifrons* occurred primarily north of *B. antiquus* and probably inhabited a more heavily wooded or forested environment. The history of bison in North America was made more complex, however, by the apparent continued immigration of Eurasian bison which probably occurred whenever a land bridge existed between Siberia and Alaska. *Bison latifrons* probably became extinct during the late Wisconsin, but *B. antiquus* survived into the Holocene, during which time it evolved into the modern species, *B. bison*, which currently has two recognized subspecies: *B. b. bison*, a grassland form, and *B. b. athabascae*, a parkland form (McDonald, 1978*a*, 1978*b*).

SUMMARY

The cumulative environmental changes of the Cenozoic, and the adaptive responses of ungulates, eventually resulted in the appearance of *Bison* during the late Pliocene. Bison dispersal and environmental changes during the Pleistocene resulted in different populations being exposed to a variety of habitats or selection regimes and differentially adapting to many of these selection regimes. This chapter has provided an outline of the chronology of the Pleistocene, patterns of vegetation and vegetation change, the evolutionary direction and ecological role of bison and its closely related genera, and the dispersal history of bison. The following chapter identifies and describes each recognized taxon of North American bison. Subsequent chapters progressively integrate expanding conceptions of these taxa with the environmental information presented in this first chapter.

CHAPTER 2

CLASSIFICATION OF NORTH AMERICAN BISON

North American bison taxonomy began formally with the tenth edition of *Systema Naturae* (Linnaeus, 1758), in which the living bison, named *Bos bison* by Linnaeus, was considered a species of cattle. The scholarly study of North American fossil bison and their evolution began a half-century later as an indirect consequence of growing scientific and popular interest in fossil and legendary proboscideans (Cuvier, 1825; Peale, 1803*a*). Rembrandt Peale, in the course of studying "mammoth" remains and legends from America, published a short paper describing a "Great Indian Buffalo," based in part on a fossilized skull fragment found near Big Bone Lick, Kentucky (Peale, 1803*b*). Following this first description of a North American fossil bison, a considerable literature developed on the taxonomy and evolution of the genus in North America.

Early attention to bison taxonomy corresponded with the early history of modern biological systematics and paleontology. Opinions of systematists differed greatly on how to delimit the group that eventually came to be recognized as *Bison*, on an acceptable taxonomic rank for the bison group, and on the taxonomic relationships among living and extinct bison of Eurasia and North America. Linnaeus (1758) classified living Eurasian bison as *Bos bonasus* and American bison as *Bos bison*. Fossil bison, known and described in the literature before 1758 (Klein, 1732), were not considered taxonomically distinct from living bison by Linnaeus in the tenth edition or by subsequent writers until 1825, when Cuvier, Harlan, and Bojanus each acknowledged the distinctiveness of modern and extinct bison. Cuvier (1825) recognized three species of bison: the living Eurasian bison (auroch or "urus"), the living American bison (*Bos americanus*), and an unnamed extinct American bison. Linnaeus (1758) used the word *bison* specifically for the living North American bison, although, interestingly, this name had been used historically by the Romans in reference to the European bison (Matthew, 1921). Cuvier (1825) considered bison, as a group, to be sufficiently distinct from other Bovidae to be subgenerically differentiated within *Bos*. Smith (1827) formally created the subgenus *Bison*. *Bison* was formally elevated to the rank of full genus by Knight (1849), who referred to living American bison as *Bison americanus*. (Earlier use of the binomial *Bison americanus* by Catesby [1754] and subsequent writers, or *Bos americanus* by Gmelin [1788], anticipated the appearance of *Bison americanus* as a taxonomic formality.) Leidy (1852*a*) was apparently the first to refer North American fossil bison to the genus *Bison*. Later, Rütimeyer raised the bison group of bovids to the supergeneric level and created the name Bisontina for this group, but neither the name nor the rank have been accepted by American taxonomists (Allen, 1876).

By late 1852 fourteen Linnaean binomials in six genera had been applied to North American bison. An additional four binomials in the genus *Taurus*, which antedated Linnaeus's tenth edition, had also been used in reference to North American bison (Allen, 1876). Leidy (1852*a*) attempted to reduce the existing confusion in bison taxonomy with the first revision of North American fossil bison based on the actual study of all available known specimens. This revision recognized *Bison* as a full genus, with two fossil species: *Bison latifrons* and *Bison antiquus*. Leidy also acknowledged the existence of a single species of living North American bison, but did not formally include this in his revision (table 5).

Twenty-four years later Allen (1876) presented a revision of all North American bison in his important monograph on bison zoogeography. This revision (table 5) was important for several reasons: (1) It was the first revision of the North American bison to include both living and extinct species; (2) it reaffirmed Leidy's (1852*a*) classification scheme by recognizing two fossil and one living species of North American bison (*B. latifrons*, *B. antiquus*, and *B. americanus*, respectively); (3) taxonomic conclusions were based, in part, on comparisons of fossil and recent skeletal materials; (4) intraspecific variability in skeletal morphology, including sexual dimorphism, was considered a necessity in the study of fossil and recent skeletal specimens; and (5) the systematic use of biometric data was employed in making species comparisons.

A period of intense taxonomic activity followed Allen's revision. The fossil record for bison substantially improved during this period with the acquisition of large numbers of specimens, particularly from the Great Plains, California, Idaho, and northwestern North America. Evolutionary theory also became generally accepted by paleontologists during this time, but direct efforts to integrate it with the rigid, traditional type-concept approach to taxonomy resulted in a proliferation of new names and considerable confusion. Open competition among a few individuals for new discoveries, acquisitions, and scientific achievements—especially the naming of new taxa—exaggerated the confusion. Lucas (1899*a*), in undertaking the first revision of North American bison since Allen (1876), attempted to "assign definite characters to the various species of bison occurring in a fossil condition in North America and to disentangle the complicated synonymy in which they have been involved" (p. 755). Lucas recognized seven species of bison in making what was a rational effort to stabilize bison taxonomy. His revision (table 5) was comprehensive and well conceived for the period in which it was written, but was, nonetheless, relatively ineffective. The most concerted surge of taxonomic splitting occurred between 1928 and 1933, during which time two new genera, eight new species, and six new subspecies were named. The first

Table 5

Names Recognized in Previous Classifications of North American Bison

Leidy (1852)	Allen (1876)	Lucas (1899)	Skinner and Kaisen (1947)
B. latifrons	*B. latifrons*	*B. latifrons*	*B. latifrons*
B. antiquus	*B. antiquus*	*B. ferox*	*B. alaskensis*
(*B. americanus*)[a]	*B. americanus* (= *B. bison*)	*B. alleni*	*B. geisti*
		B. crassicornis	*B. chaneyi*
		B. occidentalis	*B. crassicornis*
		B. antiquus	*B. alleni*
		B. bison	*B. antiquus figginsi*
			B. antiquus antiquus
			B. preoccidentalis
			B. occidentalis
			B. bison athabascae
			B. bison bison

[a] Leidy clearly recognized the existence and taxonomic distinctiveness of *B. americanus* (= *B. bison*), but did not formally include it in his systematic revision of the fossil bison.

major step toward reversing the splitting activity occurred when Frick (1937) created the subgenus *Superbison* (first mentioned incidentally in Frick, 1930) in order to differentiate longer horned from shorter horned species. A decade later Schultz and Frankforter (1946) synonymized several species of long horned bison, but named and described a new subspecies of short horned bison. The era of splitting was formally brought to an end by the Skinner and Kaisen revision (1947). This revision (table 5) was based on the actual study of most North American bison specimens and a literature review for the Eurasian bison; the authors recognized one genus, five subgenera, ten species, and four subspecies of fossil and living North American bison. This revision was important because (1) it represented by far the most detailed treatment of bison taxonomy undertaken in the history of North American systematics, (2) it was based on a much larger sample of bison skulls than any previous North American study of bison taxonomy, and (3) it eliminated many of the taxa named since the Allen (1876) and Lucas (1899*a*) revisions.

The last thirty years have been devoted more to the study of bison phylogenesis, chronology, ecology, and behavior than to pure taxonomy (McDonald, 1978*a*). Until very recently few new taxa have been named and, with the exception of one new species, all systematic names published since 1947 have been for subspecies. Recent research on several fronts has progressively shown that the Skinner and Kaisen classification is neither compatible with contemporary information on bison morphology and distribution nor with modern concepts of evolution or ecology. Thus there is a current need for a revised classification that is compatible with modern information and ideas.

RESEARCH PREMISES

Four premises were established for the research reported here: (1) The study should be based on the actual examination of specimens; (2) all pertinent specimens available for study should be examined; (3) data should be collected with specific and, where appropriate, multiple uses in mind; and (4) as many different but complementary data categories as practicable should be included in the study.

Unfortunately, much literature bearing on bison taxonomy and evolution is not based on the firsthand examination of specimens. The examination of specimens by a single person provides two advantages: (1) It maximizes the comparability of all data so collected; and (2) it provides the author with an intuitive familiarity with the complete data sample. These qualities cannot be attained through reliance on the description of specimens by others.

Ideally, all appropriate bison specimens should have been studied. In reality this was not possible, but a large and representative sample of the known specimens was studied. Bison specimens in fifty-five institutions were studied; large collections were sampled and many smaller collections were studied entirely. Certain collections, however, were partially inaccessible, notably some of the American Museum of Natural History, Los Angeles County Museum of Natural History, University of Nebraska, and University of Wyoming specimens, which were either in storage or were being studied by other researchers.

Data were sought which would permit differentiation and classification of bison and description and analysis of variation of certain characters. Primary data categories consist of (1) structural morphological data, consisting of linear and angular measurements and frequency counts, and (2) distributional data, documenting the occurrence of bison in time and space, from museum records, from the literature, and by personal inquiry.

SAMPLING AND DATA

Linear and angular measurements and frequency counts of selected morphological characters were taken from mature skulls and limb bones (humerus, radius, metacarpal, femur, tibia, metatarsal). The bison skull, as is true of most mammals, shows greater morphological variability than does the postcranial skeleton (Olsen, 1964; Yablokov, 1974). As a result of this variability, male skull characters, specifically the horn cores, have traditionally been the basis for systematic classification of the genus.

Two considerations led to the inclusion of the six limb bones in this study. First, some postcranial bones might be useful for the specific identification of skeletal remains in the absence of associated horn cores. Second, limbs constitute major functional units of bison anatomy. A systematic examination of individual and integrated limb dimensions would provide both documentation of biometric characteristics of bison limbs and insight into the nature of bison limb change through time.

Sampling Procedures

The deme—a defined interbreeding population—was the preferred population level from which samples were taken. Demic samples also permit a much more detailed examination of such ecological/evolutionary subtleties as choroclines, chronoclines, and irregular phenotypic variations that might result from differences in environmental quality, population size, or other circumstances.

Fifty randomly selected specimens of each element for each sex from each available deme was considered the ideal sample size. This sample size should yield meaningful statistical results that could be improved on only gradually with increasingly larger samples (Olson and Miller, 1958; Simpson and Roe, 1939; Yablokov, 1974). A deme-level sample of fifty was, however, rarely ever assembled, due to differences in the occurrence, preservation, discovery, retention, and/or availability of specimens; samples were assembled from collections imbalanced with regard to quantity, ontogenic age, sex, species, geologic age, and spatial provenience of specimens represented. In only a few instances were some specimens in a collection not measured because a sufficient sample size had been reached. The raw biometric data used in this study are available at the George C. Page Museum of Rancho La Brea Discoveries, Los Angeles; The Department of Paleobiology, United States National Museum, Washington, D.C.; and, in part, in McDonald (1978*a*).

Equipment and Data Recording

Most quantitative data used in this study are linear or angular measurements. The following instruments were used: (1) a 254 mm Helios dial caliper (0.1 mm graduation) for measuring dimensions consistently less than 230 mm; (2) a six-foot custom-made metal vernier caliper (1/16 inch graduation) for measurements consistently exceeding 230 mm; (3) a 7,610 mm metal tape (1 mm graduation) for measurements exceeding 1,800 mm; (4) a 1,500 mm synthetic fiber flexible tape (1 mm graduation) for curvilinear measurements; and (5) a 180° Rabone-Chesterman #200-C metal protractor (1° graduation) for measuring angles.

Measurements taken with the dial calipers were recorded to the nearest 0.1 mm; those with the vernier calipers to the nearest 0.25 inch and then converted to and recorded as millimeters; those with either tape to the nearest millimeter; and those with the protractor to the nearest degree. All measurements of any one character were taken with the specimens in as nearly as possible the same position. Estimated measurements of worn or slightly damaged specimens were recorded if the estimate was considered to be realistic and if the inclusion of the estimate in the sample was considered preferable to its omission. Less than 5 percent of all measurements are estimates.

Skull Measurements

A skull was classified either mature or immature by three criteria in decreasing order of reliability: extent of upper molar eruption and wear, the extent of horn core ossification, and the extent of frontal/frontal-parietal suture fusion. Ideally, all three criteria could be employed to determine the age of any given skull, but in many cases only part of the skull remained and aging was based upon available evidence. Specimens were considered mature if they had attained the "full maturity" toothwear stage (S-3) of Skinner and Kaisen (1947) (table 6). The extent of horn core ossification was used to determine maturity in the absence of sufficient dentition. An immature horn core is much more porous in appearance than is a mature horn core, and in males lacks the bony ridges and burrs prominent near the base of mature horn cores. The extent of frontal/frontal-parietal suture

Table 6
Superior Molar Wear and Age Classes

Age class		Molar wear stage
I	(immaturity)	M^1 erupting or in wear
A	(early adolescence)	M^2 in wear; M^3 erupting
S-1	(late adolescence)	M^1 style in wear
S-2	(early maturity)	M^2 style in full wear; M^3 style not in wear
S-3	(full maturity)	M^3 style in full wear
S-4	(old age)	M^1 and M^2 styles nearly or fully worn away; M^3 style diminishing

Source: Skinner and Kaisen, 1947.

fusion was considered a reliable indicator of relative maturity for males only. A skull was considered mature if the sagittal frontal suture was completely fused posteriorly from about midway between the planes of the orbits and bases of the horn cores, and if the frontal-parietal sutures were completely fused from the sagittal origin laterally and ventrally to below the level of the ventral horn core bases. Complete fusion of the frontal and frontal-parietal sutures does occur in females, but it is by no means ubiquitous and cannot be relied on as an indicator of relative maturity. Some borderline cases occurred where intuitive judgments of relative maturity were made, as with individuals nearing skeletal maturity or with developmental idiosyncracies. Such specimens were evaluated individually.

Mature skulls were sexed on the basis of relative robustness, degree of cranial suture fusion, and morphology of horn cores. Female skulls are smaller in size, less strongly featured, and possess less massive bones than male skulls. As noted above, the sagittal frontal and frontal-parietal sutures of female skulls are considerably less fused than are those of male skulls. Also, in most species of bison the female horn core shows no burr or rim at the dorsal base. The female horn core normally blends almost imperceptibly with its neck and the frontals, whereas the male horn core normally has a distinct burr or rim where it joins the neck. Female *B. latifrons* and *B. alaskensis*, however, do usually possess a slight burr. Two other useful criteria for sexing horn cores are the differences in spiraling and/or rotation of horn cores between the sexes. Male horn cores tend to spiral posteriorly about the longitudinal axis, if at all, whereas most female horn cores tend to have no spiral or to spiral anteriorly. Female horn cores usually (but not always) have the greatest axis of the basal diameter rotated forward; this is more variable among males.

Problems occur in the sexing of bison skulls just as they do in determining the relative maturity of skulls. Females of some taxa possess some malelike features, and it is sometimes possible to confuse a malelike female skull of one species with a male skull of a generally smaller species. This is particularly true for isolated horn cores.

Fourteen linear and three angular measurements were taken from complete skulls that were examined (table 7; figs. 7, 8). Fifteen of these were standard measurements from Skinner and Kaisen (1947), whereas two were of my own design.

Eight hundred and eleven skulls contributed biometric and/or provenience data to this study. A notable differential exists with regard to the skull characters that are represented. Generally, horn cores and the frontal-occipital region are more durable than the facial region; male skulls are more durable and more often collected than are female skulls; and geologically younger skulls are better preserved than are geologically older skulls. Also, specimens cannot be confidently identified without associated horn cores. The data are weighted, therefore, toward geologically younger male horn cores and frontal-occipitals.

Postcranial Measurements

Only mature humeri radii, metacarpi, femora, tibiae, and metatarsi were studied; an individual bone was considered mature if all epiphyses were fused. The differential rate of limb bone maturation has biased sample sizes in favor of the earlier maturing elements (tables 8–11). (Tables 9–11 indicate that there might be differences between *Bos* and *Bison*, and among *Bison*, in the maturation of limb bones. This suggests the possibility of error in estimating age

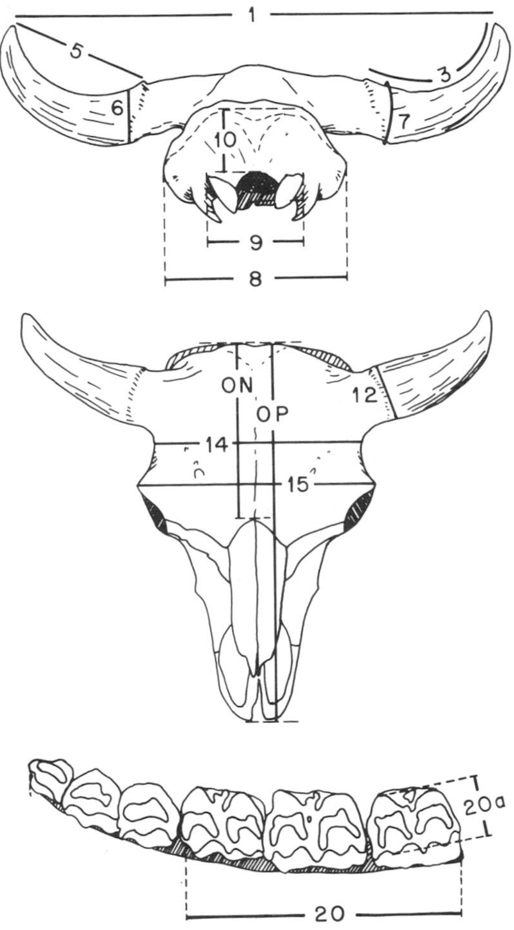

FIGURE 7 Linear measurements of skull characters used in this study. The numbers and letters identify measurements described in table 7.

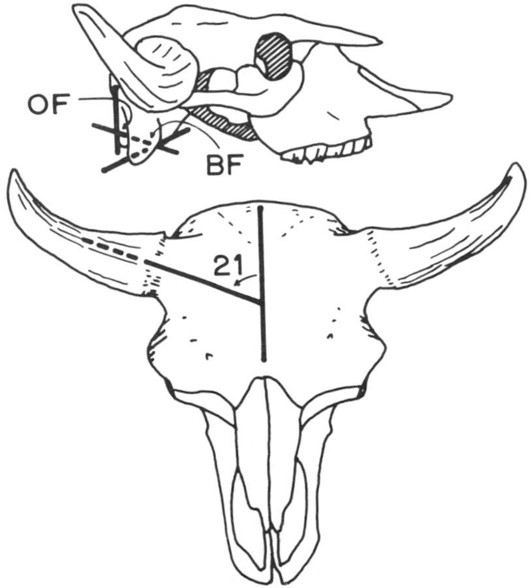

FIGURE 8 Angular measurements of skull characters used in this study. The number and letters identify measurements described in table 7.

based on epiphyseal fusion. Also, individual differences in maturation rates are an important evolutionary advantage, especially when a change in selection forces favors accelerated or decelerated maturation.)

Three characters of each of the six limb bones were measured. These measurements consisted of a modified length measurement (usually the rotational length of the bone) and the antero-posterior and transverse shaft diameters (table 12; figs. 9, 10). Heads, trochanters, tuberosities, and spines are often missing from otherwise well-preserved specimens. Consequently, the length measurements of bones whose length is determined in part by these features were modified to allow the use of specimens with partially missing extremities. Generally, minimum shaft diameters were sought for limb bones

Table 7
Skull Measurements

1	Spread of horn cores, tip to tip	The straight line distance between the tips of the horn cores.
3	Horn core length, upper curve, tip to burr	The arc from the tip to the burr of a horn core, measured along the dorsal surface of the horn core. The measurement reference point at the burr (proximal) end occurs at the rim of the burr midway between the anterior and posterior edges of the horn core.
5	Straight line distance, tip to burr, dorsal horn core	The straight line distance, or chord, between the tip and burr of a horn core, measured on the dorsal side of the horn core. The measurement reference point at the burr (proximal) end is as in number three above.
6	Dorso-ventral diameter, horn core base	The straight line distance between the rim of the ventral burr and the dorsal surface of the horn core, measured perpendicular to the antero-posterior plane of the horn core. This measurement may or may not measure the actual dorso-ventral diameter of the *burr* line, depending on the angle at which the horn core emanates from the frontal.
7	Minimum circumference, horn core base	The minimum circumference of the horn core at the proximal end, measured with the ventral point of the measured circumference tangential to the ventral burr of the horn core. This measurement may or may not measure the actual circumference of the *burr* line, depending on the angle at which the horn core emanates from the frontal.
8	Width of occipital at auditory openings	The straight line distance between the edges of the occipital at the auditory openings.
9	Width of occipital condyles	The maximum straight line distance between the lateral edges of the occipital condyles.
10	Depth, nuchal line to dorsal margin of foramen magnum	The straight line distance between the sagittal point on the nuchal line and the dorsal rim of the foramen magnum. The dorsal reference point occurs where the imaginary continuation of the nuchal line intersects the sagittal plane. (The nuchal line is not distinct in the sagittal region of the dorsal occipital.) The ventral reference point occurs at the most dorsal point along the rim of the foramen magnum.
12	Antero-posterior diameter, horn core base	The straight line distance between the anterior and posterior rims of the horn core burr, measured parallel to the antero-posterior plane of the horn core. Exaggerated ridges of bone which occur along the burr of some specimens are excluded from this measurement.
14	Least width of frontals, between horn cores and orbits	The minimum straight line distance across the frontals measured from the crotches formed between the anterior horn core surface and the posterior orbit surface.
15	Greatest width of frontals at orbits	The straight line distance across the frontals, measured from the posterior rims of the orbits.
20	M^1–M^3, inclusive alveolar length	The straight line distance along the alveolus, measured on the labial side from the anterior edge of M^1 to the posterior edge of M^3.

Table 7
Skull Measurements (Continued)

20a	M^3, maximum width, anterior cusp	The maximum transverse distance across the anterior cusp of M^3.
O-P	Distance, nuchal line to tip of premaxillae	The straight line distance from the dorsal-most point of the nuchal line to the most distal point of the premaxillae. The nuchal line reference point is as in number 10 above. The premaxillae reference point occurs where the imaginary continuation of the arc formed by the anterior edges of the premaxillae intersects the sagittal plane.
O-N	Distance, nuchal line to nasal-frontal suture	The straight line distance from the dorsal-most point on the nuchal line to where the nasal-frontal sutures meet. The nuchal line reference point is as in number 10 above.
∠21	Angle of divergence of horn cores, forward from sagittal	The acute angle separating the sagittal plane and the longitudinal axis of the horn core.
∠OF	Angle between foramen magnum and occipital planes	The angle separating the plane of the foramen magnum and the occipital plane, measured on the sagittal plane.
∠BF	Angle between foramen magnum and basioccipital planes	The angle separating the plane of the foramen magnum and the basioccipital plane, measured on the sagittal plane.

Note: All measurements except 20a, ∠OF, and ∠BF correspond to the Skinner and Kaisen (1947) system of standard measurements.

Table 8
Distribution, by Element and Sex, of Biometric Data

Element	Males	Females	Undifferentiated	Total
Skulls	619	142	1[a]	762
Limb Bones				
Humerus	103	126	11[b]	240
Radius	168	199	30[b]	397
Metacarpal	325	565	126[b]	1016
Femur	101	75	9[b]	185
Tibia	144	144	14[b]	302
Metatarsal	338	535	139[b]	1012
Total	1179	1644	329	3152
Total elements	1798	1786	330	3914

[a]This is a juvenile specimen, USNM P 11410, the holotype of *B. sylvestris*. It was included in this study because of its taxonomic significance.

[b]These are specimens from various paleontological contexts from northwestern Canada and Alaska which could not be adequately sexed. They are included here because they provide information on the total range of variation of elements from this region.

Table 9

Age of Epiphysis-Diaphysis Fusion in Limb Bones

Epiphysis	*Bos*[a] (age)	*Bison*[b] (year)	*Bison*[c] (year)
Proximal metacarpal	prenatal	(prenatal)	(prenatal)
Proximal metatarsal	(prenatal)	(prenatal)	(prenatal)
Proximal radius	1 – 1½	late 4th	2nd
Distal humerus	1½	early 4th	4th
Distal metacarpal	2 – 2½	late 4th	3rd
Distal metatarsal	....	late 4th	3rd
Distal tibia	2 – 2½	mid 4th	3rd
Proximal femur	3½	mid 5th	6th
Proximal tibia	3½ – 4	late 5th	6th
Distal femur	3½ – 4	late 5th	6th
Distal radius	3½ – 4	early 6th	5th (♂♂) 6th (♀♀)
Proximal humerus	3½ – 4	6th	6th

Sources:

[a]Getty, 1975; Sisson and Grossman, 1953;
[b]Koch, 1935;
[c]Empel and Roskosz, 1963.

Note: Parenthetical information is not provided in the original sources.

Table 10

Order of Epiphysis-Diaphysis Fusion in Cattle and Bison

	Bos[a]	*Bison*[b]	*Bison*[c]
1	Proximal metacarpal Proximal metatarsal	Proximal metacarpal Proximal metatarsal	Proximal metacarpal Proximal metatarsal
2	Proximal radius	Distal humerus	Proximal radius
3	Distal humerus	Distal tibia	Distal metacarpal Distal metatarsal Distal tibia
4	Distal metacarpal Distal metatarsal (?) Distal tibia	Distal metacarpal Distal metatarsal Proximal radius	Distal humerus
5	Proximal femur	Proximal femur	Distal radius (♂♂)
6	Proximal humerus Distal radius Distal femur	Distal femur Proximal tibia	Proximal humerus Distal radius (♀♀) Proximal femur Distal femur Proximal tibia
7		Distal radius Proximal humerus	

Sources:

[a]Getty, 1975; Sisson and Grossman, 1953;
[b]Koch, 1935;
[c]Empel and Roskosz, 1963.

Note: Numbers in the left column represent stages of development, not the age at which fusion is completed.

Table 11
Order of Limb Bone Maturation in Cattle and Bison

Bos[a]	*Bison*[b]	*Bison*[c]
Metacarpal and Metatarsal	Metacarpal and Metatarsal	Metacarpal and Metatarsal
Tibia	Tibia	Radius
Femur	Femur	Tibia
Radius	Radius	Femur
Humerus	Humerus	Humerus

Sources: [a]Getty, 1975; Sisson and Grossman, 1953; [b]Koch, 1935; [c]Empel and Roskosz, 1963.

Table 12
Limb Bone Measurements

	Humerus
1	Approximate rotational length of bone
2	Antero-posterior diameter of diaphysis at right angles to transverse minimum
3	Transverse minimum of diaphysis
	Radius
1	Approximate rotational length of bone
2	Antero-posterior minimum of diaphysis
3	Transverse minimum of diaphysis
	Metacarpal
1	Total length of bone
2	Antero-posterior minimum of diaphysis
3	Transverse minimum of diaphysis
	Femur
1	Approximate rotational length of bone
2	Antero-posterior diameter of diaphysis at right angles to the transverse minimum
3	Transverse minimum of diaphysis
	Tibia
1	Approximate rotational length of bone
2	Antero-posterior minimum of diaphysis
3	Transverse minimum of diaphysis
	Metatarsal
1	Total length of bone
2	Antero-posterior minimum of diaphysis
3	Transverse minimum of diaphysis

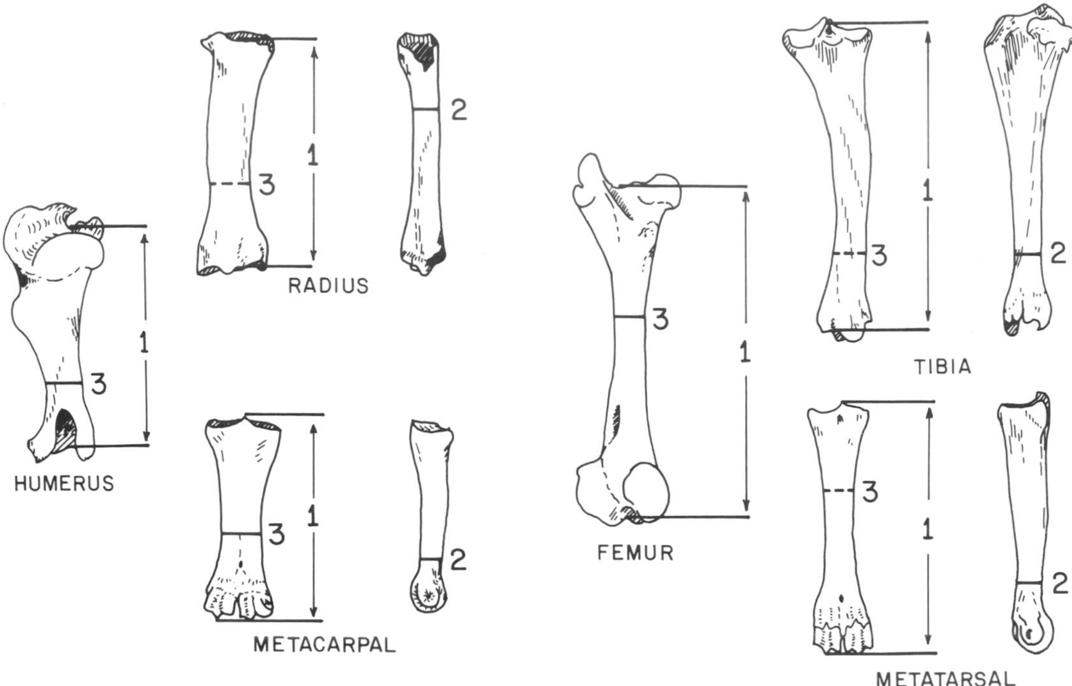

FIGURE 9 Linear measurements of front limb characters used in this study. Humerus measurement 2, not illustrated, is taken at right angles to measurement 3. The numbers identify measurements described in table 12.

FIGURE 10 Linear measurements of rear limb characters used in this study. Femur measurement 2, not illustrated, is taken at right angles to measurement 3. The numbers identify measurements described in table 12.

because these dimensions were considered potentially more meaningful in terms of analyzing the structure, function, and weight-bearing capacity of the limbs than were the midshaft diameters often used by other writers. Accordingly, the antero-posterior and transverse minima were taken for the radius, tibia, and metapodials. For the humerus and femur, however, the minimum shaft diameter was measured and the complementary diameter was measured at right angles to the minimum.

The limb bones of North American bison are sexually dimorphic. Limb bones studied were sexed by computer using the BMDP2M cluster analysis program, which evaluated and assigned cases to either sex on the basis of two characters (antero-posterior width X transverse width). Results of the cluster analyses were compared with computer-generated SPSS SCATTERGRAM scatterplots for fidelity and evaluation of assignment.

Data were collected from 3,152 limb bones (table 8). The differences in sample size between metapodials (n = 2,028) and upper limb bones (n = 1,124) are particularly important. Much of this difference can be attributed to different maturation rates for the various bones (table 11). Also important, as Guthrie (1967) has indicated, is the differential destruction of skeletal elements by natural processes. Metapodials, smaller and much more dense than the four upper limb bones, are less easily destroyed by decay or abrasion and, because there is less flesh on the foot than on the upper limb, metapodials are less likely to be destroyed or carried off by carnivores. Many of the limb bones studied came from archeological kill sites that characteristically yield fewer complete upper than lower limb bones. This is because the upper limb bones are often broken to permit access to the marrow or selectively removed from the kill site in the course of butchering and relocating the upper limbs.

Normally, no differentiation was made between right and left bones in the course of analyzing limb bone data, even though such differentiation is common practice and obviously desirable for some purposes (Grayson, 1973). In order not to reduce sample size by excluding all cases in the numerically smaller right or left group of a sample, both were included unless both right and left elements of a single individual were suspected or known, in which case only one of the pair was included (known or suspected pairs appeared only twelve times). This practice substituted one bias for another, but both were functions of systematic data compilation and evaluation and the one used provided more data than the other.

Distribution Data

Most specimens included in this study have provenience data. The geologic age of specimens was also determined when possible, but this information is less accurate than is provenience. (Absolutely dated specimens are listed in Appendix 2. Relatively dated specimens are discussed below.)

CLASSIFICATION REASONING AND FORMAT

Certain problems had to be resolved before this revision could be completed. Among the more important of these were (1) the recognition of taxonomic ranks, (2) the recognition and handling of morphological variability among specimens studied, and (3) the selection and weighting of taxonomically important characters.

Taxonomic Categories

Bison is retained as a full genus within the family Bovidae and is considered generically separate from the closely related genus *Bos*. Recently, some writers have considered, implied, or urged the synonymy of *Bison* and *Bos* (e.g., Bhambhani and Kuspira, 1969; Stormont, Miller, and Suzuki, 1961; M. Wilson, 1974*b*, 1975). Other writers have opposed, or showed reason for opposing, this synonymy (e.g., Basrur and Moon, 1967; Krasinska and Pucek, 1967; Sartore et al., 1969; Shaw and Patel, 1962; Wilber and Gorski, 1955).

Mayr (1969) states that a genus should be composed of a single species or a monophyletic group of species which can be separated from other genera by a decided gap, should occupy a distinctive niche or adaptive zone, should possess general internal similarity, and should satisfy certain practical requirements. *Bison* and *Bos*, as currently conceived, can be readily separated by differences in the skull structure, which have been conserved within each group. They have also evolved different distribution, population, and ecological patterns.

Subgenera are not recognized in this classification. The five recognized species and the simple patterns of inferred phylogenesis constitute much too little information to necessitate the retention of the five subgenera recognized by Skinner and Kaisen (1947).

Five species of bison are recognized in this classification. Species are considered to have evolved relatively rapidly, by standards of geologic time, while adapting to new environments, and to have thereafter remained relatively stable (morphologically, behaviorally, and ecologically) throughout their history. This view of species and speciation conforms to the "punctuated equilibrium" concept advocated by Eldredge and Gould (1972:84):

> Paleontology's view of speciation has been dominated by the picture of "phyletic gradualism." It holds that new species arise from the slow and steady transformation of entire populations. Under its influence, we seek unbroken fossil series linking two forms by insensible gradation as the only complete mirror of Darwinian process; we ascribe all breaks to imperfections in the record. . . .
>
> The history of life is more adequately represented by a picture of "punctuated equilibria" than by the notion of phyletic gradualism. The history of evolution is not one of stately unfolding, but a story of homeostatic equilibria, disturbed only "rarely" (i.e., rather often in the fullness of time) by rapid and episodic events of speciation.

Acceptance of "punctuated equilibrium" as the primary process of speciation is not to say that cumulative change did not occur among bison. Acceptance does imply, however, that any period of morphological (or other specifically differentiating) change was relatively brief and amounted to a transformation from less fit to more fit phenotypes as a result of a restructured selection regime. The distance and rate of species differentiation (i.e., adaptation to a new environment) would be proportional to the distance and rate of change between the old and new selection regimes. Thus, a gradual transition from one environment to another would theoretically result in gradual adaptation among bison, whereas a rapid transition would result in rapid adaptation. The main point here is that adaptation occurred rapidly, but only to the extent required by the new environment. Ideally, distinct morphological (and behavioral and ecological) gaps should exist among species adapted to different environments.

The dynamic environments of late Quaternary North America and Eurasia allowed or caused substantial shifts in population numbers and distributions of Holarctic bison. As a result, occasional interbreeding among contemporary species probably took place in certain zones and at certain times. Interbreeding and secondary intergradation were insufficient, however, to produce morphologically distinct and adaptively stable populations, or to erode the specific distinctiveness of established species. "An amount of gene flow between species that does not lead to the breakdown of the differences between them is compatible with their being properly described as reproductively isolated." (Raup and Stanley, 1978:102).

Specific status is therefore retained for the five species recognized in this classification, even though there is morphological evidence that interbreeding probably existed among some of the species. (Hybridization is discussed more fully in chapter 6.) The retention of specific, rather than subspecific, status conveys more biological meaning and is potentially more useful for further study of the nature of natural selection and adaptation among bison.

In contrast, when a distinctive morphologic character cline can be demonstrated among contiguous populations of bison, the populations at either end of these clines may be considered conspecific. There has recently appeared a tendency among some students of late Quaternary megafauna to regard late Pleistocene and Holocene taxa of the same phyletic lineage as conspecifics (e.g., Flerow, 1965, 1967, 1971; Fuller and Bayrock, 1965; Harris and Mundel, 1974; Sher, 1974; M. Wilson, 1974*a*, 1974*b*, 1975). Most of these authors consider extant and shorter horned late Pleistocene-early Holocene bison to be only subspecifically distinct. The phyletic continuity implied by this taxonomic lumping is not being contested here; modern North American bison probably did evolve directly from late Pleistocene North American bison. Distinct differences in the morphology and distribution of recent and fossil bison can be demonstrated, however, and each existed in a significantly different environment in which selection regimes favored substantially different phenotypes (see chapter 5). In this classification late Pleistocene-early Holocene short horned and extant bison are considered to be specifically distinct, a decision that gives greater biological meaning to the status of each and allows more freedom to work with the respective taxa as cohesive, functional population units.

Sources of Morphologic Variation

This classification is based on the analysis of qualitative and quantitative variation among bison crania. Three possible sorts of variation were considered in the analyses of these data: (1) continuous variation (one mode of normal variation with populations); (2) hybridization; and (3) genetic drift and its consequences. Continuous variation of characters was assumed for each canalized taxon (Dodson, 1975; Mayr, 1969; Salthe, 1975; Simpson and Roe, 1939; Yablokov, 1974) (see chapter 3). Hybridization was considered a possible source of variation in view of the environmental dynamics of the late Quaternary and the periodic breakdown of geographic isolation of populations. Empirical data—skulls possessing characters of both North American and Eurasian autochthons—corroborate the occurrence of probable hybrid zones and periods (see chapter 6). Several deviant phenotypes, whose abnormality could not be attributed to either hybridization or normal continuous variation, occurred among the skulls studied, suggesting the periodic occurrence of small inbreeding populations in which recessive characters, normally with probably neutral or moderate selective value, were expressed in phenotypes. Extreme cases of prolonged inbreeding may, however, have resulted in the appearance of debilitating characters.

Selection of Taxonomic Characters

Male horn cores have traditionally been considered the most useful and dependable skeletal element with which to classify bison. Only rarely has another skeletal element been given higher taxonomic priority. Notably, Hay (1914) and Figgins (1933) created new taxa on the basis of variations in dentition, but these taxa and dentition based taxonomy were short-lived. Statements are often voiced that postcranial characters can probably be used to classify bison and/or to specifically identify individual or groups of postcranial bones. However, no classification of bison based on postcranial characters, or suggestions of how to attempt such a classification, have been put into writing.

After studying over 800 skulls and 3,100 limb bones, I reached the conclusion that horn cores are taxonomically the most useful and reliable character complex in the bison skeleton. The use of horn cores as the basis for both alpha and beta taxonomy has, however, generated controversy. Guthrie (1966*a*) has argued that (1) horn cores continue to grow throughout the life of the individual, thus annual increments were added to the core; (2) variation in the quality and quantity of diet produced variation in the size, shape, pattern, and rate of horn (and horn core?) growth; (3) genotypic differences can produce variation in the size, shape, pattern and rate of horn (and horn core?) growth; (4) sexual dimorphism would tend to increase between-sex variation; and (5) accidents to horns could influence the size, shape, pattern, and rate of horn development.

Guthrie's criticism of the taxonomic utility of horn cores reflects a point of view in which subtle quantitative differences among individual horn cores are of presumed taxonomic significance. The nature and magnitude of variation with which Guthrie concerns himself is, in my mind, insufficient to transcend the morphological gaps which separate species. Horn cores are important ranking, defensive and offensive organs that function in the determination of relative individual fitness within a deme and within the greater megafaunal community (Estes, 1974; Geist, 1966; Lott, 1974, pers. comm., July 5, 1977; Schaller, 1967; Walther, 1974; pers. observ.). They are therefore subject to natural selection. Furthermore, horn cores do show canalization of diagnostic characters. Lastly, many of Guthrie's assertions are unsupportable or based on unacceptable methodologies. The most important example of this, insofar as this study is concerned, is the claim that significant growth occurs after the horn core has reached maturity. Most writers who have studied the problem of bovid horn core growth believe that little or no growth (i.e., increase in size) occurs after the horn core has reached maturity (Allen, 1876; Fuller, 1959; Grigson, 1974, 1976; Hammond, 1955; Skinner and Kaisen, 1947). My observations agree. Even allowing for some growth after maturity, however, the change would be slight and insignificant insofar as the taxonomic validity of the horn core complex is concerned.

In addition to horn cores, several skeletal complexes were studied, including limb bones, the occipital region, the frontal region, and dentition. Limb bones and dentition were easily rejected as taxonomically useful elements: they both show too little variation. Frontal and occipital complexes show more tendency to cluster into meaningful phena than do limb bones or dentition, but they cluster into substantially less meaningful phena than do horn core complexes. The horn core complex emerged as the most biologically meaningful taxonomic characters when time-space-habitat relations were considered and when allowances were made for the consequences of hybridization and genetic drift.

Each bison species has a unique, differentiating set of horn core characters. Each character complex includes the same considerations of shape, orientation, and size. The horn core of any species can, therefore, be identified using the same procedure. No single criterion is diagnostic, but when the entire complex is considered, specific identification is readily possible. The horn core complex of characters is listed in table 13, and illustrated in figures 11 and 12.

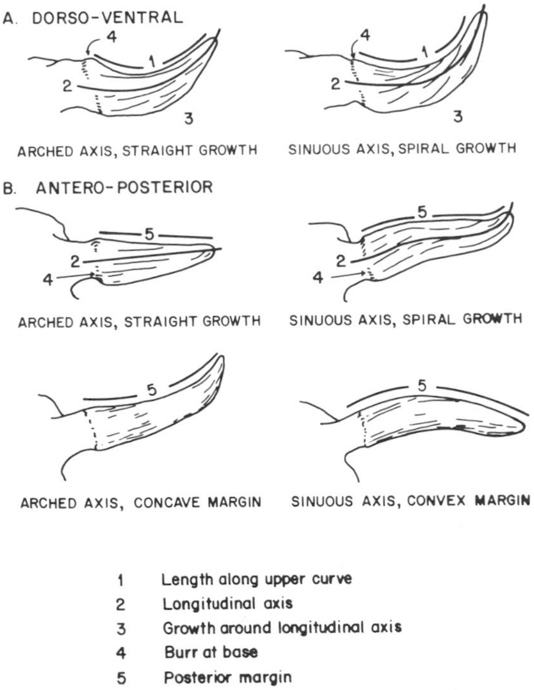

FIGURE 11 Axes, margins, and surfaces of the horn core character complex locating characters used in determining the identity of specimens (see also fig. 12, tables 13, 43).

Table 13

The Horn Core Character Complex

Shape
Cross section of core at base — symmetrical or asymmetrical about the dorso-ventral axis?
Posterior margins — straight, sinuous, concave, or convex?
Cross section of core at tip — cordiform, triangular, circular, or elliptical?
Configuration of longitudinal axis — straight, arched, or sinuous?
Growth pattern around longitudinal axis of core — spiraled or straight?
Orientation
Relation of antero-posterior plane of horn core to plane of frontals — parallel (i.e., not rotated), rotated forward, or rotated backward?
Size
Length of horn core along upper curve, tip to burr.

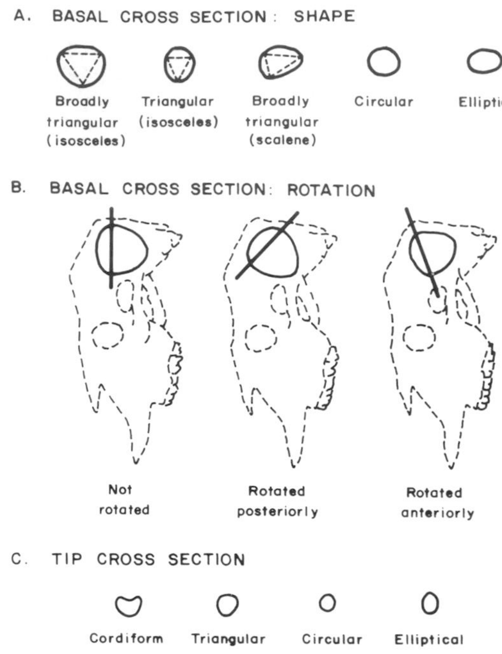

FIGURE 12 Cross section shapes and rotation positions of the horn core character complex describing characters used in determining the identity of specimens (see also fig. 11; tables 13, 43).

Revision Format

In this classification each species and subspecies is introduced by its scientific name and author. This is followed by (1) a bibliographic citation of its original (type) description; (2) information about the origin and current location of the holotype; (3) the principle synonymy; (4) a statement of the content of the taxon; (5) a diagnosis; (6) a differential diagnosis; (7) a biometric summary; (8) a list of specimens referred to the taxon; (9) a description; (10) a review of known spatial and temporal distribution; and (11) a discussion of taxonomic, distributional, and other relevant considerations.

Only the original description of each valid name is cited, except in those cases where preliminary announcement of the name preceded description, or vice versa, in which cases both are cited. The principal synonymy includes all synonyms recognized as available by articles 10–20 of the *International Code of Zoological Nomenclature* (1964 revised edition) and names created for hybrid individuals, following Mayr's (1969) suggestion for their practical handling. The diagnosis lists morphological characters that distinguish the taxon being described, and which can be used to determine the taxonomic identity of a specimen. The differential diagnosis lists morphological characters that differentiate other taxa from the taxon being described. Formal diagnoses and differential diagnoses have not been used before in North American bison systematics. All diagnoses and differential diagnoses presented in this revision are, therefore, original. The description presents more complete, but not necessarily diagnostic, information about the morphological characters of a taxon.

CLASSIFICATION OF THE NORTH AMERICAN BISON

The genus *Bison* is represented in North America by five species. Three of these species appear to have been North American autochthons (i.e., they evolved their specific characteristics in North America) whereas the other two species appear to have been Eurasian autochthons. This classification is accordingly divided into two phyletic groups: a North American lineage and a Eurasian lineage. The North

American lineage is well represented by specimens and has hence been well studied. The Eurasian lineage, however, is well studied only so far as it is represented by specimens found in North America. Only a few specimens from Eurasian localities, now in North American museums, were available for study.

Table 14 lists the world bison taxa recognized in this study. Only those taxa marked with an asterisk were studied in sufficient detail to be systematically included in this revision. The synonymy of many Eurasian taxa implied in this list is based on inferences made from photographs or drawings of types and referred specimens. This method is admittedly inadequate for properly evaluating the respective specimens, but some reorganization of the lengthy Eurasian synonymy was necessary to bring it into working alignment with the North American classification presented below. The taxa listed in table 14 may be taken as my current opinion on world bison classification, subject to revision after the Eurasian bison materials have been studied in detail.

Description of the Taxa

Genus *Bison* (Smith, 1827)

Original Description:

Charles Hamilton Smith. 1827. "Supplement to the Order Ruminantia" in Georges Cuvier, *The Animal Kingdom* (London: Geo. B. Whittaker), vol. 4, pp. 398–404 and vol. 5, pp. 373–374.

Type Species:

Bison bison (Linnaeus, 1758).

Table 14

Bison Taxa Recognized in This Study

North American Autochthons
Bison latifrons (Harlan, 1825)*
Bison antiquus Leidy, 1852*
Bison antiquus antiquus Leidy, 1852*
Bison antiquus occidentalis (Lucas, 1898)*
Bison bison (Linnaeus, 1758)*
Bison bison bison (Linnaeus, 1758)*
Bison bison athabascae Rhoads, 1897*
Eurasian Autochthons
Bison sivalensis Lydekker, 1878
Bison priscus (Bojanus, 1827)*
Bison alaskensis Rhoads, 1897*
Bison schoetensacki Freudenberg, 1910[a]
Bison bonasus (Linnaeus, 1758)
Bison bonasus bonasus (Linnaeus, 1758)
Bison bonasus caucasicus (Turkin and Satunin, 1904)[b]

[a] As cited in Skinner and Kaisen, 1947;
[b] As cited in Bohlken, 1967;
*These taxa are systematically included in this revision.

Type Locality:
Canadian River Valley, eastern New Mexico, United States.

Type Specimen:
None. Description of type species by Francisco Hernandez, 1651, *Rerum Medicarum Novae Hispaniae Thesaurus* (Rome, Vitalis Mascardi), p. 587 and addendum (*Historiae Animalium et Mineralium Novae Hispaniae*), chapter 30, p. 10 (plate 1).

Principal Synonymy:
Bos Linnaeus, 1758, in part
Urus Bojanus, 1827, in part
Harlanus Owen, 1847
Simobison Hay and Cook, 1930
Stelabison Figgins, 1933

Diagnosis:
Horn cores more or less triangular in cross section, sometimes elliptical, rarely circular; extend outward laterally from frontals forward of occipital plane. Parietals situated on or about same plane as frontals. Parietal plane forms obtuse angle with occipital plane. Frontals flat to convex between horn cores; never bossed; quadrate. Occipital broad. Orbits tubular; protrude antero-laterally. Nasals broad; short; triangular. Premaxilla triangular. Nasal process of premaxilla does not reach nasals.

Differential diagnosis (from *Bos*):
Bos horn cores commonly circular to subcircular; extend outward laterally from frontals tangential to occipital plane well back from orbits. Frontals concave to convex; longer than wide; occasionally bossed between horn cores. Parietals reduced; form part of occipital plane. Frontal plane forms acute angle with parieto-occipital plane. Orbits less tubular; directed more forward; protrude less than in bison. Nasals narrow; long; rectangular. Premaxilla rectangular. Nasal process of premaxilla may or may not reach nasals.

Description:
The genus *Bison* is systematically placed in the order Artiodactyla (even-toed ungulates), suborder Pecora (true ruminants), family Bovidae (hollow horned), tribe Bovini (oxenlike). Bison are, therefore, cattlelike, hollow horned ruminants with cloven hooves. They are rather massively built with large but narrow bodies, short thick necks, large heads, and smooth curved horns. All known bison are normally brown in color, ranging from true brown to very dark black-brown. The pelage is longer and woolly over the front half of the body and shorter and smoother over the rear half.

The bison skull can be divided into four topographic regions: the frontals, the horn cores, the occipital, and the face. The frontals are broad, roughly square but somewhat wider than long. The frontals and parietals form most of the upper surface of the bison skull, exclusive of the horn cores; they are generally convex but in some species are nearly flat. The horn cores emanate from the frontals in a lateral or postero-lateral direction, and from a slightly downward (= subhorizontal) direction to a slightly upward (= suprahorizontal) direction. The horn cores extend away from the cranium for most of their length; most curve upward and some curve inward at the distal end. The occipital is broad and low, semi-elliptical in shape, and has a surface that ranges from relatively flat in some species to rather concave and rugose in others. The external occipital protuberance and nuchal line are well developed, especially in larger species. The occipital condyles are prominent; their long axes are normally ca. 180° apart and thus form a straight line. Bison have thirty-two teeth arranged according to the formula

$$I\,\frac{0}{3}\;C\,\frac{0}{1}\;PM\,\frac{3}{3}\;M\,\frac{3}{3}.$$

Molar teeth are hypsodontal and selenodont, with crescentic fossae. Labial ridges and lingual intercusp styles appear on maxillary molars; these are reversed on lower molars. Premolars are less complex than molars; incisors and canines are simple. The face is relatively short and broad, generally triangular in shape. Nasals are broader at the nasal-frontal suture and normally taper gradually toward the muzzle. The maxilla and premaxilla also taper toward the muzzle. The orbits are well developed, are anterolaterally directed and protrude strongly in males, less in females. Empel (1962) and Flerow (1965) have described the bison skull in technical detail.

Bison were the dominant high latitude ruminants of the late Pleistocene of Eurasia and the dominant ruminants of late Pleistocene and Holocene North America. Bison appear to have been most numerous and to have reached their greatest densities in arctic steppe tundra and temperate grasslands, especially in the Holocene grasslands of midlatitude North America. Bison were, however, browsers as well as grazers. Some species appear to have been adapted to woodland or forest environments.

Distribution:

The genus *Bison* is known from throughout most of Europe, Asia, and North America (fig. 13). Bison first appears in the fossil record during the late Pliocene of temperate Asia as a constituent of the middle Villafranchian fauna. During the middle Pleistocene bison occurred from Europe east through northern Asia into Beringia. Bison first appeared in the fossil record of sub-Beringian North America in deposits of Illinoian age. Most bison are known from the late Pleistocene of North America and higher latitude Eurasia, and the Holocene steppes and adjacent forests of Eurasia and North America.

Discussion:

Taxonomy: What is now the genus *Bison* was originally (i.e., 1758) subsumed within Linnaeus's genus *Bos*. Bojanus (1827) created the genus *Urus* to accommodate extinct bison. Charles Hamilton Smith used *Bison* in a subgeneric sense to accommodate the bison group. Charles Knight appears to have been the first to give *Bison* generic rank. Subsequent nineteenth-century American and German writers, especially, accepted the generic rank and concept of *Bison* and thereby effected its establishment in scientific literature and its taxonomic recognition. *Bos* remained in use for some time, however, as a close rival genus in which the bison group was frequently placed (e.g., Blake, 1898*a*, 1898*b*; Cope, 1895; Lydekker, 1912). *Harlanus* was created to accommodate a supposed tapiroid pachyderm from Georgia (Owen, 1847), but the type for this genus was subsequently found to be a partial bison ramus containing three well-worn molars and a fragmentary premolar (Leidy, 1854).

The genus *Simobison* was created to accommodate bison supposedly characterized by extremely short faces, flat horn cores situated well in front of the occiput, flattened crania, orbits situated well forward, and arched rami. The type specimen (CoMNH VP 574) was a very badly crushed and incomplete skull of a young adult male. "The nasal bones were not found and a considerable part of the maxillae in front of the orbits was in a decayed condition; also the greater part of the premaxillae is missing" (Hay and Cook, 1930:24). The horn core tips were also missing.

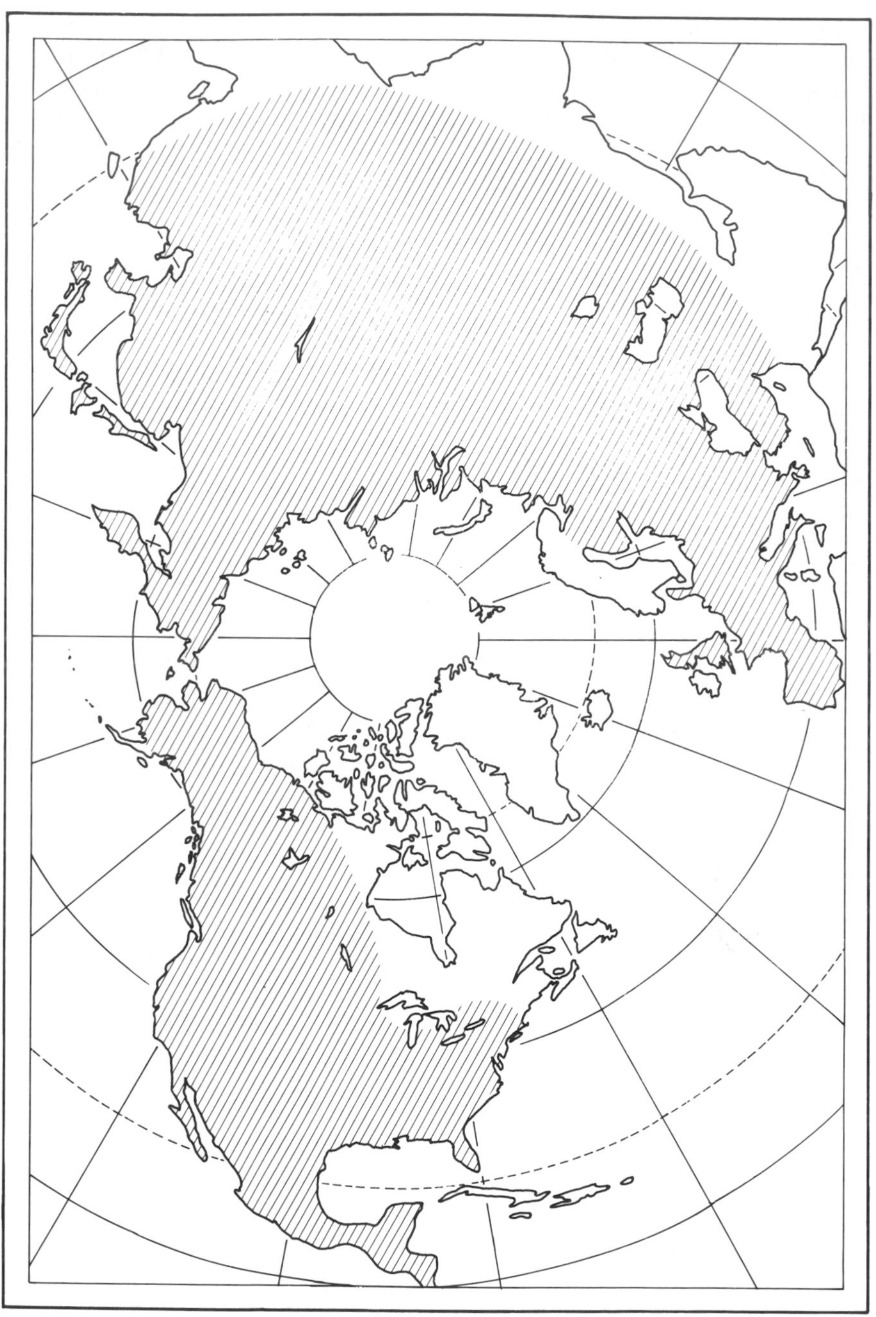

FIGURE 13 World distribution of the genus *Bison* during the Quaternary.

The *Simobison* type specimen was poorly and disproportionately restored. The horn cores are artificially flattened, lowered, lengthened, and misdirected, and the cranium is unnaturally flattened. As Skinner and Kaisen (1947) pointed out, the face of the restored specimen has been fitted with rami from a somewhat younger animal (as confirmed by differences in tooth eruption and wear between the upper and lower dentition). The face was apparently sculptured to fit the jaw of a younger, smaller animal, and this created the disproportionate features used by Hay and Cook to establish the new genus. Even after allowing for distortion by crushing, imaginative restoration and ontogenic age of the specimen, this skull is identifiable as *Bison antiquus*.

Figgins (1933) created the genus *Stelabison* to accommodate "all races" of bison possessing external pillars on molar teeth. "A character so markedly at variance to its nearly total absence in other races of *Bison* makes advisable a new genus" (p. 18). *Stelabison* was described from two isolated maxillary molars found in Texas (CoMNH VP 1363, 1365) and a third molar with partial skull from the Yukon Territory (AMNH VP F:AM13721). Teeth in bison have been considered, with substantial empirical justification, to show little variation among normal populations of bison taxa. Teeth do, however, exhibit certain variations, perhaps resulting from genetic drift, one of which may be the presence of labial styles (see chapter 6). Such a deviant character is not a normal morphological character of a viable, adapted population. Because the character used to establish this genus has no primary taxonomic significance, *Stelabison* is synonymized with *Bison*.

Type species and type locality: Linnaeus (1758) recognized both the American bison (as *Bos bison*) and the European bison (as *Bos bonasus*) in the tenth edition of *Systema Naturae*. Additionally, both had first been listed in Linnaeus's sixth edition of *Systema Naturae* (Thomas, 1911). The type description, however, for *B. bison* (Hernandez, 1651) antedates the type description for *B. bonasus* (Ray, 1693), so *B. bison* has priority as the type species for the genus.

Hernandez apparently had not actually seen free living American bison before writing his description. He did, however, implicitly base his description of American bison on descriptions of the "cattle of the Quivira region" provided by members of the Francisco Vazquez de Coronado expedition, even though several other eyewitness accounts of bison were available to him (Brand, 1967; Reed, 1952, 1955). Evidence and reason direct that the Coronado expedition accounts should be recognized as the source for Hernandez's description of *B. bison* (plate 1).

Thomas (1911) identified "Mexico" as the type locality of *B. bison*, recognizing that sixteenth-century "Mexico" (= New Spain) included much of what is now the southwestern United States. After reviewing several Spanish colonial accounts of bison, Reed (1952, 1955) concluded that Europeans first saw bison in the southern or southeastern United States, and he identified this region as the type locality of the species. First sighting, however, does not in itself convey type locality status for a subsequently named

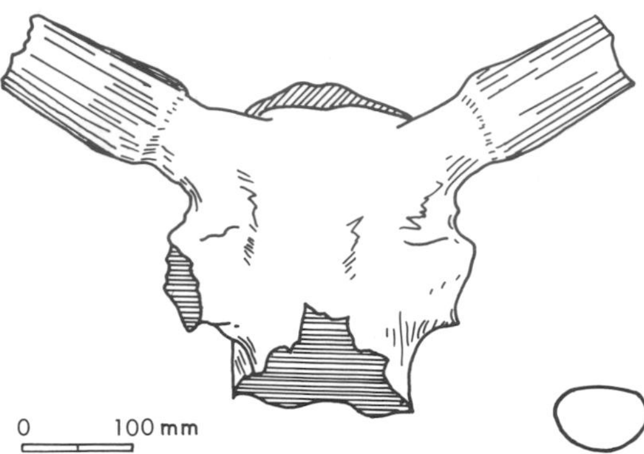

FIGURE 14 The holotype of *Bison sivalensis* has formerly been considered to be a female skull, but the large size of the horn cores relative to the frontals, the distinct dorsal burr, the dorsal groove (described by Lydekker), and the compressed isosceles triangle cross section shape of the base all suggest that this is a male skull (after the original engraving in Lydekker, 1878).

and described taxon. Hershkovitz (1957) recognized this flaw in Reed's reasoning and designated central Kansas as "the precise type locality" for *B. bison*, properly equating central Kansas with the Quivira of the Coronado expedition.

Quivira, however, cannot be accepted as the *B. bison* type locality. The main force of the Coronado expedition first saw bison in the Canadian River Valley of eastern New Mexico, and encountered them regularly thereafter until Coronado divided his force in what is probably now the eastern Texas panhandle. The main force stayed south of the Canadian River, and the advance force proceeded north toward Quivira. The advance force saw bison until it was within three or four days of the westernmost village of Quivira (i.e., the first village of Quivira they reached). Coronado's force did not see, or at least did not report seeing, bison in or near Quivira, even though this force traveled throughout the region for nearly a month (Bolton, 1949; Hodge and Lewis, 1907; Winship, 1896). Considering the Coronado expedition's experience with bison, and accepting the reports of members of this expedition as the source of Hernandez's information on *B. bison*, the type locality is here redesignated as the Canadian River Valley of eastern New Mexico.

Early Eurasian Bison: *Bison* possibly evolved from *Leptobos* in temperate eastern Eurasia during the late Pliocene, although current research on late Tertiary bovids from China has produced an as yet undescribed taxon that might be an even more likely ancestor of *Bison* (Guthrie, oral comm., September 2, 1978). The earliest and most primitive species in the genus, *B. sivalensis*, is known from two specimens: the holotype, from Upper Siwalik deposits near Pinjor, India (Lydekker, 1878; Skinner and Kaisen, 1947), and Skull B of the Nihowan fauna (Teilhard de Chardin & Piveteau, 1930) from about 150 km west of Peking, China (figs. 14, 15). The latter has been the

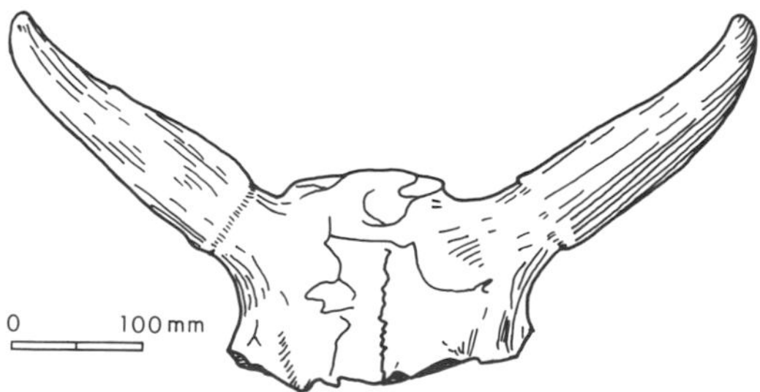

FIGURE 15 Skull B of the Nihowan, China, bison series was originally identified as a female *Bison paleosinensis* but this specimen is here referred to *Bison sivalensis* and is tentatively identified as a male skull (after a photograph in Teilhard de Chardin and Piveteau, 1930).

syntype for *B. paleosinensis*, but it is here referred to *B. sivalensis*. Both the Siwalik and Nihowan specimens have previously been considered female skulls (Skinner and Kaisen, 1947; Teilhard de Chardin and Piveteau, 1930), but both are here identified as male skulls on the basis of several horn core characters, including the proportionately larger core-to-skull ratio, the presence of a dorsal burr, the shape of the basal core in cross section, and, in the Nihowan specimen, the distal groove and posterior deflection of the tip and large size vis-à-vis Skull B. The horn cores of *B. sivalensis* differ little in size from those of *Leptobos* (table 15). The *B. sivalensis* horn cores are posteriorly placed and oriented and the frontal-parietal surface is narrower and more rounded, characters more suggestive of *Leptobos* than of later *Bison*.

The occurrence of *B. sivalensis* in roughly contemporaneous (i.e., middle or late Villafranchian) early Pleistocene deposits of southern and eastern Asia suggests that this species was reasonably well canalized and widely dispersed at that time. Other occurrences of *Bison* are reported from additional middle and late Villafranchian faunas of midlatitude Eurasia, particularly eastern Asia, which indicates that the genus, at least, was differentiated and

Table 15

Comparative Biometrics: Leptobos and Early Eurasian Bison

Standard measurement	*Leptobos*[a]	*Bison sivalensis*[b]	*Bison sivalensis*[c]	*Bison priscus*[d]
Spread of horn cores, tip to tip	....	....	(545 mm)	(536 mm)
Horn core length, upper curve, tip to burr	....	....	235 mm	300 mm
Straight line distance, tip to burr, dorsal horn core	....	....	220 mm	200 mm
Dorso-ventral diameter, horn core base	74 mm	66 mm	....	(76 mm)
Minimum circumference, horn core base	245 mm	254 mm	....	(245 mm)
Width of occipital at auditory openings	....	(194 mm)	210 mm	230 mm
Width of occipital condyles	....	99 mm	100 mm	102 mm
Depth, nuchal line to dorsal margin of foramen magnum	....	(86 mm)	....	91 mm
Antero-posterior diameter, horn core base	80 mm	86 mm	(70 mm)	(80 mm)
Least width of frontals, between horn cores and orbits	....	221 mm	220 mm	234 mm
Greatest width of frontals at orbits	....	(242 mm)	....	....
Distance, nuchal line to nasal-frontal suture	....	(169 mm)	(165 mm)	(175 mm)

Sources: Skinner and Kaisen, 1947; Teilhard de Chardin and Piveteau, 1930.
Note: Parenthetical data are estimates by Skinner and Kaisen, 1947.

[a] YPM VP 24541, apparently from Upper Siwalik deposits, India;
[b] *B. sivalensis* holotype, from Upper Siwalik deposits near Pinjor, India;
[c] *B. sivalensis* (ref.), formerly syntype for *B. paleosinensis*, from Nihowan fauna about 150 km west of Peking, China;
[d] *B. priscus* (ref.), formerly lectotype for *B. paleosinensis*, from Nihowan fauna about 150 km west of Peking, China.

widely dispersed at the time (table 16). *B. sivalensis* probably represents an early, and possibly middle, Pleistocene forest/woodland bison adapted to seral openings and edges where both grazing and browsing were possible. This possible adaptive mode would have made possible the widespread distribution of relatively small overall numbers of *B. sivalensis* throughout forested or wooded eastern Asia. Such a distribution pattern would have resulted in *B. sivalensis* being positioned geographically to give rise to both the Eurasian and the North American lineages of later Pleistocene *Bison*.

B. paleosinensis has formerly been considered a later Pliocene-early Pleistocene species, but it is here considered later Villafranchian in age (Aigner, 1972; Kurten, 1968) and synonymized with *B. priscus*. The holotype

Table 16

Early Eurasian Bison Localities

Locality	Bison reported[a]	Assigned age	Source
Kuruksay, Tadzhik Depression, USSR	*Bison* sp.	1.95 – 1.79 MYA	Ranov and Davis, 1979
Yushe, Shansi, China	*B. priscus* = *B. paleosinensis*	Lower Pleistocene	Aigner, 1972
Nihowan, Sankanho Valley, China	*B. priscus* = *B. paleosinensis*	Lower Pleistocene	Aigner, 1972
Lantian 64098, Shensi, China	*B. priscus* = *B. paleosinensis* and *Bison* sp.	Lower Pleistocene	Aigner, 1972
Ch'angchiap'o, Shensi, China	*B. priscus* = *B. paleosinensis* and *Bison* sp.	Lower Pleistocene	Aigner, 1972
Tangshan, Hopei, China	*B. priscus* = *B. paleosinensis*	Lower Pleistocene	Aigner, 1972
Huituipo, Shensi, China	*B. priscus* = *B. paleosinensis*	Lower Pleistocene	Aigner, 1972
Lingyi, Shansi, China	*Bison* sp.	Lower Pleistocene	
Upper Siwaliks, India (Pinjor phase)	*B. sivalensis*	Pre-Mindel, possibly post-Tiglian	Aigner, 1972; Skinner and Kaisen, 1947
Episcopia, Europe	*B. schoetensacki*	Waalian	Kurten, 1968
Gombasek, Europe	*B. schoetensacki*	Gunz II	Kurten, 1968
Hundsheim, Europe	*B. priscus*	Gunz II	Kurten, 1968
Süssenborn, Europe	*B. priscus*	Gunz II	Kurten, 1968
K'oho, Shansi, China	*Bison* sp.	Middle Pleistocene	Aigner, 1972
Choukoutien I, Peiching, China	*Bison* sp.	Middle Pleistocene	Aigner, 1972
Mauer, Europe	*B. schoetensacki*	Cromerian	Kurten, 1968
Mosbach I, Europe	*B. priscus*	Mindel	Kurten, 1968
Koneprusy, Europe	*B. schoetensacki*	Mindel	Kurten, 1968
Mosbach II, Europe	*B. schoetensacki*	Mindel	Kurten, 1968
Martinet Quarry, Europe	*Bison* sp.	Middle Pleistocene (probably Mindel)	Delpech *et al*, 1978
Ubeidiya, Jordan Valley, Israel	*Bison* sp.	Mindel	Horowitz, 1975
Tadzhik Depression, USSR	*Bison* sp.	Mindel	Ranov and Davis, 1979

[a]The species presented here are those reported in the sources. The criteria for species identification are variable and not entirely clear in all cases. A standardized, systematic evaluation of all specimens that are specifically identified here would probably result in several changes.

(formerly lectotype) of *B. paleosinensis* is only slightly larger than *Leptobos* and *B. sivalensis* (table 15), but it does show significant morphological differences that set it apart from *B. sivalensis*: the broader and less-rounded parieto-frontal region, forward placement of the horn cores, and possession of the horn core character complex diagnostic of middle and late Pleistocene Eurasian bison. The holotype of *B. paleosinensis* probably represents the early stage of *B. priscus* evolution (figs. 16, 17). The cooccurrence of two species of bison in the Nihowan fauna is explained by the fact that this fauna probably represents an unusually long period of accumulation (Aigner, 1972).

In summary, *B. sivalensis* is considered the primitive species of bison which evolved from *Leptobos* or a closely related Bovini genus during the late Pliocene. This species was probably adapted to forest or woodland openings or edges, was apparently widely dispersed throughout southern and eastern Asia during the early Pleistocene, and probably gave rise to the later lineages of both Eurasian and North American bison.

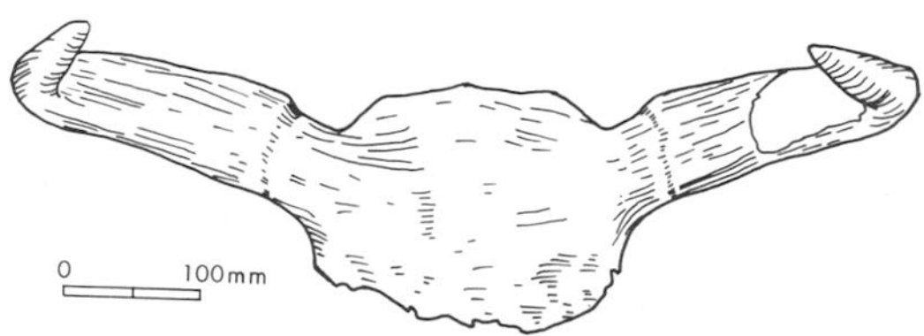

FIGURE 16 Skull A of the Nihowan, China, bison series was made the lectotype for *Bison paleosinensis* by Skinner and Kaisen (1947), but this specimen is here referred to *Bison priscus* (after a photograph in Teilhard de Chardin and Piveteau, 1930).

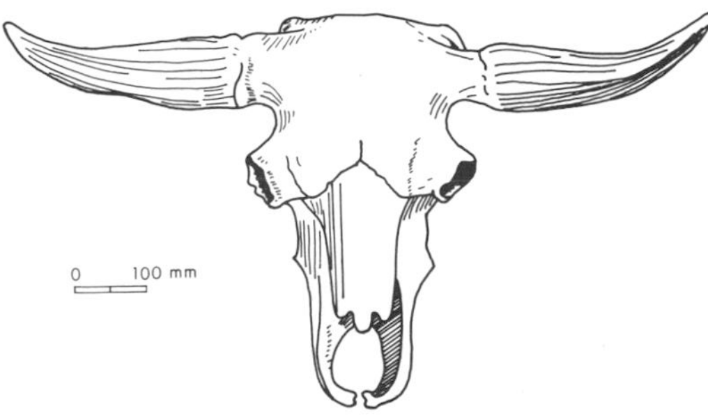

FIGURE 17 The figured type of *Bison priscus*, redrawn from Cuvier (1825).

Autochthonous North American Species

I. *Bison latifrons* (Harlan, 1825)

Original Description:

Rembrandt Peale, 1803*b*. "Account of Some Remains of a Species of Gigantic Oxen Found in America and Other Parts of the World." *Philosophical Magazine,* vol. 15, pp. 325–327. (Description).

Richard Harlan. 1825. *Fauna Americana: Being a Description of the Mammiferous Animals Inhabiting North America* (Philadelphia: Anthony Finley), p. 273. (Name applied).

Type Locality:

Creek bed 12–14 miles north of Big Bone Lick, Kentucky.

Type Location:

Department of Geology, Academy of Natural Sciences of Philadelphia, Philadelphia, Pennsylvania.

Type Specimen:

ANSP G 12993—partial male skull consisting of occipital, partial frontal, and proximal right horn core (plate 2).

Principal Synonymy:

Bison ferox Marsh, 1877
Bison alleni Marsh, 1877
Bos crampianus Cope, 1895
Bos arizonica Blake, 1898*a*
Bison regius Hay, 1914

Bison angularis Figgins, 1933
Bison rotundus Figgins, 1933

Species Content:

B. latifrons was the late-middle (?) and late Pleistocene forest/woodland bison of North America.

Diagnosis:

Male horn cores broadly triangular (isosceles) in cross section and nearly symmetrical about dorso-ventral axis at base; tip cordiform to triangular in cross section; posterior margin straight; growth mildly spiraled around arched longitudinal axis; antero-posterior plane nearly parallel with frontal plane; length along upper curve greater than 500 mm (plates 3, 4).

Female horn cores circular to elliptical in cross section; tip circular to elliptical in cross section; posterior margin straight; growth straight to mildly spiraled around slightly arched longitudinal axis; plane of greatest diameter rotated forward from vertical; slight dorsal burr at base; length along upper curve greater than 500 mm (plate 5).

Differential Diagnosis:

Male *B. antiquus* and *B. bison* horn cores triangular to circular in cross section at base; length less than 500 mm on upper curve. *B. a. occidentalis* horn core tips circular to elliptical in cross section; longitudinal axis spiraled. *B. alaskensis* and *B. priscus* horn cores asymmetrical at base about dorso-ventral axis; posterior margin sinuous; longitudinal axis spiraled; antero-posterior plane usually rotated forward from frontal plane; *B. priscus* horn core less than 500 mm along upper curve. Female horn cores of all other taxa less than 500 mm along upper curve. *B. alaskensis* and *B. priscus* horn cores normally sinuous or convex along posterior edge.

Biometric Summary:

Skulls, table 17
Limb bones, table 18

Referred Specimens:

Table 19

Description:

B. latifrons was the largest species of bison ever to live. The horn cores of male *B. latifrons* are longer and proportionately more slender than in any other species of North American bison. The horn cores emanate horizontally from the frontals at an angle of about 70° to 80° forward from the sagittal plane. The horn cores are broadly triangular (isosceles) in cross section at the base and taper gradually toward the tip. The entire core curves gradually upward; the distal curve is neither recurved nor posteriorly deflected. A prominent dorsal groove usually extends over the distal one third of the core. The tip of the core is normally above the frontal plane and behind the occipital plane. The sheath was apparently recurved, but not posteriorly deflected (plate 6). The frontals are relatively flat. The orbits are high, directed forward, and protrude less than in any other species of North American bison. The nasals are relatively high and narrow, especially above the maxilla-nasal suture, which gives the nasals a more rectangular shape than is found in other species. The occipital is more concave than in other bison, a result of both the prominent nuchal line/external occipital protuberance and the exaggerated protrusion of the occipital condyles. The articular surface of the occipital condyles is more frequently and strongly flanged along the ventro-lateral margin than in any other species.

Female horn cores are long and relatively slender. They emanate from the frontals at a lesser angle forward from the sagittal plane than do the horn cores of males. The horn cores are circular to elliptical in cross section at the base, and they taper gradually toward the tip. The core has some, but very little, upward curvature. A prominent dorsal burr appears at the base of the horn core—a characteristic unique among females to

B. latifrons. A dorsal groove may or may not be present on the distal core. The tip of the core is above the frontal plane and behind the occipital plane.

The frontals are mildly crenulated to concave transversely, usually being somewhat convex at the horn core base and the sagittal suture line, but concave between these regions. The orbits are relatively high and directed forward, but do not protrude strongly. The nasals are relatively high and rectangular. The occiput is more concave than in the females of the other species because of the relatively prominent nuchal line/external occipital protuberance. The occipital condyles do not protrude as strongly in the females as in the males. The ventro-lateral margins of the occipital condyles are not usually flanged.

B. latifrons is the least well known of all species of North American bison. Females are especially poorly known; no complete skull of a female exists. The description of the species, therefore, and of the female sex especially, is the most in need of supplementary information and most susceptible to modification of all North American bison species descriptions.

Distribution:

B. latifrons is known from throughout much of the United States, particularly from the Great Plains, the Great Basin, coastal California, and Florida (fig. 18). No *B. latifrons* is known from Canada, and the species is rare in

Table 17

Summary of *Bison latifrons* Skull Biometrics

Standard measurement	n	Range	Mean ± SE	σ	V
Males					
Spread of horn cores, tip to tip	19	1445 – 2235 mm	1789.1 ± 48.0 mm	209.1 mm	11.7
Horn core length, upper curve, tip to burr	25	551 – 1090 mm	876.0 ± 28.3 mm	141.4 mm	16.1
Straight line distance, tip to burr, dorsal horn core	20	529 – 979 mm	805.4 ± 26.2 mm	117.3 mm	14.6
Dorso-ventral diameter, horn core base	34	107 – 178 mm	144.9 ± 2.8 mm	16.6 mm	11.4
Minimum circumference, horn core base	36	408 – 669 mm	489.2 ± 8.7 mm	52.4 mm	10.7
Width of occipital at auditory openings	14	287 – 343 mm	322.9 ± 4.6 mm	17.3 mm	5.3
Width of occipital condyles	17	140 – 179 mm	159.5 ± 2.5 mm	10.1 mm	6.4
Depth, nuchal line to dorsal margin of foramen magnum	12	109 – 141 mm	125.5 ± 3.5 mm	12.2 mm	9.7
Antero-posterior diameter, horn core base	37	137 – 226 mm	164.7 ± 3.1 mm	18.6 mm	11.3
Least width of frontals, between horn cores and orbits	19	299 – 406 mm	355.1 ± 7.4 mm	32.2 mm	9.1
Greatest width of frontals at orbits	13	352 – 444 mm	407.8 ± 7.5 mm	27.1 mm	6.6
M^1–M^3, inclusive alveolar length	5	100 – 110 mm	103.9 ± 1.9 mm	4.2 mm	4.0
M^3, maximum width, anterior cusp	4	29.4 – 34.7 mm	31.5 ± 1.1 mm	2.3 mm	7.2
Distance, nuchal line to tip of premaxillae	2	681 – 751 mm	716.0 ± 35.0 mm	49.5 mm	6.9
Distance, nuchal line to nasal-frontal suture	10	295 – 356 mm	326.7 ± 6.5 mm	20.7 mm	6.3
Angle of divergence of horn cores, forward from sagittal	16	62° – 84°	77.3° ± 1.3°	5.4°	7.0
Angle between foramen magnum and occipital planes	8	118° – 142°	128.5° ± 3.2°	9.1°	7.1
Angle between foramen magnum and basioccipital planes	9	118° – 145°	127.4° ± 2.7°	8.2°	6.4
Females					
Spread of horn cores, tip to tip	1	....	1238.0 mm	....	....
Horn core length, upper curve, tip to burr	3	519 – 653 mm	564.0 ± 44.5 mm	77.0 mm	13.7
Straight line distance, tip to burr, dorsal horn core	2	506 – 620 mm	563.0 ± 57.0 mm	80.6 mm	14.3
Dorso-ventral diameter, horn core base	9	89 – 119 mm	105.9 ± 3.0 mm	9.1 mm	8.6
Minimum circumference, horn core base	9	287 – 371 mm	324.9 ± 8.8 mm	26.4 mm	8.1
Width of occipital at auditory openings	3	235 – 281 mm	260.3 ± 13.5 mm	23.4 mm	9.0
Width of occipital condyles	3	135 – 155 mm	146.7 ± 6.0 mm	10.4 mm	7.1
Depth, nuchal line to dorsal margin of foramen magnum	3	95 – 115 mm	107.0 ± 6.1 mm	10.6 mm	9.9
Antero-posterior diameter, horn core base	9	92 – 119 mm	103.0 ± 3.2 mm	9.7 mm	9.4
Least width of frontals, between horn cores and orbits	5	276 – 306 mm	285.2 ± 5.4 mm	12.2 mm	4.3
Greatest width of frontals at orbits	2	327 – 335 mm	331.0 ± 4.0 mm	5.7 mm	1.7
M^1–M^3, inclusive alveolar length	1	....	106.0 mm	....	....
M^3, maximum width, anterior cusp	1	....	29.4 mm	....	....
Distance, nuchal line to tip of premaxillae	0	....	....	....	....
Distance, nuchal line to nasal-frontal suture	1	....	257.0 mm	....	....
Angle of divergence of horn cores, forward from sagittal	3	....	78.0°	....	....
Angle between foramen magnum and occipital planes	1	....	124.0°	....	....
Angle between foramen magnum and basioccipital planes	1	....	119.0°	....	....

Table 18

Summary of *Bison latifrons* Limb Biometrics

Standard measurement	n	Range	Mean ± SE	σ	V
Males					
Humerus:					
Approximate rotational length of bone	7	379 – 418 mm	401.7 ± 5.7 mm	15.0 mm	3.7
Antero-posterior diameter of diaphysis	14	73 – 86 mm	79.8 ± 1.0 mm	3.9 mm	4.9
Transverse minimum of diaphysis	14	60 – 72 mm	66.9 ± 1.0 mm	3.7 mm	5.5
Radius:					
Approximate rotational length of bone	9	362 – 393 mm	383.2 ± 3.3 mm	9.8 mm	2.6
Antero-posterior minimum of diaphysis	11	41 – 49 mm	45.5 ± 0.8 mm	2.7 mm	5.9
Transverse minimum of diaphysis	11	65 – 82 mm	75.6 ± 1.3 mm	4.3 mm	5.7
Metacarpal:					
Total length of bone	16	234 – 264 mm	248.6 ± 2.1 mm	8.3 mm	3.3
Antero-posterior minimum of diaphysis	19	33 – 40 mm	36.3 ± 0.5 mm	2.1 mm	5.9
Transverse minimum of diaphysis	18	54 – 74 mm	65.7 ± 1.2 mm	4.9 mm	7.5
Femur:					
Approximate rotational length of bone	3	479 – 483 mm	481.7 ± 1.3 mm	2.3 mm	0.5
Antero-posterior diameter of diaphysis	5	61 – 70 mm	65.8 ± 1.5 mm	3.4 mm	5.2
Transverse minimum of diaphysis	5	61 – 65 mm	63.0 ± 0.8 mm	1.9 mm	3.0
Tibia:					
Approximate rotational length of bone	9	441 – 479 mm	485.1 ± 3.8 mm	11.3 mm	2.4
Antero-posterior minimum of diaphysis	13	47 – 54 mm	49.6 ± 0.6 mm	2.3 mm	4.5
Transverse minimum of diaphysis	13	65 – 75 mm	68.4 ± 1.0 mm	3.4 mm	5.0
Metatarsal:					
Total length of bone	19	286 – 314 mm	303.6 ± 1.6 mm	6.8 mm	2.2
Antero-posterior minimum of diaphysis	19	37 – 44 mm	40.9 ± 0.4 mm	1.7 mm	4.1
Transverse minimum of diaphysis	19	48 – 58 mm	52.6 ± 0.7 mm	3.0 mm	5.7
Females					
Humerus:					
Approximate rotational length of bone	1	....	345.0 mm	....	...
Antero-posterior diameter of diaphysis	5	66 – 75 mm	72.0 ± 1.5 mm	3.5 mm	4.8
Transverse minimum of diaphysis	5	55 – 62 mm	57.2 ± 1.3 mm	2.9 mm	5.2
Radius:					
Approximate rotational length of bone	5	344 – 366 mm	359.2 ± 4.0 mm	8.9 mm	2.5
Antero-posterior minimum of diaphysis	10	35 – 38 mm	36.3 ± 0.3 mm	0.9 mm	2.6
Transverse minimum of diaphysis	8	55 – 67 mm	60.3 ± 1.6 mm	4.6 mm	7.6
Metacarpal:					
Total length of bone	9	238 – 259 mm	248.1 ± 2.1 mm	6.3 mm	2.5
Antero-posterior minimum of diaphysis	11	30 – 34 mm	32.2 ± 0.4 mm	1.3 mm	3.9
Transverse minimum of diaphysis	11	49 – 59 mm	52.6 ± 0.9 mm	3.0 mm	5.7
Femur:					
Approximate rotational length of bone	3	415 – 446 mm	428.7 ± 9.1 mm	15.8 mm	3.7
Antero-posterior diameter of diaphysis	4	56 – 58 mm	57.3 ± 0.5 mm	1.0 mm	1.7
Transverse minimum of diaphysis	4	51 – 56 mm	53.5 ± 1.0 mm	2.1 mm	3.9
Tibia:					
Approximate rotational length of bone	2	409 – 410 mm	409.5 ± 0.5 mm	0.7 mm	0.2
Antero-posterior minimum of diaphysis	3	41 – 42 mm	41.3 ± 0.3 mm	0.6 mm	1.4
Transverse minimum of diaphysis	3	....	56.0 mm	....	...
Metatarsal:					
Total length of bone	3	280 – 289 mm	285.3 ± 2.7 mm	4.7 mm	1.7
Antero-posterior minimum of diaphysis	4	35 – 38 mm	36.5 ± 0.6 mm	1.3 mm	3.5
Transverse minimum of diaphysis	4	39 – 41 mm	39.8 ± 0.5 mm	1.0 mm	2.4

Note: This table includes limb bones both known and assumed to represent *B. latifrons*. Four articulated skeletons of mature *B. latifrons* have yielded limb bones with the skull, which is necessary for positive identification (table 45). Most limb bones in the above sample are, however, isolated specimens found on the American Falls Reservoir (Idaho) beaches. Both *B. latifrons* and *B. alaskensis* skulls have been found on these beaches, with those of *B. latifrons* being much more common (cf. tables 19 and 41). The majority of these limb bones are, therefore, assumed to represent *B. latifrons*, although some *B. alaskensis* limb bones are possibly included. Stevens (1978) has described the American Falls sample of limb bones.

Table 19

Bison latifrons: Referred Specimens

Institution and identification number			Sex	Provenience
AMNH	VP	6840	♂	Brush Creek, Brown County, Ohio
AMNH	VP	14346	♂	Hoxie, Sheridan County, Kansas
AMNH	VP	20074	♂	Near Independence School, along Missouri River, North Dakota
AMNH	VP	26828	♂	Bradenton Canal, Manatee County, Florida
AMNH	VP	A*	♂	Near Raton, Colfax County, New Mexico
ANSP	G	3	♂	Near Wellington, Sumner County, Kansas
ANSP	G	12993	♂	12-14 miles north of Big Bone Lick, Boone County, Kentucky
BM		20706[a]	♂	Near San Felipe, Austin County, Texas
CoMNH	VP	1164	♂	Near Sutton, Clay County, Nebraska
CoMNH	VP	1187	♂	Near Dorchester, Saline County, Nebraska
CoMNH	VP	1208	♂	Near Nada, Texas
CoMNH	VP	1364	♂	Near Waco, McLennan County, Texas
CoMNH	VP	1603	♂	Hardesty, Texas County, Oklahoma
FMNH	P	14636	♂	Gage, Ellis County, Oklahoma
IMNH	VP	27[b]	♂	American Falls Reservoir, southeast Idaho
IMNH	VP	1392	♂	American Falls Reservoir, southeast Idaho
IMNH	VP	1924	♂	American Falls Reservoir, southeast Idaho
IMNH	VP	2225	♂	American Falls Reservoir, southeast Idaho
IMNH	VP	2252	♂	American Falls Reservoir, southeast Idaho
IMNH	VP	2662	♂	American Falls Reservoir, southeast Idaho
IMNH	VP	6377-258	♀	American Falls Reservoir, southeast Idaho
IMNH	VP	6377-259	♂	American Falls Reservoir, southeast Idaho
IMNH	VP	6377-286	♂	American Falls Reservoir, southeast Idaho
IMNH	VP	6377-741	♂	American Falls Reservoir, southeast Idaho
IMNH	VP	15218	♀	American Falls Reservoir, southeast Idaho
IMNH	VP	16710	♂	American Falls Reservoir, southeast Idaho
IMNH	VP	17212	♂	American Falls Reservoir, southeast Idaho
IMNH	VP	17403	♂	American Falls Reservoir, southeast Idaho
IMNH	VP	17536	♀	American Falls Reservoir, southeast Idaho
IMNH	VP	24566	♂	American Falls Reservoir, southeast Idaho
IMNH	VP	25494	♂	Bishop Company Gravel Pit, Rupert, Minidoka County, Idaho
IMNH	VP	26594	♀	American Falls Reservoir, southeast Idaho
IMNH	VP	26602[b]	♂	American Falls Reservoir, southeast Idaho
IMNH	VP	26604	♂	American Falls Reservoir, southeast Idaho
IMNH	VP	A*	♂	American Falls Reservoir, southeast Idaho
IMNH	VP	B*	♂	American Falls Reservoir, southeast Idaho
IMNH	VP	C*	♂	American Falls Reservoir, southeast Idaho
IMNH	VP	D*	♂	American Falls Reservoir, southeast Idaho
IMNH	VP	E*	♀	American Falls Reservoir, southeast Idaho
IMNH	VP	F*	♀	American Falls Reservoir, southeast Idaho (?)
ISM		300JS77[c]	♂	Jones Spring, Hickory County, Missouri
KU	VP	201	♂	25 miles southeast of Coldwater, Comanche County, Kansas
KU	VP	12170	♂	Hoxie, Sheridan County, Kansas
LACM	VP	15223	♂	Costeau Pit, Orange County, California
LACM	VP	15273	♂	Costeau Pit, Orange County, California
LACM	VP	15362	♂	Costeau Pit, Orange County, California
LACM	VP	15461	♂	Costeau Pit, Orange County, California
LACM	VP	16060	♂	Costeau Pit, Orange County, California
LACM	VP	19055	♀	Costeau Pit, Orange County, California

Table 19

Bison latifrons: Referred Specimens (Continued)

Institution and identification number			Sex	Provenience
LACM	VP	HC6013	♂	Rancho la Brea, Los Angeles County, California
MRG		A*	♂	Lago de Chapala, Jalisco
MU		A[d]*	♂	Near Munster, Cooke County, Texas
PPHM		2315-1	♂	Waters Ranch, near Lipscomb, Lipscomb County, Texas
SDSM		5889[e]	♂	Near Midland, Haakon County, South Dakota
SM		A*	♂	Forney Dam, Collin County, Texas
TMM		2228	♂	Texas (?)
TMM		2561	♂	Texas (?)
TMM		2563	♂	Eastland Pit, McLennan County, Texas
TMM		A*	♂	Texas (?)
UCMP		4067	♂	MacArthur, Shasta Couty, California
UCMP		39220	♂	San Miguel Hill, Contra Costa County, California
UCMP		42157	♂	Near Susanville, Lassen County, California
UCMP		77458	♂	Lake Heinz, Monterey County, California
UCMP		A*	♂	Secret Valley, Lassen County, California
UF	VP	2263	♂	Bradenton 51st Street, Manatee County, Florida
UF	VP	2264	♂	Bradenton 51st Street, Manatee County, Florida
UF	VP	2622	♂	Bradenton 51st Street, Manatee County, Florida
UF	VP	3558	♂	Bradenton 51st Street, Manatee County, Florida
UF	VP	7559	♂	Haile Pit VIII-A, Alachua County, Florida
UF	VP	16600	♂	Waccasasso River, Levy County, Florida
UF	VP	16681[f]	♂	North Havanna Road, Sarasota County, Florida
UMMP		29560	♂	Jinglebob Pasture, XI Ranch, Meade County, Kansas
UMMP		33510	♂	Iowa Park, Wichita County, Texas
UNSM	VP	1-12-12-41[ag]	♂	15 miles west of Stockton, Rooks County, Kansas
UNSM	VP	1115	♂	Kansas
UNSM	VP	30324	♂	Near Giltner, Hamilton County, Nebraska
UNSM	VP	30359	♀	Near Trenton, Hitchcock County, Nebraska
UNSM	VP	30361	♂	Geist Gravel Pit, near Scott City, Scott County, Kansas
UNSM	VP	30956	♂	Pickle Farm Pit, Lancaster County, Nebraska
UNSM	VP	55181[g]	♂	Near Naponee. Franklin County, Nebraska
UNSM	VP	60071	♂	Nebraska (?)
USGS		D901-4, P327	♀	Near Acequia, Minidoka County, Idaho
USGS		D901-A*	♂	Near Acequia, Minidoka County, Idaho
USNM	P	1171	♂	Withlacoochie River, Marion County, Florida
USNM	P	5318	♂	Minidoka, Minidoka County, Idaho
USNM	P	25912	♂	Near Canon City, Fremont County, Colorado
UU	VP	7015[h]	♂	Near Park City, Summit County, Utah
WTSU		A[i]*	♀	Lake Meredith, Potter County, Texas
YPM	VP	10910	♂	Niobrara River, Nebraska
YPM	VP	11911	♂	Blue River, Riley County, Kansas
Lost Specimen		A[j]	♂ (?)	Near Beeville, Bee County, Texas
Lost Specimen		B[k]	♂ (?)	Near Greaterville, Pima County, Arizona

Sources of supplementary data:

[a]Skinner and Kaisen, 1947;
[b]John A. White, field notes, August 9, 1975;
[c]Jeffrey J. Saunders, written comm., May 7, 1979;
[d]Dalquest, 1961*a*;
[e]Green and Martin, 1960;
[f]Robertson, 1974;
[g]Schultz and Hillerud, 1977;
[h]Miller, 1976;
[i]Anderson, 1977;
[j]Hay, 1924;
[k]Blake, 1898*a*.
*No identification number was found for these specimens.

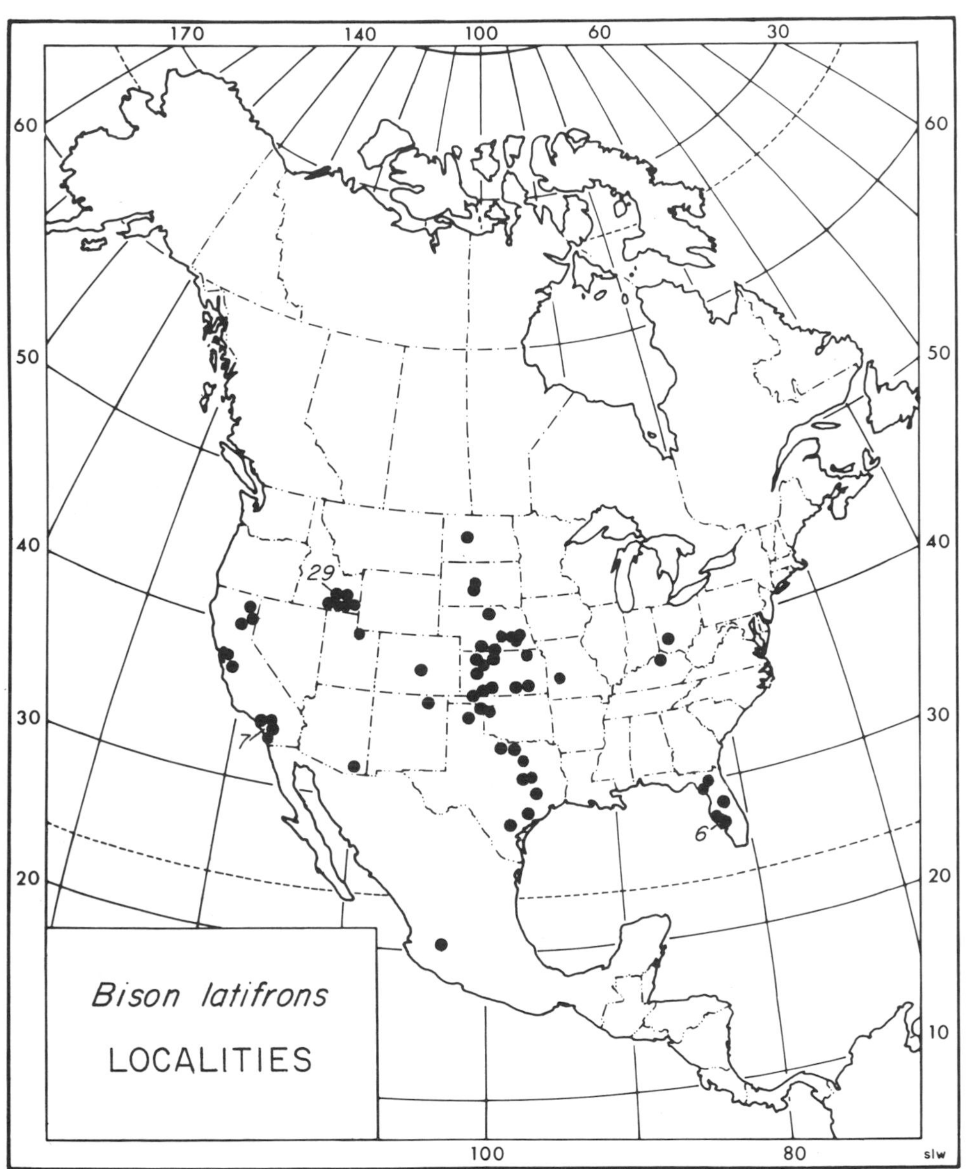

FIGURE 18 The distribution of *Bison latifrons* specimens used in this study. The primary *Bison latifrons* range appears to have formed a crescent from the Pacific Coast of California, across the northern Great Basin, onto the Great Plains, and south onto the coastal plain of Texas. A major outlier existed in Florida. The species' interglacial range probably extended further north than is indicated here, but evidence (possibly destroyed by glacial activity) to support this is not available. The densely forested southeast and the *Bison antiquus*-occupied southwest have yielded conspicuously few *Bison latifrons* specimens. Numbers indicate total specimens from clustered localities.

Table 20

Geologic Ages Reported for *Bison latifrons* Specimens

Locality	Reported age	Dating method	Source
Lago de Chapala, Jalisco	Late Pleistocene	Biostratigraphy	Downs, 1958
	30,000 BP ±	Radiocarbon	F. R. Solorzano, oral comm., July 14, 1976
Costeau Pit, California	Sangamon – early Wisconsin	Biostratigraphy	Miller, 1971
	> 42,000 BP	Radiocarbon	Miller, 1971
Rancho la Brea, California	< 40,000 – > 12,000 BP	Radiocarbon	UCLA-773C*, UCLA-1292B*
Canon City, Colorado	Sangamon	Geostratigraphy	Lewis, 1970; Scott and Lindvall, 1970
	ca. 160,000 BP	Radiometric	Scott and Lindvall, 1970
	Pre-Sangamon	Geostratigraphy	Schultz and Hillerud, 1977
Haile VIII-A, Florida	Sangamon	Geostratigraphy	Robertson, 1974
Bradenton Canal, Florida	Sangamon	Biostratigraphy	Robertson, 1974
	Wisconsin (?)	Geostratigraphy	Robertson, 1974
Acequia, Idaho	Late Pleistocene	Geostratigraphy	Site report (D901), USGS/P&S/Denver
American Falls Reservoir, Idaho	Illinoian	Geostratigraphy	Hopkins, Bonnichson, and Fortsch, 1969
	Pre-Illinoian – early Illinoian	Geostratigraphy	Schultz and Hillerud, 1977
	> 32,000 BP	Radiocarbon	W-358*
	31,000 – 21,500 BP	Radiocarbon	McDonald and Anderson, 1975
Comanche County, Kansas	Illinoian – Wisconsin	Geostratigraphy	Hibbard, 1955a
Jinglebob Pasture, Kansas	Sangamon (?)	Geostratigraphy and biostratigraphy	Hibbard, 1955a
	Yarmouth – Illinoian	Geostratigraphy	Schultz and Hillerud, 1977
Scott City, Kansas	Illinoian – Wisconsin	Geostratigraphy	Hibbard, 1955a
	Yarmouth – Illinoian	Geostratigraphy	Schultz and Hillerud, 1977
Wellington, Kansas	Illinoian – Wisconsin	Geostratigraphy	Hibbard, 1955a
Dorchester, Nebraska	Yarmouth – Illinoian	Geostratigraphy	Schultz and Hillerud, 1977
Giltner, Nebraska	Yarmouth – Illinoian	Geostratigraphy	Schultz and Hillerud, 1977
Lancaster County, Nebraska	Yarmouth – Illinoian	Geostratigraphy	Schultz and Hillerud, 1977
Naponee, Nebraska	Yarmouth – Illinoian	Geostratigraphy	Schultz and Hillerud, 1977
Sutton, Nebraska	Yarmouth – Illinoian	Geostratigraphy	Schultz and Hillerud, 1977
Trenton, Nebraska	Yarmouth – Illinoian	Geostratigraphy	Schultz and Hillerud, 1977
Midland, South Dakota	Illinoian	Geostratigraphy	Green, 1962
	Illinoian or before	Geostratigraphy	Schultz and Hillerud, 1977
Clear Creek, Texas	28,840 BP	Radiocarbon	SM-534*
Forney Dam, Texas	Sangamon	Geostratigraphy	B. Slaughter, written comm., May 19, 1978
Good Creek, Texas	Sangamon	Geostratigraphy and ecological analogs	Dalquest, 1962
Lipscomb County, Texas	Late Pleistocene	Geostratigraphy and biostratigraphy	Schultz and Lansdown, 1972
	Yarmouth – Illinoian	Geostratigraphy	Schultz and Hillerud, 1977
Wichita County, Texas	Late Pleistocene	Geostratigraphy	UMMP records
Summit County, Utah	Late Sangamon – early Wisconsin	Biostratigraphy	Miller, 1976
	> 40,000 BP	Radiocarbon	Miller, 1976

*Denotes radiocarbon date laboratory number.

Mexico and the arid southwestern United States. The distribution pattern for this species suggests that it evolved in mid-latitude North America, and that it was probably a browser-grazer primarily adapted to forest openings or woodlands (but not closed forests).

Hibbard (1955*a*) reviewed the geologic age of *B. latifrons* and concluded that no specimen could be considered older than the Illinoian. Although Hibbard's conclusion was not unanimously accepted (Schultz and Hillerud, 1977; Schultz, Tanner and Martin, 1972), there is no convincing evidence that *B. latifrons* appeared before the Illinoian. More recent opinions and radiometric evidence support an Illinoian appearance for the species (Green, 1962; Guthrie, 1970; Scott and Lindvall, 1970). Radiocarbon dates indicate that *B. latifrons* probably lived until near or after the onset of the late Wisconsin glacial stade. The geologic age of *B. latifrons* appears to have extended from the Illinoian to the late Wisconsin (table 20), with maximum numbers and greatest distribution probably occurring during the Sangamon interglacial when suitable habitat was at its maximum.

Discussion:

Synonymy: *B. latifrons* is the largest North American fossil bison and, as such, presents a greater range of variation in size than other species. This fact probably accounts for much of the synonymy of the species. *B. latifrons* was poorly known when Marsh (1877) described *B. ferox*, based on a medial horn core section (YPM VP 10910) found somewhere along the Niobrara or Loup rivers in Nebraska (Schultz and Frankforter, 1946). Now, with a much larger sample available, and with allowances for intraspecific variation, YPM VP 10910 is easily recognized as a *B. latifrons* horn core, especially in its cross section shape, posterior profile, fluting, and dorsal groove characteristics.

B. alleni was originally (and poorly) described by Marsh (1877) from a nearly complete horn core (YPM VP 11911) found along the Blue River near Manhattan, Kansas. Marsh provided measurements for the specimen, including greatest and least basal diameters of 140 mm and 110 mm, respectively, and remarked that this specimen had more curved horns than did the *B. ferox* holotype. He did not mention that a portion of the horn core base was missing, a condition that may have resulted in an unrealistically shortened dorso-ventral diameter, which in turn exaggerated the degree of dorso-ventral compression. YPM VP 11911 is more strongly curving than most *B. latifrons*, but curvature is not specifically diagnostic (table 13). The holotype of *B. alleni* (plate 7) possesses all of the diagnostic characters of *B. latifrons* possible in the preserved material: a broadly triangular (isosceles) basal cross section, a cordiform tip, a straight posterior margin, and slightly spiraled growth around an arched axis.

Many specimens of *B. alaskensis* (= *B. chaneyi*, *B. aguascalentensis*) have been identified as, or confused with, *B. alleni*, a result of exaggerated emphasis on the basal compression data reported by Marsh (1877). The working concept of *B. alleni* morphology is, as a result, much different than the facts represented by the holotype.

Bos crampianus was based on morphological characters only slightly different from *B. latifrons* found in two partial horn cores and a partial skull (ANSP G 3) from Wellington, Kansas

(Cope, 1895). Diagnostic features are clearly *B. latifrons*.

Bos arizonica was initially described as a large horned ox (Blake, 1898*a*), but in a subsequent note, Blake (1898*b*) tacitly acknowledged the likelihood that his *Bos arizonica* was a giant bison. The holotype of this species was never figured, and it has since been lost. The measurements Blake provided are not diagnostic, but if they were reasonably accurate the extent of dorso-ventral compression in the holotype (152 mm antero-posteriorly, 140 mm dorso-ventrally) suggests that the specimen was more likely *B. latifrons* than the other possibility, *B. alaskensis*.

Hay (1914:192) differentiated *B. regius* as "a species related to *B. latifrons*, but having the horn-cores relatively longer, slenderer, and more strongly curved. Teeth with the enamel of the 'lakes' furnished with reentering folds." Hay's comments on horn core morphology are diagnostically meaningless. The reentering folds noted in the upper molars of the holotype are not unique to this specimen or to large bison in general, and are certainly not taxonomically meaningful characters.

The holotype of *B. angularis* (CoMNH VP 1164) is perhaps the most unusual specimen of *B. latifrons* on record. Figgins (1933:23) justified specific differentiation on the bases of "flattened and depressed [frontals]; occipital high and angular; foramen magnum broad and angular to an unique degree; horn-cores little curved, thrusting backwards and upwards; terminal portions of horn-cores deeply and broadly flattened on upper surface." The horn cores do emanate unusually high on the frontals and do extend rearward and upward to an unusual degree. The other criteria Figgins considered as differentiating are not unique to CoMNH VP 1164 among *B. latifrons*. Even the horn core abnormalities are not sufficently extreme to preclude identification of this specimen as *B. latifrons*.

Minor characters of individual variation only were used by Figgins to establish the distinctiveness of *B. rotundus*. The holotype of this species (CoMNH VP 1187) possessed "horncores intermediate in curvatures between those of *B. regius* and *B. latifrons* but tending to the former" (Figgins, 1933:24).

Geologic age of *B. latifrons:* The appearance and duration of *B. latifrons* is extremely controversial. The oldest current view is that the species migrated into North America from Eurasia during or after late Aftonian time and lived until late Yarmouth or early-to-middle Illinoian time, after which the species became extinct (Schultz and Frankforter, 1946; Skinner and Kaisen, 1947). This view was originally allied with the principle of progressive horn core diminution which maintained that the earliest North American bison had the largest horns (Allen, 1876). More recent opinions maintain that *B. latifrons* probably first appeared during the Illinoian and became extinct during or immediately after the Illinoian (Guthrie, 1970; Hibbard, 1955*a*).

A critical review of estimates of recent authors of the appearance of *B. latifrons* indicates that Schultz and his colleagues (Schultz and Frankforter, 1946; Schultz and Hillerud, 1977; Schultz, Tanner, and Martin, 1972) and Skinner and Kaisen (1947) consider the species to have originated before the

Illinoian (table 10). Skinner and Kaisen did not critically discuss the provenience of individual specimens. Schultz and Hillerud (1977) have presented the most thorough discussion to date of the provenience of individual specimens, but these authors consider only fourteen specimens (cf. table 19), which they consider critical to an understanding of *B. latifrons*. All specimens considered were assigned pre-Illinoian or Illinoian ages. The provenience of four specimens in their sample is, however, uncertain, according to published records or those in the University of Nebraska State Museum. The provenience of six of the remaining specimens considered by Schultz and Hillerud has been studied by other researchers (Green, 1962; Hibbard, 1955*a*, 1970; Hopkins, Bonnichsen, and Fortsch, 1969; Lewis, 1970; Schultz and Lansdown, 1972; Scott and Lindvall, 1970). Each of the six has been assigned an Illinoian or more recent age by these researchers. A radiometric date of 160,000 BP places a *B. latifrons* skull from Canon City, Colorado, in the late Illinoian, and represents the strongest available support for definitely assigning the species to the Illinoian. Finite radiocarbon dates associated with *B. latifrons* from Rancho La Brea, California; American Falls Reservoir, Idaho; Lago de Chapala, Jalisco; and Clear Creek, Texas (table 20) strongly suggest that the species survived until just before, if not actually into, the late Wisconsin stade. The geographically dispersed nature of these dates also indicates that middle and late(?) Wisconsin *B. latifrons* were widespread.

The geologic age of *B. latifrons*, then, appears to have extended from at least the Illinoian to the middle, and possibly late, Wisconsin. The species could have existed in central North America earlier than the Illinoian, but convincing evidence to support this is not yet available. Hillerud's claim (oral presentation, 1978 AMQUA, September 2) that a pre-Illinoian *B. latifrons* had been unearthed by Péwé on the Seward Peninsula of Alaska was erroneous (Péwé, written comm., August 22, 1979). The most recent radiocarbon dates indicate that *B. latifrons* survived until at least the period 30,000–20,000 BP, and it is not unreasonable to suspect that the species could have survived into the late Wisconsin.

Distribution: *B. latifrons* is known best from the Great Plains, the Great Basin, and the Pacific Coast, and the species was probably most concentrated throughout this region. Numerous remains of other species have been found in southwestern North America, but only two *B. latifrons* are known from this region and one of them (from southern Arizona: Blake, 1898*a*, 1898*b*; now lost) is only tentatively identified as *B. latifrons*. The other specimen, from Lago de Chapala, could represent an aberrant population or could simply reflect the lack of paleontological research in the northern part of Mexico where, perhaps, other specimens of *B. latifrons* await discovery. The paucity of specimens from the eastern United States probably reflects the low carrying capacity of the natural closed forests of this region. Florida, which probably had more open vegetation, was inhabited by *B. latifrons*. *B. latifrons* also probably occurred in northern North America during interglacial periods, but, if so, subsequent glacial activity has apparently destroyed or obscured the fossil evidence.

II. *Bison antiquus* Leidy, 1852

IIa. *Bison antiquus antiquus* Leidy, 1852

Original Description:

Joseph Leidy. 1852*b*. (Report on Bison from Big Bone Lick and Vicinity). *Proceedings, Academy of Natural Sciences of Philadelphia*, vol. 6, p. 117. (Name proposed).

Joseph Leidy. 1852*a*. "Memoir on the Extinct Species of American Ox." *Smithsonian Contributions to Knowledge*, vol. 5, art. 3. (Description).

Type Locality:

Big Bone Lick, Kentucky.

Type Location:

Department of Geology, Academy of Natural Sciences, Philadelphia.

Type Specimen:

ANSP G 12990—proximal and medial right horn core with partial frontal (plate 8).

Principal Synonymy:

Bos scaphoceras Cope, 1895
Bison californicus Rhoads, 1897
Bison pacificus Hay, 1927
Bison bison antiquus Wilson, 1974*b*

Species Content:

B. antiquus was the late Pleistocene-early and middle Holocene savanna/steppe bison of North America.

Subspecies Content:

B. a. antiquus was the savanna bison of late Pleistocene North America, and the conservative subspecies of savanna/steppe bison of early and middle Holocene North America.

Diagnosis:

Male horn cores triangular (isosceles) in cross section and symmetrical about dorso-ventral axis at base; tip cordiform or triangular in cross section; posterior margin straight; growth straight along arched longitudinal axis; antero-posterior plane nearly parallel with frontal plane; length along upper curve greater than 200 mm, less than 400 mm; greater than 5,000–4,000 BP.

Female horn cores circular to elliptical in cross section at base; symmetrical about greatest diameter at base; tip elliptical in cross section; posterior margins straight; growth straight along arched longitudinal axis; plane of greatest diameter rotated forward from frontal plane; length along upper curve greater than 125 mm, less than 300 mm.

Differential Diagnosis:

Male *B. a. occidentalis* tip circular or elliptical in cross section; growth straight or spiraled around arched or sinuous longitudinal axis; antero-posterior plane may be rotated rearward from frontals; less than 11,000 BP. *B bison* tip circular or elliptical in cross section; antero-posterior plane usually rotated rearward from frontal plane; less than 5,000–4,000 BP. *B. latifrons* horn cores broadly triangular in cross section at base; length along upper curve greater than 500 mm. *B. alaskensis* and *B. priscus* horn cores asymmetrical in cross section about dorso-ventral axis at base; posterior margin sinuous; growth spiraled about sinuous longitudinal axis; antero-posterior plane usually rotated forward from frontal plane.

Female *B. a. occidentalis* horn cores very similar to *B. a. antiquus* but, if attached to frontals, *B. a. occidentalis* cores are directed more posteriorly and upward; less strongly arched. *B. bison* length along upper curve less than 200 mm; less than 5,000–4,000 BP. *B. latifrons* cores with slight spiral growth around mildly arched longitudinal axis; dorsal basal burr present; length along upper curve greater than 500 mm; basal circumference greater than 300 mm. *B. priscus* with sinuous or convex posterior margin. *B. alaskensis* with sinuous or convex posterior margin; length along upper curve greater than 250 mm.

Biometric Summary:

Skulls, table 21.
Limb bones, table 22.

Referred Specimens:
Table 23.
Description:
B. a. antiquus is the smaller, shorter horned bison of late Pleistocene-early and middle Holocene North America (fig. 19; plate 9). The length of horn cores along the upper curve is about equal to, but normally less than, their basal circumference. They emanate subhorizontally or, less often, horizontally from the frontals at an angle of about 75°–80° forward from the sagittal plane. This species has the most laterally directed of all male bison horn cores. The cores are triangular (isosceles) in cross section at the base, where the antero-posterior and dorso-ventral diameters are about equal. They taper gradually toward the tip. The cores are not strongly curved; most of the upward curvature normally occurs in the distal one-half of each core. A prominent dorsal groove extends over the distal 10 percent to 20 percent of the core. The tip is characteristically thick or heavy, is normally near the frontal plane, and is well in front of the occipital plane. As a result of these characters, the horn cores of *B. a. antiquus* males are both the lowest in relation to the frontals and the most forward oriented of all the bison taxa. Sheath characteristics of *B. a. antiquus* are not known, but comparison of the growth pattern and orientation of

Table 21

Summary of *Bison antiquus antiquus* Skull Biometrics

Standard measurement	n	Range	Mean ± SE	σ	V
Males					
Spread of horn cores, tip to tip	27	765 – 1067 mm	870.0 ± 13.7 mm	71.0 mm	8.2
Horn core length, upper curve, tip to burr	32	203 – 364 mm	279.2 ± 6.2 mm	35.1 mm	12.6
Straight line distance, tip to burr, dorsal horn core	30	185 – 330 mm	249.7 ± 5.3 mm	29.2 mm	11.7
Dorso-ventral diameter, horn core base	39	81 – 126 mm	101.9 ± 1.6 mm	9.7 mm	9.5
Minimum circumference, horn core base	38	233 – 392 mm	324.4 ± 5.3 mm	32.6 mm	10.0
Width of occipital at auditory openings	20	251 – 318 mm	287.9 ± 4.2 mm	18.6 mm	6.5
Width of occipital condyles	20	132 – 161 mm	143.7 ± 1.9 mm	8.4 mm	5.9
Depth, nuchal line to dorsal margin of foramen magnum	18	94 – 134 mm	111.6 ± 2.2 mm	9.2 mm	8.3
Antero-posterior diameter, horn core base	41	76 – 129 mm	105.6 ± 1.9 mm	12.2 mm	11.5
Least width of frontals, between horn cores and orbits	25	276 – 352 mm	314.7 ± 3.9 mm	19.4 mm	6.2
Greatest width of frontals at orbits	15	338 – 400 mm	371.3 ± 4.6 mm	18.6 mm	4.8
M^1–M^3, inclusive alveolar length	2	105.2 – 106 mm	105.6 ± 0.4 mm	0.6 mm	0.5
M^3, maximum width, anterior cusp	2	29.6 – 30.0 mm	29.8 ± 0.2 mm	0.3 mm	0.9
Distance, nuchal line to tip of premaxillae	1	....	629.0 mm	....	....
Distance, nuchal line to nasal-frontal suture	14	260 – 314 mm	283.9 ± 4.9 mm	18.4 mm	6.5
Angle of divergence of horn cores, forward from sagittal	22	72° – 86°	79.2° ± 1.0°	4.8°	6.1
Angle between foramen magnum and occipital planes	17	113° – 136°	125.4° ± 1.6°	6.6°	5.3
Angle between foramen magnum and basioccipital planes	17	107° – 132°	115.6° ± 1.7°	7.1°	6.1
Females					
Spread of horn cores, tip to tip	20	524 – 802 mm	669.5 ± 15.9 mm	70.9 mm	10.6
Horn core length, upper curve, tip to burr	26	145 – 253 mm	202.9 ± 5.1 mm	26.0 mm	12.8
Straight line distance, tip to burr, dorsal horn core	25	136 – 234 mm	190.1 ± 5.0 mm	25.1 mm	13.2
Dorso-ventral diameter, horn core base	26	53 – 79 mm	66.5 ± 1.2 mm	6.0 mm	9.1
Minimum circumference, horn core base	24	172 – 241 mm	209.0 ± 3.3 mm	16.2 mm	7.7
Width of occipital at auditory openings	13	221 – 264 mm	246.2 ± 3.2 mm	11.6 mm	4.7
Width of occipital condyles	13	116 – 149 mm	133.2 ± 2.5 mm	9.2 mm	6.9
Depth, nuchal line to dorsal margin of foramen magnum	11	86 – 109 mm	100.7 ± 1.9 mm	6.1 mm	6.1
Antero-posterior diameter, horn core base	26	54 – 75 mm	65.8 ± 1.1 mm	5.4 mm	8.3
Least width of frontals, between horn cores and orbits	21	238 – 303 mm	262.8 ± 3.4 mm	15.8 mm	6.0
Greatest width of frontals at orbits	16	289 – 341 mm	314.6 ± 3.7 mm	14.8 mm	4.7
M^1–M^3, inclusive alveolar length	6	92.8 – 99.0 mm	96.2 ± 1.0 mm	2.4 mm	2.5
M^3, maximum width, anterior cusp	7	26.2 – 30.1 mm	28.8 ± 0.6 mm	1.6 mm	5.4
Distance, nuchal line to tip of premaxillae	4	567 – 587 mm	577.0 ± 5.4 mm	10.9 mm	1.9
Distance, nuchal line to nasal-frontal suture	16	219 – 276 mm	248.1 ± 3.4 mm	13.8 mm	5.5
Angle of divergence of horn cores, forward from sagittal	18	69° – 88°	82.1° ± 1.3°	5.3°	6.5
Angle between foramen magnum and occipital planes	10	114° – 131°	123.8° ± 1.5°	4.8°	3.8
Angle between foramen magnum and basioccipital planes	10	106° – 125°	115.0° ± 2.1°	6.6°	5.8

Table 22

Summary of *Bison antiquus antiquus* Limb Biometrics

Standard measurement	n	Range	Mean ± SE	σ	V
Males					
Humerus:					
Approximate rotational length of bone	1	....	339.0 mm	....	....
Antero-posterior diameter of diaphysis	3	68 – 73 mm	69.7 ± 1.6 mm	2.8 mm	4.1
Transverse minimum of diaphysis	3	58 – 59 mm	58.7 ± 0.2 mm	0.5 mm	0.9
Radius:					
Approximate rotational length of bone	4	335 – 351 mm	343.0 ± 3.8 mm	7.7 mm	2.2
Antero-posterior minimum of diaphysis	4	33 – 35 mm	33.7 ± 0.4 mm	0.9 mm	2.8
Transverse minimum of diaphysis	4	53 – 56 mm	54.5 ± 0.6 mm	1.2 mm	2.3
Metacarpal:					
Total length of bone	28	211 – 238 mm	222.0 ± 1.2 mm	6.6 mm	3.0
Antero-posterior minimum of diaphysis	28	28 – 33 mm	30.7 ± 0.2 mm	1.2 mm	4.0
Transverse minimum of diaphysis	28	47 – 61 mm	52.3 ± 0.6 mm	3.5 mm	6.6
Femur:					
Approximate rotational length of bone	4	414 – 448 mm	429.0 ± 7.0 mm	14.0 mm	3.2
Antero-posterior diameter of diaphysis	5	52 – 62 mm	56.2 ± 1.8 mm	4.2 mm	7.4
Transverse minimum of diaphysis	5	57 – 62 mm	58.8 ± 0.9 mm	2.1 mm	3.6
Tibia:					
Approximate rotational length of bone	9	393 – 430 mm	412.0 ± 4.9 mm	14.7 mm	3.5
Antero-posterior minimum of diaphysis	12	34 – 49 mm	42.5 ± 0.8 mm	2.9 mm	6.9
Transverse minimum of diaphysis	12	54 – 64 mm	59.0 ± 0.9 mm	3.2 mm	5.4
Metatarsal:					
Total length of bone	22	251 – 287 mm	273.9 ± 1.7 mm	8.3 mm	3.0
Antero-posterior minimum of diaphysis	22	32 – 40 mm	35.2 ± 0.4 mm	2.3 mm	6.7
Transverse minimum of diaphysis	22	37 – 47 mm	40.9 ± 0.5 mm	2.7 mm	6.7
Females					
Humerus:					
Approximate rotational length of bone	9	311 – 357 mm	343.5 ± 4.6 mm	13.8 mm	4.0
Antero-posterior diameter of diaphysis	10	55 – 68 mm	62.9 ± 1.3 mm	4.2 mm	6.7
Transverse minimum of diaphysis	10	45 – 56 mm	50.6 ± 0.9 mm	3.1 mm	6.1
Radius:					
Approximate rotational length of bone	8	310 – 351 mm	329.0 ± 5.2 mm	14.8 mm	4.5
Antero-posterior minimum of diaphysis	8	28 – 37 mm	32.8 ± 1.2 mm	3.5 mm	10.7
Transverse minimum of diaphysis	8	43 – 59 mm	51.1 ± 2.0 mm	5.7 mm	11.2
Metacarpal:					
Total length of bone	40	206 – 230 mm	219.5 ± 1.0 mm	6.9 mm	3.1
Antero-posterior minimum of diaphysis	40	25 – 32 mm	28.6 ± 0.2 mm	1.5 mm	5.5
Transverse minimum of diaphysis	40	38 – 51 mm	45.0 ± 0.5 mm	3.5 mm	7.8
Femur:					
Approximate rotational length of bone	11	378 – 430 mm	400.7 ± 4.5 mm	15.2 mm	3.8
Antero-posterior diameter of diaphysis	14	47 – 53 mm	49.9 ± 0.5 mm	2.1 mm	4.3
Transverse minimum of diaphysis	14	45 – 55 mm	51.1 ± 0.7 mm	2.9 mm	5.7
Tibia:					
Approximate rotational length of bone	20	365 – 405 mm	390.2 ± 2.2 mm	10.2 mm	2.6
Antero-posterior minimum of diaphysis	19	36 – 46 mm	40.1 ± 0.5 mm	2.4 mm	6.1
Transverse minimum of diaphysis	19	49 – 57 mm	52.5 ± 0.5 mm	2.2 mm	4.1
Metatarsal:					
Total length of bone	37	252 – 286 mm	266.0 ± 1.4 mm	8.9 mm	3.2
Antero-posterior minimum of diaphysis	37	28 – 36 mm	32.8 ± 0.3 mm	2.4 mm	7.3
Transverse minimum of diaphysis	37	29 – 41 mm	35.4 ± 0.5 mm	3.3 mm	9.5

Table 23

Bison antiquus antiquus: Referred Specimens

Institution and identification number			Sex	Provenience
AMNH	VP	22755	♂	Bradenton Canal, Manatee County, Florida
AMNH	VP	31070	♂	Beaumont, Riverside County, California
AMNH	VP	42885	♂	Comondu, Baja California
ANSP	G	297	♂	San Francisco, San Francisco County, California
ANSP	G	10226	♂	Clovis area, New Mexico
ANSP	G	12990	♂	Big Bone Lick, Boone County, Kentucky
BeMNH		603B	♂	Nye, Wisconsin
BeMNH		3339	♂	Melrose Bog, Minnesota
CAS		15502	♂	Gottville, Siskiyou County, California
CM		10195	♀	Rancho la Brea, Los Angeles County, California
CoMNH	VP	1642	♂	Arikaree River, Colorado
DM		Bu6	♂	Near Drumheller, Alberta
EC		A[a]*	♂	Near Vincennes, Indiana
FMNH	P	M702	♂	Arroyo las Positas, Alameda County, California
IG		49-76[b]	♂	Barranca de Acatlan, Tequixquiac, Mexico
IMNH	VP	1838	♂	Hop-Strawn Gravel Pit, Power County, Idaho
IMNH	VP	17867	♂	Near McCammon, Bannock County, Idaho
KU	VP	10297	♂	Lawrence, Douglas County, Kansas
LACM	VP	1244	♀	Rancho la Brea, Los Angeles County, California
LACM	VP	1245	♀	Rancho la Brea, Los Angeles County, California
LACM	VP	1246	♀	Rancho la Brea, Los Angeles County, California
LACM	VP	1252	♀	Rancho la Brea, Los Angeles County, California
LACM	VP	1255	♀	Rancho la Brea, Los Angeles County, California
LACM	VP	1256	♀	Rancho la Brea, Los Angeles County, California
LACM	VP	6000	♂	Rancho la Brea, Los Angeles County, California
LACM	VP	19058	♂	McKittrick Tar Seep, Kern County, California
LACM	VP	28678	♀	China Lake, California
LACM	VP	98596	♂	McKittrick Tar Seep, Kern County, California
LACM	VP	HC6001	♀	Rancho la Brea, Los Angeles County, California
LACM	VP	HC6002	♀	Rancho la Brea, Los Angeles County, California
LACM	VP	HC6003	♀	Rancho la Brea, Los Angeles County, California
LACM	VP	HC6004	♀	Rancho la Brea, Los Angeles County, California
LACM	VP	HC6005	♀	Rancho la Brea, Los Angeles County, California
LACM	VP	HC6006	♀	Rancho la Brea, Los Angeles County, California
LACM	VP	HC6007	♂	Rancho la Brea, Los Angeles County, California
LACM	VP	HC6008	♀	Rancho la Brea, Los Angeles County, California
LACM	VP	HC6009	♀	Rancho la Brea, Los Angeles County, California
LACM	VP	HC6010	♀	Rancho la Brea, Los Angeles County, California
LACM	VP	HC6011	♀	Rancho la Brea, Los Angeles County, California
LACM	VP	HC6012	♀	Rancho la Brea, Los Angeles County, California
LACM	VP	RLB913	♂	Rancho la Brea, Los Angeles County, California
LACM	VP	RLB1007	♂	Rancho la Brea, Los Angeles County, California
LACM	VP	A*	♀	McKittrick Tar Seep, Kern County, California
LACM	VP	B*	♂	White River, Tulare County, California
LACM	VP	C*	♂	Costeau Pit, Orange County, California
MMMN	VP	W597	♂	Treesbank, Manitoba
MMMN	VP	A*	♂	Near Prairie Grove, Manitoba
MNA		P1.90	♂	Near Badger Springs, Arizona
MRG		A*	♂	Lago de Chapala, Jalisco
NMC	VP	12442	♂	Athabasca, Alberta
PMA	P	69.20.1	♂	Athabasca, Alberta

Table 23
Bison antiquus antiquus: Referred Specimens (Continued)

Institution and identification number			Sex	Provenience
PMA	P	71.6.1	♂	Clover Bar Bridge, near Edmonton, Alberta
RO		5[c]	♂	Wacissa River, Florida
SM		60003/1.49	♂	Lugow Pit, near Dallas, Texas
SM		60003/2.50	♂	Lugow Pit, near Dallas, Texas
SMP		157-56-z	♂	Near Carlisle, Iowa
TMM		937-764a	♂	Blackwater Draw, Roosevelt County, New Mexico
TMM		937-764b	♂	Blackwater Draw, Roosevelt County, New Mexico
TMM		957-A*	♂	Onion Creek Gravel Pit, Travis County (?), Texas
TMM		1860[d]	♀	Tedford Farm, San Patricio County, Texas
TMM		30967-423	♂	Tedford Farm, San Patricio County, Texas
TMM		A*	♂	Archeological Site 41VV162A, Val Verde County, Texas
TMM		B*	♂	Near mouth of Little Brazos River, Texas
TTU	VP	465	♂	Weatherly Lake, Texas
UA1		600	♂	Clover Gravel Bar, near Edmonton, Alberta
UA1		619	♂	Twin Bridges Gravel Pit, Alberta
UALP		2538	♂	Charlie Day Spring, Coconino County, Arizona
UALP		10358[e]	♂	Brophy Cienega, near Elgin, Arizona
UCMP		19478[f]	♂	John Day Region, Oregon
UCMP		21151[f]	♀	Rancho la Brea, Los Angeles County, California
UCMP		21154[f]	♂	Rancho la Brea, Los Angeles County, California
UCMP		21186	♀	Rancho la Brea, Los Angeles County, California
UCMP		21190	♀	Rancho la Brea, Los Angeles County, California
UCMP		36662	♂	Doolan Canyon, Alameda County, California
UCMP		36960	♂	Homogera, El Salvador
UCMP		37615	♀	Burke Ranch, Contra Costa County, California
UCMP		38341a	♀	Delta Mendota 15, San Joaquin County, California
UCMP		38341b	♂	Delta Mendota 5, San Joaquin County, California
UCMP		41164	♂	Positas Arroyo General, California
UCMP		43689	♂	Lower Yukon Valley, Alaska
UCMP		64688	♂	Tule Springs #2, Clark County, Nevada
UCMP		A*	♀	Rancho la Brea, Los Angeles County, California
UF	VP	10001	♂	Aucilla River, Florida
UF	VP	11861	♂	Wacissa River, Florida
UF	VP	15078	♂	Florida
UF	VP	19376	♂	Ichetucknee River, Florida
UF	VP	SF-I	♂	Santa Fe River I, Florida
UF	VP	A*	♂	Sloth Hole, Aucilla River, Florida
UMMP		18209	♂	Near Yarmany Station, Colorado
UNSM	VP	30312	♂	Devils Gap, Dawson County, Nebraska
UNSM	VP	30795	♂	Cass County, Nebraska
UNSM	VP	UIowa-350	♂	Iowa (?)
UNSM	VP	A*	♀	Nebraska (?)
UP	G	A*	♂	Nicaragua
USNM	A	DH18-1	♂	Dutton Archeological Site, Yuma County, Colorado
USNM	A	DK40-4	♂	Dutton Archeological Site, Yuma County, Colorado
USNM	P	7254	♀	Klamath River, near Oak Bar, California
USNM	P	8523	♂	Near Walla Walla, Walla Walla County, Washington
USNM	P	10544	♀	Sagamore Mine, Riverton, Crow Wing County, Minnesota
USNM	P	13683	♂	Lea County, New Mexico
USNM	P	A*	♂	Valsequillo, Puebla
USNM	P	B*	♂	Valsequillo, Puebla

Table 23
Bison antiquus antiquus: Referred Specimens (Continued)

Institution and identification number		Sex	Provenience
USNM	P C*	♂	Valsequillo, Puebla
USNM	P D*	♂	Valsequillo, Puebla
USNM	P E*	♂	Valsequillo, Puebla
WPBSM	A[c]*	♂	West Palm Beach, Palm Beach County, Florida
YPM	VP 10602	♀	Rancho la Brea, Los Angeles County, California

Sources of supplementary data:
[a]Middleton and Moore, 1900;
[b]Hibbard, 1955*b*;
[c]Robertson, 1974;
[d]Lundelius, 1972;
[e]Tom Stafford, written comm., September 6, 1979;
[f]Chandler, 1916;
*No identification number was found for these specimens.

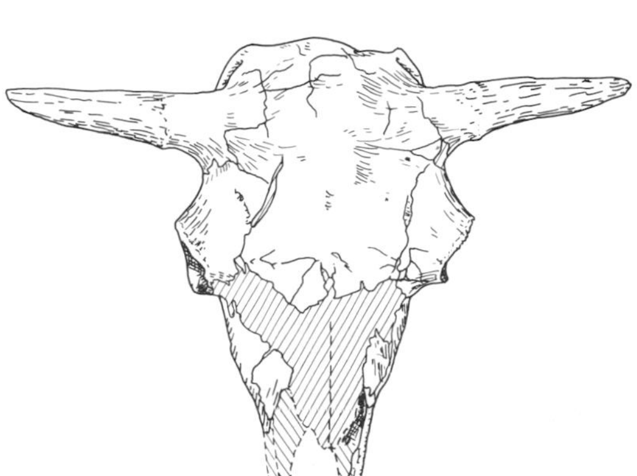

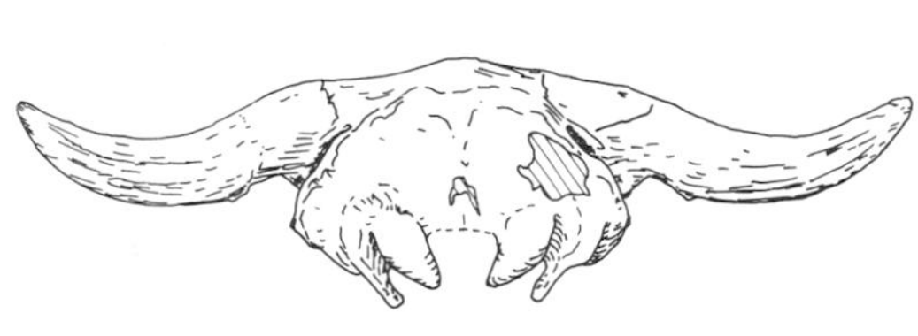

FIGURE 19 One of the most complete known female *Bison antiquus antiquus* skulls (TMM 1860) is from the Ingleside fauna, found in San Patricio County, Texas (after figure in Lundelius, 1972).

B. a. antiquus horn cores with those of other taxa suggests that the sheath tips were not (or were perhaps slightly) posteriorly directed and probably were situated in front of the occipital plane.

The frontals are mildly domed. The orbits are normally lower than in other late Pleistocene bison, but are directed relatively forward and protrude more than in *B. latifrons*. The nasals are low, broad, and triangular. The occipital is moderately concave to nearly flat. The nuchal line/external occipital protuberance is prominent, but the occipital condyles do not protrude outward from the occipital plane as prominently or as frequently as in *B. latifrons* and the Pleistocene Eurasian bison. The ventro-lateral margin of the condyles is occasionally, but not typically, flanged.

Female horn cores are shorter and somewhat more slender than male horn cores, and are more robust than all other female horn cores except *B. latifrons*. Horn cores emanate subhorizontally from the frontals and are more laterally directed than in any other taxon. The horn cores are circular to elliptical in cross section at the base, and they taper gradually toward the tip. The tip is near the frontal plane and forward of the occipital plane.

The frontals are crenulated transversely, being convex near the horn cores and the sagittal line, but concave between these regions. The orbits are high, directed forward, and are relatively more robust than the orbits of other female bison. The occipital is relatively flat. The nuchal line/external occipital protuberance is well developed, but the occipital condyles do not protrude far beyond the occipital plane. The ventrolateral margin of the occipital condyles is normally not flanged, but occasionally a flange is present.

Distribution:

B. a. antiquus, known from throughout most of North America, is the most widely distributed of all North American bison (fig. 20). The primary Pleistocene range of the taxon appears to have been the southwestern quarter of the continent, with a significant outlier in Florida. During the late Wisconsin the range of *B. a. antiquus* appears to have expanded, especially its northern edge. Subsequently, the whole range shifted northward during the early and middle Holocene, away from the southwest onto the northern plains of central North America (see chapter 7). This distribution pattern suggests that *B. a. antiquus* first evolved in southwestern North America, probably as a grazer/browser primarily adapted to a savanna or wooded steppe environment.

The antiquity of *B. a. antiquus* is not clearly established. Hibbard (1955*b*) and Miller (1971) reported *B. a. antiquus* (as conceived herein) from Upper Becerra (Tequixquiac) and Costeau Pit faunas, respectively. Both faunas were assigned to the Sangamon. The Upper Becerra horizon, however, has been assigned to the late Wisconsin by other writers (de Terra, Romero, and Stewart, 1949; Wormington, 1957), and Miller apparently based his determination of geologic age on a combination of a radiocarbon age of the fauna in excess of 40,000 BP and biostratigraphic reasoning. *B. antiquus* is present in the Valsequillo fauna from Puebla, which might be as old as the Illinoian but which is probably no older than 30,000 BP based on radiocarbon dates (Steen-McIntyre, Fryxell, and Malde, 1973; Szabo, Malde, and Irwin-Williams, 1969). Hansen and Begg (1970) report uranium and actinium dates for two sites containing bison (cf. *B. antiquus*) near Sacramento, California, ranging from 186,000 BP to 45,000 BP. Considering the possibility of Sangamon ages for any or all of these faunas, the widespread occurrence of *B. a. antiquus* during the Wisconsin, and the probability that speciation and dispersal of a savanna-adapted taxon would have been more likely during an interglacial than a glacial period, a Sangamon appearance for *B. a. antiquus* is tentatively accepted here. Table 24 lists geologic ages reported for faunas which include *B. a. antiquus*.

B. antiquus was a monotypic taxon until the very late Wisconsin-early Holocene, at which time environmental changes brought about changes in the population dynamics, numbers, and distribution of the species. These changes were accompanied by morphological divergences among local populations. Those *B. antiquus* populations that antedated this period of divergence, and those that remained relatively unchanged morphologically during the period of divergence, are herein recognized as *B. a. antiquus*, the nominate and conservative subspecies. Those *B. antiquus* populations that became morphologically differentiated from earlier phenotypes of the taxon are recognized as *B. a. occidentalis*. Following this relatively brief period of allopatry and morphological divergence (lasting from about 11,000 BP to about 9,500 BP), range shifts brought about a probable fusion (i.e., sympatry and hybridization) of the two subspecies. This period

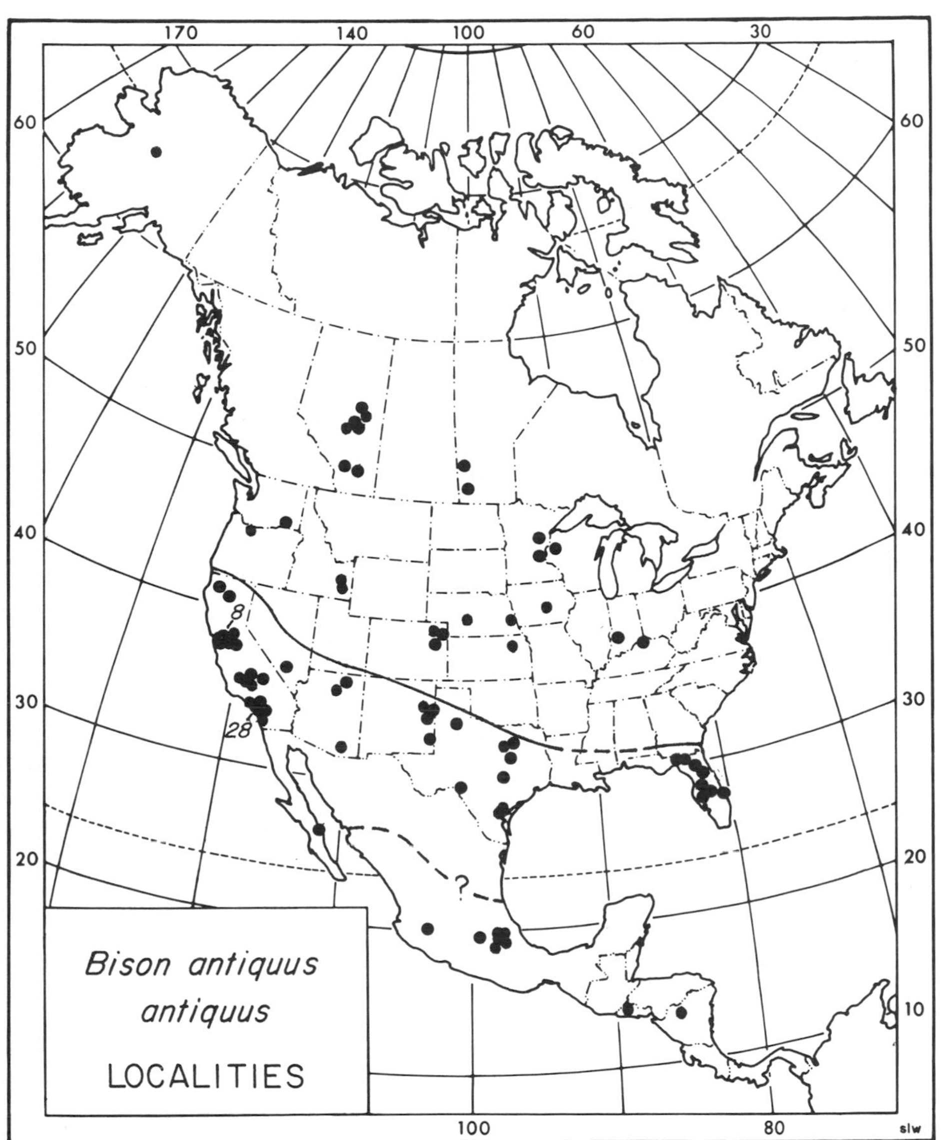

FIGURE 20 The distribution of *Bison antiquus antiquus* specimens used in this study. The primary range of this taxon was apparently what is now the southwestern United States and northern Mexico. A major outlier existed in Florida. These areas are bounded by the solid and broken lines. Sporadic populations occurred southward into Central America as far as Nicaragua. Most occurrences north of the primary range probably date from the late-Wisconsin to middle-Holocene period, and represent a range shift made possible largely by the extinction of *Bison latifrons* and the withdrawal of continental and cordilleran glaciers (see figs. 97–103). Numbers indicate total specimens from clustered localities.

of coalescence lasted from about 9,500 BP to about 5,000–4,000 BP, after which time a new phenotype—*B. bison*—appeared. The appearance and radiation of *B. bison* is recognized here as the termination of *B. antiquus*.

B. a. antiquus probably reached maximum population numbers and densities during the Sangamon interglacial and Wisconsin interstades, when temperate savanna and grassland habitats were most widespread. Maximum distribution (i.e., greatest range size) probably occurred during the late Wisconsin when northward dispersals into the former range of *B. latifrons*, perhaps as far as Beringia, appear to have taken place.

Discussion:

Validity of the name *B. antiquus*: *B. antiquus* was described by Leidy (1852*a*) on the basis of a partial skull from Big Bone Lick, Kentucky (plate 8)—a specimen here considered to represent a hybrid individual (see chapter 6). The name *B. antiquus* has since been used widely, and now almost without exception, with reference to the late Pleistocene shorter horned bison of the southwestern United States and adjoining regions. Article 1 of the *International Code of Zoological Nomenclature* (1964 revised edition) excludes names given to hybrids. Such names are, therefore, technically unavailable (Mayr, 1969:349). Stability of nomenclature, however, is of paramount importance. A change of names from *B. antiquus* to the senior synonym, *B. scaphoceras* (which has been synonymized with *Ovis scaphoceras* since 1899) would be extremely impractical. The widespread use of *B. antiquus*, therefore, argues for its de facto retention as the preferred name for the late Pleistocene shorter horned American bison. Further support for the retention of *B. antiquus* comes from the inferred phyletic ancestry of the holotype; this specimen shows characters of both *B. antiquus* (antero-posterior plane parallel to frontal plane; near lateral

Table 24

Geologic Ages Reported for *Bison antiquus antiquus* Specimens

Locality	Reported age	Dating method	Source
Lago de Chapala, Jalisco	Late Pleistocene	Biostratigraphy	Downs, 1958
	30,000 BP ±	Radiocarbon	F. R. Solorzano, oral comm., July 14, 1976
Tequixquiac, Mexico	Wisconsin	Geostratigraphy	Hibbard, 1955b
Valsequillo, Puebla	>35,000 – 9,150 BP	Radiocarbon	W-1898*, W-1896*
Costeau Pit, California	Sangamon – early Wisconsin	Biostratigraphy	Miller, 1971
	>42,000 BP	Radiocarbon	Miller, 1971
Davis and Teichert Sites, California	186,000 – 45,000 BP	Uranium; actinium	Hansen and Begg, 1970
McKittrick Tar Seep, California	38,000 BP	Radiocarbon	UCLA-728*
Rancho la Brea, California	<40,000 – >12,000 BP	Radiocarbon	UCLA-773C*, UCLA-1292B*
Aucilla River, Florida	Wisconsin	Geostratigraphy	Robertson, 1974
Bradenton Canal, Florida	Sangamon	Biostratigraphy	Robertson, 1974
Santa Fe River, Florida	Wisconsin	Geostratigraphy	Robertson, 1974
Wacissa River, Florida	Early Holocene	Biostratigraphy	Robertson, 1974
Wakulla River, Florida	Wisconsin	Geostratigraphy	Robertson, 1974
West Palm Beach, Florida	21,150 BP	Radiocarbon	Robertson, 1974
Big Bone Lick, Kentucky	17,000 – 10,600 BP ±	Radiocarbon	W-1358*, W-1617*
Blackwater Draw, New Mexico			
Gray Sand Horizon	Late Wisconsin	Geostratigraphy	Hester, 1972
Brown Sand Horizon	11,600 – 11,040 BP	Radiocarbon	Hester, 1972
Lugow Pit, Texas	Mid-Wisconsin	Geostratigraphy	Slaughter, 1966
San Patricio County, Texas	Early Wisconsin	Geostratigraphy	Lundelius, 1972

*Denotes radiocarbon date laboratory number.

direction of horn core; base isosceles triangle in cross section) and *B. priscus* (sinuous posterior profile), but those of the former are more pronounced. An appeal has been filed with the International Commission of Zoological Nomenclature to validate the name and traditional usage of *B. antiquus*.

Synonymy: *Bos scaphoceras* was described from a proximal and medial horn core (plate 10) from Nicaragua (Cope, 1895). Lucas (1899*a*) considered this specimen to be the horn core of a sheep and referred to it as *Ovis scaphoceras*. Subsequent students of bison have not studied the specimen, but the triangular cross section, dorsal groove, basal burr, and growth pattern of the core are all characteristically *Bison*, not *Ovis*. The configuration of the core is somewhat abnormal for bison, but is within the range of variation of other Pleistocene *B. antiquus* horn cores from Puebla, Michoacan, northern California, and Florida, all of which may owe their differences to genetic drift or inbreeding. Bison populations in tropical America were probably quite small and sporadic in occurrence and genetic drift or inbreeding is, under such circumstances, theoretically likely. I therefore synonymize *B. scaphoceras* with *B. antiquus*.

Rhoads (1897) described *B. californicus* from a partial skull (ANSP G 297) containing horn cores and frontals found in the Pilarcitos Valley near San Francisco, California. This specimen is *B. antiquus*, morphologically and with provenience more typical of the taxon than is the holotype.

B. pacificus was described by Hay (1927) on the basis of a distal horn core fragment found near Gottville, Siskiyou County, California. This specimen is identical to the tips of horn cores in *B. antiquus* and, consequently, *B. pacificus* is synonymized with that species.

Wilson (1974*b*) proposed *B. bison antiquus* as a replacement for *B. antiquus* (i.e., for *B. a. antiquus* of this classification). Pleistocene and late Holocene bison are morphologically different and existed under significantly different selection regimes. In my opinion, they are, therefore, better and more meaningfully understood as different species.

IIb. *Bison antiquus occidentalis* (Lucas, 1898)

Original Description:
Frederic A. Lucas. 1898. (The fossil bison of North America). *Science*, n.s., vol. 8, p. 678.
Frederic A. Lucas. 1899*b*. "The Characters of *Bison occidentalis*, the fossil bison of Kansas and Alaska." *Kansas University Quarterly*, ser. A, vol. 8, pp. 17–18. (Description).
Type Locality:
Fort Yukon, Alaska.
Type Location:
Department of Paleobiology, United States National Museum, Washington, D.C.
Type Specimen:
USNM P 4157—frontals, occipital, complete right and proximal two-thirds of left horn cores (plate 11).
Principal Synonymy:
Bison kansensis McClung, 1904
Bison texanus Hay and Cook, 1928
Bison taylori Hay and Cook, 1928
Bison figginsi Hay and Cook, 1928
Simobison figginsi Hay and Cook, 1930
Bison oliverhayi Figgins, 1933
Stelabison occidentalis Figgins, 1933
Stelabison occidentalis francisi Figgins, 1933.

Bison bison occidentalis Fuller and Bayrock, 1965

Subspecies Content:

B. a. occidentalis was the early and middle Holocene steppe bison of North America, a taxon characterized by great morphological variability resulting from extreme population fluctuations, rapidly and extensively changing environments, and a reorganized selection regime.

Diagnosis:

Male horn cores triangular (isosceles) in cross section and symmetrical about dorso-ventral axis at base; tip circular or elliptical in cross section; posterior margin straight or concave; growth straight or spiraled around arched or sinuous longitudinal axis; antero-posterior plane rotated rearward from frontal plane; length along upper curve greater than 175 mm, less than 400 mm; less than 11,000 BP, greater than 5,000–4,000 BP (plate 11).

Female horn cores circular to elliptical in cross section at base; symmetrical about greatest and least diameters at base; tip circular to elliptical in cross section; posterior margin straight; growth straight along mildly arched longitudinal axis; plane of greatest diameter usually rotated forward from frontal plane; length along upper curve greater than 150 mm, less than 250 mm (plate 12).

Differential Diagnosis:

Male *B. a. antiquus* tips cordiform or triangular in cross section; posterior margin straight; growth straight along arched longitudinal axis; antero-posterior plane parallel with frontal plane. *B. bison* growth nearly or clearly straight about arched longitudinal axis. *B. latifrons* horn core broadly triangular in cross section at base; cordiform or triangular in cross section at tips; posterior margin straight; longitudinal axis arched; length along upper curve greater than 500 mm. *B. priscus* and *B. alaskensis* horn cores asymmetrical about dorso-ventral axis; tips cordiform or triangular in cross section; posterior margins sinuous; growth spiraled around sinuous longitudinal axis; antero-posterior plane rotated forward from frontals; length along upper curve greater than 275 mm. Females as with *B. a. antiquus* females. Specimens resembling *B. a. antiquus* but greater than 11,000 BP are probably *B. a. antiquus*.

Biometric Summary:

Skulls, table 25.

Limb bones, table 26.

Referred Specimens:

Table 27.

Description:

B. a. occidentalis is a highly variable taxon. Male horn cores are moderately short and compact, but exhibit a relatively wide range of variation in length, basal circumference, and orientation. The length along the upper curve is about equal to the circumference of the core at the base. The horn cores emanate horizontally or suprahorizontally from the frontals at an angle of about 65° to 75° forward from the sagittal plane. The cores are triangular (isosceles) in cross section at the base, where the antero-posterior and dorso-ventral diameters are about equal. They taper gradually toward the tip. The cores vary greatly in curvature; generally, they curve upward in a continuous arc that begins at or near the base. Primitive *B. a. occidentalis*, however, often had long, slender, and less curved horn cores. The core is usually rotated rearward from the vertical, and the distal core is normally deflected to the rear. A dorsal groove is normally not present on the distal core. The horn core tip is much thinner and more pointed than that of *B. a. antiquus* and other late Pleistocene bison; it normally is situated above the frontal plane and behind the occipital plane. Sheath characteristics are not known for *B. a. occidentalis*, but the growth pattern and orientation of the horn core of this taxon suggest that the sheath tips were deflected to the rear and probably occurred behind the occipital plane.

The frontals vary from flat to moderately domed. The orbits vary from high to low, but are more frequently low than high; domed frontals and low orbits are usually correlated. Nasals are low, broad and triangular. The occipital is relatively flat. The nuchal line/external occipital protuberance is prominent, but less so than in earlier bison. The occipital condyles protrude from the occipital plane less strongly than in pre-Holocene bison. The ventro-lateral margins of the occipital condyles are occasionally but not normally flanged. The *B. a. occidentalis* skull is less robust overall than that of any other late Pleistocene bison.

Female *B. a. occidentalis* horn cores are short and slender, and much less robust than female *B. a. antiquus* horn cores. The horn cores emanate horizontally from the frontals at an angle of about 70° to 80° forward from the sagittal plane. The horn cores are circular to elliptical in cross section at the base, and taper gradually toward the tip. The cores curve upward gradually (less strongly than in *B. a. antiquus*). The tip is near (usually

Table 25

Summary of *Bison antiquus occidentalis* Skull Biometrics

Standard Measurement	n	Range	Mean ± SE	σ	V
Males					
Spread of horn cores, tip to tip	77	626 – 1055 mm	779.3 ± 8.8 mm	76.9 mm	9.9
Horn core length, upper curve, tip to burr	86	186 – 392 mm	277.8 ± 4.2 mm	39.1 mm	14.1
Straight line distance, tip to burr, dorsal horn core	81	175 – 350 mm	248.1 ± 3.5 mm	31.8 mm	12.8
Dorso-ventral diameter, horn core base	85	70 – 114 mm	94.6 ± 0.9 mm	8.4 mm	8.9
Minimum circumference, horn core base	89	237 – 355 mm	300.3 ± 2.9 mm	27.7 mm	9.2
Width of occipital at auditory openings	61	238 – 294 mm	262.0 ± 1.7 mm	13.2 mm	5.0
Width of occipital condyles	71	111 – 151 mm	135.0 ± 0.9 mm	7.7 mm	5.7
Depth, nuchal line to dorsal margin of foramen magnum	57	89 – 120 mm	104.0 ± 0.9 mm	7.0 mm	6.7
Antero-posterior diameter, horn core base	91	77 – 120 mm	98.8 ± 1.1 mm	10.0 mm	10.1
Least width of frontals, between horn cores and orbits	74	261 – 348 mm	296.6 ± 2.0 mm	16.8 mm	5.7
Greatest width of frontals at orbits	64	311 – 394 mm	348.0 ± 2.1 mm	16.7 mm	4.8
M^1–M^3, inclusive alveolar length	3	90 – 102 mm	97.3 ± 3.7 mm	6.4 mm	15.6
M^3, maximum width, anterior cusp	2	27.8 – 29.1 mm	28.4 ± 0.7 mm	0.9 mm	3.2
Distance, nuchal line to tip of premaxillae	25	511 – 606 mm	564.3 ± 5.0 mm	24.8 mm	4.4
Distance, nuchal line to nasal-frontal suture	57	233 – 287 mm	259.5 ± 1.6 mm	12.3 mm	4.7
Angle of divergence of horn cores, forward from sagittal	62	63° – 83°	72.1° ± 0.7°	5.2°	7.2
Angle between foramen magnum and occipital planes	57	110° – 142°	129.6° ± 1.0°	7.3°	5.7
Angle between foramen magnum and basioccipital planes	56	98° – 126°	113.4° ± 0.7°	5.6°	4.9
Females					
Spread of horn cores, tip to tip	13	556 – 749 mm	614.4 ± 14.2 mm	51.3 mm	8.4
Horn core length, upper curve, tip to burr	19	165 – 235 mm	197.1 ± 4.4 mm	19.0 mm	9.6
Straight line distance, tip to burr, dorsal horn core	16	154 – 212 mm	182.9 ± 3.5 mm	14.0 mm	7.7
Dorso-ventral diameter, horn core base	25	54 – 71 mm	60.4 ± 1.0 mm	4.8 mm	8.0
Minimum circumference, horn core base	24	168 – 219 mm	191.5 ± 2.7 mm	13.4 mm	7.0
Width of occipital at auditory openings	11	208 – 237 mm	217.3 ± 2.4 mm	8.0 mm	3.7
Width of occipital condyles	12	115 – 140 mm	126.8 ± 2.0 mm	7.0 mm	5.5
Depth, nuchal line to dorsal margin of foramen magnum	11	78 – 99 mm	86.9 ± 1.8 mm	5.9 mm	6.8
Antero-posterior diameter, horn core base	27	52 – 73 mm	60.3 ± 1.1 mm	5.7 mm	9.3
Least width of frontals, between horn cores and orbits	19	214 – 262 mm	238.2 ± 2.8 mm	12.3 mm	5.2
Greatest width of frontals at orbits	12	276 – 310 mm	292.3 ± 3.2 mm	10.9 mm	3.7
M^1–M^3, inclusive alveolar length	4	95 – 101 mm	97.3 ± 1.4 mm	2.9 mm	3.0
M^3, maximum width, anterior cusp	4	22.4 – 28.1 mm	24.3 ± 1.3 mm	2.6 mm	10.8
Distance, nuchal line to tip of premaxillae	8	503 – 557 mm	523.5 ± 6.3 mm	17.8 mm	3.4
Distance, nuchal line to nasal-frontal suture	15	200 – 246 mm	227.3 ± 3.5 mm	13.7 mm	6.0
Angle of divergence of horn cores, forward from sagittal	15	68° – 84°	75.0° ± 1.1°	4.4°	5.9
Angle between foramen magnum and occipital planes	9	122° – 140°	131.7° ± 2.4°	7.1°	5.4
Angle between foramen magnum and basioccipital planes	9	103° – 119°	110.2° ± 1.8°	5.4°	4.9

Table 26

Summary of *Bison antiquus occidentalis* Limb Biometrics

Standard measurement	n	Range	Mean ± SE	σ	V
Males					
Humerus:					
Approximate rotational length of bone	39	326 – 402 mm	356.9 ± 3.3 mm	21.2 mm	5.9
Antero-posterior diameter of diaphysis	41	58 – 76 mm	66.4 ± 0.7 mm	4.5 mm	6.8
Transverse minimum of diaphysis	41	48 – 66 mm	55.0 ± 0.6 mm	4.2 mm	7.7
Radius:					
Approximate rotational length of bone	82	306 – 381 mm	343.8 ± 1.6 mm	14.7 mm	4.2
Antero-posterior minimum of diaphysis	83	30 – 39 mm	35.5 ± 0.2 mm	2.1 mm	5.9
Transverse minimum of diaphysis	83	52 – 69 mm	58.5 ± 0.3 mm	3.3 mm	5.6
Metacarpal:					
Total length of bone	139	199 – 240 mm	219.8 ± 0.7 mm	8.6 mm	3.9
Antero-posterior minimum of diaphysis	143	26 – 35 mm	30.1 ± 0.1 mm	1.8 mm	6.0
Transverse minimum of diaphysis	142	40 – 61 mm	51.4 ± 0.3 mm	3.7 mm	7.2
Femur:					
Approximate rotational length of bone	38	362 – 448 mm	403.7 ± 2.9 mm	18.0 mm	4.4
Antero-posterior diameter of diaphysis	40	44 – 62 mm	50.0 ± 0.5 mm	3.2 mm	6.5
Transverse minimum of diaphysis	40	44 – 55 mm	49.3 ± 0.3 mm	2.3 mm	4.7
Tibia:					
Approximate rotational length of bone	63	372 – 464 mm	403.5 ± 2.2 mm	17.6 mm	4.3
Antero-posterior minimum of diaphysis	64	34 – 45 mm	38.9 ± 0.2 mm	2.2 mm	5.8
Transverse minimum of diaphysis	64	49 – 61 mm	53.5 ± 0.3 mm	3.0 mm	5.6
Metatarsal:					
Total length of bone	150	248 – 300 mm	270.3 ± 0.8 mm	10.0 mm	3.7
Antero-posterior minimum of diaphysis	161	29 – 38 mm	33.1 ± 0.1 mm	1.8 mm	5.5
Transverse minimum of diaphysis	156	32 – 48 mm	39.2 ± 0.2 mm	2.9 mm	7.5
Females					
Humerus:					
Approximate rotational length of bone	45	286 – 395 mm	321.7 ± 3.1 mm	21.2 mm	6.6
Antero-posterior diameter of diaphysis	77	44 – 67 mm	54.1 ± 0.5 mm	4.5 mm	8.3
Transverse minimum of diaphysis	76	35 – 44 mm	44.5 ± 0.4 mm	3.6 mm	8.2
Radius:					
Approximate rotational length of bone	121	284 – 372 mm	315.2 ± 1.4 mm	16.1 mm	5.2
Antero-posterior minimum of diaphysis	119	25 – 37 mm	29.8 ± 0.1 mm	2.1 mm	7.0
Transverse minimum of diaphysis	119	37 – 57 mm	47.3 ± 0.3 mm	3.4 mm	7.2
Metacarpal:					
Total length of bone	323	194 – 233 mm	211.2 ± 0.4 mm	7.4 mm	3.5
Antero-posterior minimum of diaphysis	331	22 – 32 mm	26.5 ± 0.1 mm	1.5 mm	5.8
Transverse minimum of diaphysis	331	32 – 49 mm	40.3 ± 0.1 mm	2.9 mm	7.3
Femur:					
Approximate rotational length of bone	28	388 – 397 mm	367.6 ± 3.2 mm	17.1 mm	4.6
Antero-posterior diameter of diaphysis	36	35 – 54 mm	43.6 ± 0.6 mm	3.6 mm	8.3
Transverse minimum of diaphysis	36	39 – 48 mm	43.3 ± 0.3 mm	2.3 mm	5.3
Tibia:					
Approximate rotational length of bone	71	311 – 419 mm	362.2 ± 2.2 mm	18.8 mm	5.2
Antero-posterior minimum of diaphysis	75	29 – 39 mm	33.5 ± 0.2 mm	2.1 mm	6.4
Transverse minimum of diaphysis	75	40 – 54 mm	46.1 ± 0.3 mm	3.0 mm	6.6
Metatarsal:					
Total length of bone	297	230 – 284 mm	259.9 ± 0.5 mm	9.6 mm	3.6
Antero-posterior minimum of diaphysis	297	25 – 35 mm	30.2 ± 0.1 mm	1.9 mm	6.2
Transverse minimum of diaphysis	297	25 – 39 mm	32.1 ± 0.1 mm	2.6 mm	8.2

Table 27

Bison antiquus occidentalis: Referred Specimens

Institution and identification number		Sex	Provenience
AMNH	VP F:AM14479	♂	Near Great Falls, Cascade County, Montana
AMNH	VP F:AMA508-5331	♂	Near Fairbanks, Alaska
AMNH	VP F:AM23347	♂	Near Silverton, Briscoe County, Texas
AMNH	VP F:AMAINS-749	♂	Soles Ranch, Sheridan County, Nebraska
AMNH	VP F:AMAINS-825-479-12	♂	Pine Creek, Brown County, Nebraska
AMNH	VP F:AMBB-1931	♂	Salt Creek, Hudspeth County (?), Texas
AMNH	VP F:AMCRO-79-2220	♂	Swartout, Texas
AMNH	VP F:AMFOL-16	♀	Folsom, Union County, New Mexico
AMNH	VP F:AMA*	♂	Folsom, Union County, New Mexico
AMNH	VP F:AMB*	♂	Folsom, Union County, New Mexico
ANSP	G 13731	♀	Near Clovis, New Mexico
ANSP	G 13750	♀	Near Clovis, New Mexico
BeMNH	198	♂	Minneapolis, Hennepin County, Minnesota
BeMNH	450	♂	Hennepin County, Minnesota
BeMNH	603a	♂	Nye, Wisconsin
BeMNH	603e	♂	Nye, Wisconsin
BeMNH	723	♂	Minneapolis, Hennepin County, Minnesota
BeMNH	738	♂	Taconite, Minnesota
BeMNH	3173	♂	Minneapolis, Hennepin County, Minnesota
BeMNH	3567	♂	Minneapolis, Hennepin County, Minnesota
BeMNH	3569	♀	Minneapolis, Hennepin County, Minnesota
CoMNH	VP 574	♂	Lone Wolf Creek, near Colorado, Mitchell County, Texas
CoMNH	VP 629	♂	Michies, Dawson County, Texas
CoMNH	VP 630	♂	Michies, Dawson County, Texas
CoMNH	VP 1236	♂	Folsom, Union County, New Mexico
CoMNH	VP 1240	♂	Folsom, Union County, New Mexico
FMNH	P 26249	♂	Near Columbus, Platte County, Nebraska
IMNH	VP 6377-109	♂	Pocatello, Bannock County, Idaho
IMNH	VP A*	♂	Wasden Archeological Site, Idaho
KU	VP 9905	♂	Kaw River, Near Morris, Kansas
KU	VP A*	♂	Near Twelve Mile Creek, Logan County, Kansas
MMMN	VP 129	♂	Betula Lake, White Shell Park, Manitoba
MMMN	VP 222	♀	Treesbank, Manitoba
MMMN	VP 411	♂	Russell, Manitoba
MMMN	VP 613	♂	Near Piney Fort, Carberry Sand Hills, Manitoba
MPM	VP 3323	♂	Interstate Bog, Wisconsin
NDSU	G 165[a]	♂	Spring Creek, near Zap, Mercer County, North Dakota
NDSU	G 166[a]	♂	Spring Creek, near Zap, Mercer County, North Dakota
NMC	VP 2242	♂	Upper Dominion Creek, Yukon Territory
NMC	VP 2813	♂	Arden, Manitoba
NMC	VP 8186	♂	Near Drumheller, Alberta
NMC	VP 17306	♂	Quartz Creek, Yukon Territory
NMC	VP 17525	♂	Edmonton, Alberta
NMC	VP 17530	♂	Rumsey, Alberta
PMA	P 68.2.131	♀	Duffield, Alberta
PMA	P 68.2.148	♀	Duffield, Alberta
PMA	P 68.2.150	♂	Duffield, Alberta
PMA	P 68.2.151	♂	Duffield, Alberta
PMA	P 68.2.195	♂	Duffield, Alberta
PMA	P 68.2.624	♂	Duffield, Alberta
PMA	P 69.17.19	♂	Duffield, Alberta

Table 27

Bison antiquus occidentalis: Referred Specimens (Continued)

Institution and identification number		Sex	Provenience
PMA	P 69.17.20	♀	Duffield, Alberta
PMA	P 71.6.5.	♂	Near Leduc, Alberta
PMA	P 74.16.2	♂	Duffield, Alberta
PPHM	2659-1	♂	Near Crowell, Hardeman County, Texas
PPHM	A*	♂	Rex Rodgers Archeological Site, Briscoe County, Texas
PPHM	B*	♀	Rex Rodgers Archeological Site, Briscoe County, Texas
SIU	P 199[b]	♂	Sheridan Dam, Pennington County, South Dakota
SM	62176	♂	Texas
SM	A*	♂	Texas
SMNH	P1240a	♂	South of Outlook, Saskatchewan
SMNH	P1240b	♂	South of Outlook, Saskatchewan
SMNH	P1240c	♂	South of Outlook, Saskatchewan
SMP	21-67-z	♂	Northwest Iowa
SMP	43-51-z	♂	Cherokee County, Iowa
SMP	48-59-z	♂	Magnusson Farm, near Washta, Iowa
SMP	153-56-z	♂	Polk County, Iowa
SMP	155-56-z	♂	Polk County, Iowa
SMP	156-56-z	♂	Polk County, Iowa
SMP	157-56-z	♂	Polk County, Iowa
SMP	A*	♂	Clay County, Iowa
SMP	B*	♂	Plymouth County, Iowa
SMP	C*	♂	Simonsen Archeological Site, Cherokee County, Iowa
SPSM	P 67.1.34	♀	Minnesota
SUI	34815[c]	♂	Hughes Bog, near Marion, Linn County, Iowa
SUI	38959	♂	Dows Bog, Iowa
TMM	725-A*	♀	Plainview Archeological Site, Texas
TMM	725-B*	♂	Plainview Archeological Site, Texas
TMM	802-1	♂	Near Brownfield, Terry County, Texas
TMM	892-2E	♂	Lubbock Lake Archeological Site, Lubbock County, Texas
TMM	892-3E	♀	Lubbock Lake Archeological Site, Lubbock County, Texas
TMM	892-46	♂	Lubbock Lake Archeological Site, Lubbock County, Texas
TMM	892-77	♂	Lubbock Lake Archeological Site, Lubbock County, Texas
TMM	A*	♂	Bonfire Shelter, Val Verde County, Texas
TTU	A 1975/18	♂	Canyon Lake, near Lubbock, Lubbock County, Texas
TTU	A 2299	♂	Lubbock Lake Archeological Site, Lubbock County, Texas
TTU	A 2300	♂	Lubbock Lake Archeological Site, Lubbock County, Texas
TTU	A 11099	♂	Lubbock Lake Archeological Site, Lubbock County, Texas
UAI	620	♂	Dunregan, Alberta
UCy	A OS-340	♂	Athabasca, Alberta
UCy	G 2691[d]	♂	Milan, near Three Hills, Alberta
UCy	G 3417[d]	♂	Milan, near Three Hills, Alberta
UCM	A 10694	♂	Olsen-Chubbock Archeological Site, Cheyenne County, Colorado
UCM	A 10829	♀	Olsen-Chubbock Archeological Site, Cheyenne County, Colorado
UCM	A 10866	♂	Olsen-Chubbock Archeological Site, Cheyenne County, Colorado
UCM	A A*	♀	Olsen-Chubbock Archeological Site, Cheyenne County, Colorado
UNSM	VP 1-31-11-31	♂	Fort Robinson, Hall County, Nebraska
UNSM	VP 2-10-6-32SP	♂	Scottsbluff Archeological Site, Scottsbluff County, Nebraska
UNSM	VP 201-58	♀	Nebraska (?)
UNSM	VP 491-10-6-32SP	♀	Nebraska (?)
UNSM	VP 1196-52	♀	Nebraska (?)
UNSM	VP 6207-72	♂	Near Concord, Dixon County, Nebraska

Table 27
Bison antiquus occidentalis: Referred Specimens (Continued)

Institution and identification number		Sex	Provenience
UNSM	VP 30362	♂	Custer County, Nebraska
UNSM	VP 30784	♀	Cuming County, Nebraska
UNSM	VP 30809	♀	Scottsbluff County, Nebraska
UNSM	VP 30811	♀	Scottsbluff County, Nebraska
UNSM	VP A[e]*	♂	Lipscomb Archeological Site, Lipscomb County, Texas
UNSM	VP B[e]*	♂	Meserve Archeological Site, Hall County, Nebraska
UNSM	VP C[e]*	♂	Scottsbluff Archeological Site, Scottsbluff County, Nebraska
USGS	18-93	♂	Olsen-Chubbock Archeological Site, Cheyenne County, Colorado
USGS	A*	♀	Olsen-Chubbock Archeological Site, Cheyenne County, Colorado
USNM	P 2325	♂	Talto River, Alaska
USNM	P 4157	♂	Fort Yukon, Alaska
USNM	P 5513	♂	Pelly River, Yukon Territory
USNM	P 10541	♂	Sagamore Mine, Riverton, Crow Wing County, Minnesota
USNM	P 10545	♂	Sagamore Mine, Riverton, Crow Wing County, Minnesota
USNM	P 10546	♂	Sagamore Mine, Riverton, Crow Wing County, Minnesota
USNM	P 11164	♀	Minturn Strip Mine, Campbell County, Wyoming
USNM	P 16860	♀	Sagamore Mine, Riverton, Crow Wing County, Minnesota
USNM	P 19111	♂	South of Devils Lake, North Dakota
USNM	P 19112	♂	South of Devils Lake, North Dakota
USNM	P A*	♂	Lindenmeier Archeological Site, Colorado
USNM	P B*	♂	Lindenmeier Archeological Site, Colorado
USNM	P C*	♂	Valsequillo, Puebla
UW	8	♂	Hawken II Archeological Site, Wyoming
UW	60218	♂	Hawken I Archeological Site, Wyoming
UW	60220	♂	Hawken I Archeological Site, Wyoming
UW	60352	♀	Hawken I Archeological Site, Wyoming
UW	60369	♀	Hawken I Archeological Site, Wyoming
UW	B0006	♂	Kerr-McGee Pit, Shirley Basin, Wyoming
UW	B1998	♀	Hawken II Archeological Site, Wyoming
UW	CKH-303-60219	♂	Hawken I Archeological Site, Wyoming
UW	CKH-351-60221	♂	Hawken I Archeological Site, Wyoming
UW	CKH-352-60222	♂	Hawken I Archeological Site, Wyoming
UW	CKH-355-A*	♂	Hawken I Archeological Site, Wyoming
UW	CKH-60217	♂	Hawken I Archeological Site, Wyoming
UW	NC2326	♂	Casper Archeological Site, Natrona County, Wyoming
UW	NC2560	♀	Casper Archeological Site, Natrona County, Wyoming
UW	A*	♀	Hawken II Archeological Site, Wyoming
UW	B*	♀	Casper Archeological Site, Natrona County, Wyoming
UW	C*	♀	Casper Archeological Site, Natrona County, Wyoming
UW	D*	♂	Finley Archeological Site, Wyoming
UW	E*	♂	Hawken II Archeological Site, Wyoming
WM	Z 550[f]	♂	Fort Peck, Montana
WTSU	P180-1	♂	Palo Duro Canyon, Texas

Sources of supplementary data:
[a]Brophy, 1965;
[b]Galbreath and Stein, 1962;
[c]Hall, 1972;
[d]Shackleton and Hills, 1977;
[e]Hillerud, 1970;
[f]Rasmussen, 1974.
*No identification number was found for these specimens.

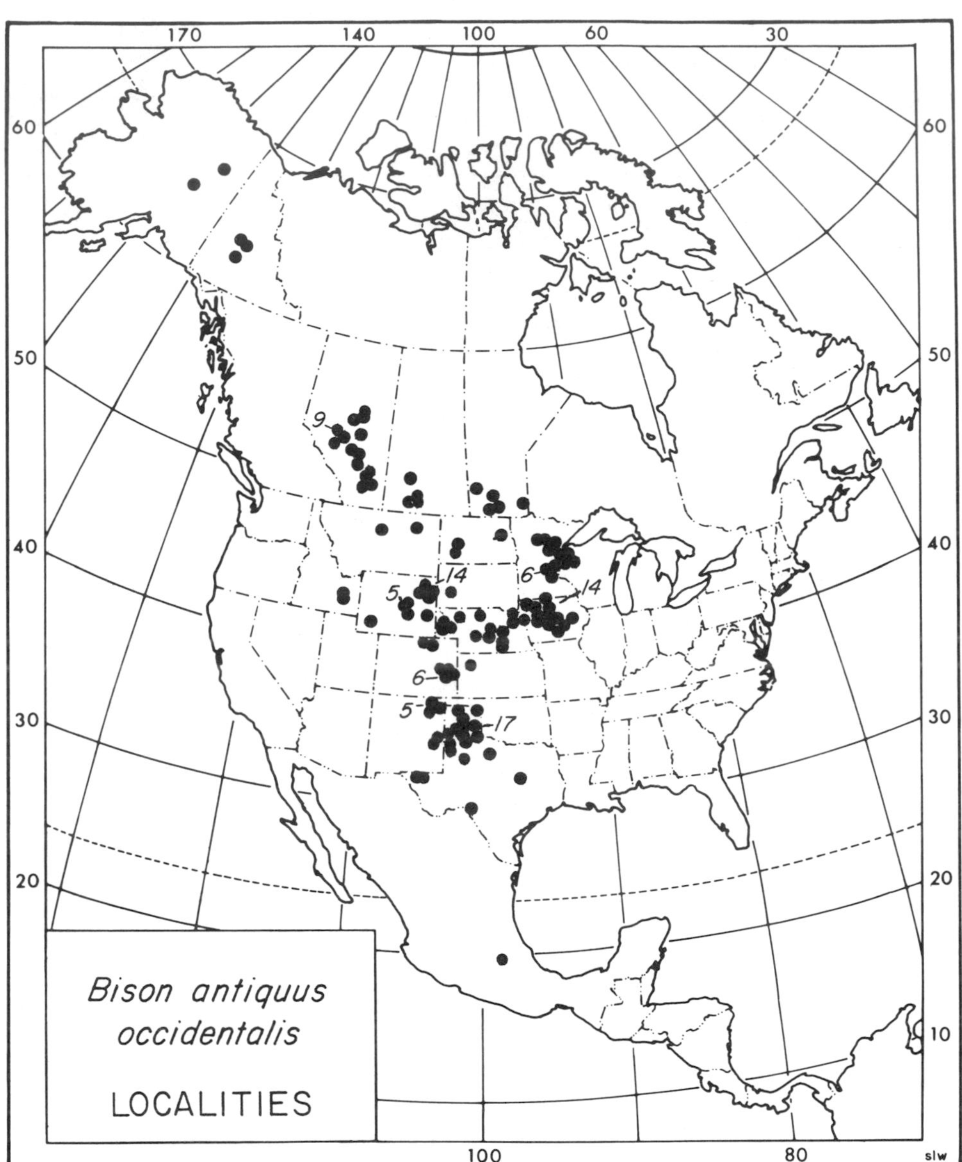

FIGURE 21 The distribution of *Bison antiquus occidentalis* specimens used in this study. *Bison antiquus occidentalis* apparently originated in the western Great Plains about 11,000 years ago, and from there spread northward and eastward, following the expanding grassland and its prairie margin, until becoming anchored on the northern plains about 7,000 years ago. Some populations passed through the boreal forest filter and reached Alaska and northwest Canada. *Bison antiquus occidentalis,* which evolved into *Bison bison* about 5,000–4,000 years ago, was primarily restricted to the central grassland region, although some populations existed far to the south and north. Numbers indicate total specimens from clustered localities.

just above) the frontal plane and near or just behind the occipital plane.

The frontals of female *B. a. occidentalis* are crenulated transversely, being convex near the horn cores and the sagittal suture line, and concave between these regions. The orbits vary from relatively low to relatively high; they are directed forward and protrude less than in earlier bison. The occipital is relatively flat. The nuchal line/external occipital protuberance is prominent, but less so than in female *B. a. antiquus*. The occipital condyles do not protrude from the occipital plane. The ventro-lateral margins of the occipital condyles are normally not flanged. The female *B. a. occidentalis* skull is less robust than that of any pre-Holocene female bison.

Distribution:

B. a. occidentalis is known from an area extending from the Valsequillo area of Puebla north to the Yukon Valley of Alaska. Most records are from the Great Plains and adjacent prairies, with fewer records from the northeastern Great Basin and the Yukon Valley (fig. 21). Specimens of *B. a. occidentalis* have not been dated earlier than about 10,990 BP nor later than about 5,440 BP (table 28). *B. a. occidentalis* spans a very brief period of geologic time and appears to represent the adaptive transition from *B. a. antiquus* to *B. bison*. The temporal and spatial distribution record for this taxon suggests that it evolved in midlatitude North America during very late Wisconsin time, and that it was probably a grazer/browser primarily associated with open grasslands and prairies. *B. a. occidentalis* probably attained maximum distribution, densities, and overall population size during the early middle Holocene, between about 9,500 BP and 6,500 BP (i.e., following the wave of late Wisconsin megafaunal extinctions and preceding the Altithermal/Atlantic maximum; see chapter 7).

Table 28

Radiocarbon Ages Reported for *Bison antiquus occidentalis* Localities

Locality	Radiocarbon age	Laboratory number
Duffield, Alberta	8,150 – 7,350 BP	S-106, S-107
Russell, Manitoba	6,320 BP	GSC-280
Valsequillo, Puebla	>35,000 – 9,150 BP	W-1898, W-1896
Lindenmeier, Colorado	10,990 BP*	I(UW)-141, I-622
Olsen-Chubbock, Colorado	10,150 BP	A-744
Wasden, Idaho (level 16)	7,100 BP	M-1853
Hughes Bog, Iowa	5,640 BP	ISGS-56
Simonsen, Iowa	8,430 BP	I(UW)-79
Zap, North Dakota	7,840 BP	I-2536
Zap, North Dakota	5,440 BP	W-1537
Bonfire Shelter, Texas	10,083 BP*	Tx-153, Tx-657, Tx-658
Lubbock Lake, Texas (Folsom level)	10,155 BP*	C-558, L-283G, SMU-292
Plainview, Texas	9,800 BP	L-303
Rex Rodgers, Texas	9,391 BP	SMU-274
Casper, Wyoming	9,945 BP*	RL-125, RL-208
Hawken, Wyoming	6,370 BP*	RL-185, RL-437
North Walker Pit, Wyoming	9,050 BP	RL-454

Note: This list includes only radiocarbon dates representing specimens that could be identified by horn cores.
*Denotes the arithmetic average of two or more dates.

Discussion:

Concept of the subspecies: *B. a. occidentalis* was the dominant early and middle Holocene bison of central and northwestern America. This subspecies has been the object of considerable confusion owing in part to its great variability and in part to its supposed Eurasian or Beringian origin (see, e.g., Geist, 1971*a*; Guthrie, 1970; Skinner and Kaisen, 1947, Wilson, 1969, 1974*a*, 1974*b*). The results of this study indicate, however, that *B. a. occidentalis* probably evolved directly from *B. a. antiquus* in the midlatitude grasslands of North America. The origin, variability, and dispersal history of this taxon can be attributed to (1) genetic drift and/or inbreeding in small populations isolated as a result of excessive human predation (ca. 11,000–9,500 BP); which (2) subsequently rebounded (after ca. 9,500 BP) and coalesced with other similarly drifted populations; and (3) with relict *B. a. antiquus* populations; while (4) simultaneously experiencing newly ordered selection pressures; in (5) a rapidly and extensively changing environment (see chapters 6, 7).

Synonymy: *B. kansensis* was described (McClung, 1904) on the basis of a partial horn core and adjacent skull found near North Lawrence, Kansas. The holotype (KU VP 388) differs only slightly from the holotype of *B. a. occidentalis*, and *B. kansensis* is thus synonymized with *B. a. occidentalis*.

B. texanus, *B. taylori*, and *B. figginsi* were briefly described by Hay and Cook (1928) and later described more fully by the same authors (1930). *B. texanus*, based on a skull (CoMNH VP 629) from Michies, Dawson County, Texas, was specifically differentiated by strongly curved horn cores. The authors considered *B. texanus* to be related to *B. occidentalis*. *B. taylori*, described from a partially crushed skull (CoMNH VP 1236) from the Folsom Bison Quarry, Union County, New Mexico, was specifically distinguished by its broad forehead, its narrow muzzle, and its widely spreading and drooping horn cores. *B. figginsi* (= *Simobison figginsi*) was discussed above under genus *Bison*. The holotype, a badly crushed and incomplete skull (CoMNH VP 574) with associated skeleton, was found on Lone Wolf Creek at Colorado, Mitchell County, Texas. All three are synonymized with *B. a. occidentalis*, because they are based either on minor and taxonomically insignificant individual variations or distorted reconstructions of badly damaged specimens.

B. oliverhayi, also from the Folsom Bison Quarry, was described from a badly crushed and restored skull (CoMNH VP 1240). Specific status was based on the supposed narrow skull of this specimen (Figgins, 1933), but it too falls within the expected range of individual variation for *B. a. occidentalis*.

Stelabison has been synonymized with *Bison* earlier in this chapter. *S. occidentalis francisi* was given taxonomic distinctiveness on the basis of a single, abnormal tooth (CoMNH VP 1363) from Hearne, Robertson County, Texas, but is here synonymized with *B. a. occidentalis*. Although positive identification of the tooth is impossible, the facts that (1) it was found in a region inhabited by *B. a. occidentalis*; (2) it appears to be of very late Pleistocene-early Holocene age (on the basis of preservation); and (3) it possesses abnormalities relatively common among very late Pleistocene-early Holocene bison teeth make this the most reasonable synonymy.

Fuller and Bayrock (1965) proposed the name *B. b. occidentalis* as a replacement for *B. occidentalis*, reasoning that *B. b. occidentalis* was found in western Canada during the late Wisconsin and, thereafter, evolved into *B. b. athabascae*. There is no sound evidence that *B. a. occidentalis* was found in western Canada during the late Wisconsin (table 28). According to Fuller and Bayrock, *B. occidentalis* would be more meaningfully classified if it were placed in the same species as *B. b. athabascae*. My reasons for recognizing *B. a. occidentalis* as a subspecies of *B. antiquus*, rather than of *B. bison*, were established earlier.

III. *Bison bison* (Linnaeus, 1758)

IIIa. *Bison bison bison* (Linnaeus, 1758)

Original Description:
Francisco Hernandez. 1651. *Rerum Medicarum Novae Hispaniae Thesaurus.* (Rome, Vitalis Mascardi), p. 587 and addendum (*Historiae Animalium et Mineralium Novae Hispaniae*), Chapter 30, p. 10. (Description).
Carolus Linnaeus. 1758. *Systema Naturae per Regna Tria Naturae.* (Stockholm: Laurentii Salvii), tenth edition, t. 1, pp. 71–72. (Name applied.)

Type Locality:
Canadian River Valley, eastern New Mexico.

Type Location:
None.

Type Specimen:
None. Hernandez's (1651) description is the Linnaean type for *B. bison.*

Principal Synonymy:
Bison sylvestris Hay, 1915
Bison americanus pennsylvanicus Shoemaker, 1915
Bison bison oregonus Bailey, 1932
Bison bison septemtrionalis Figgins, 1933
Bison bison haningtoni Figgins, 1933

Species Content:
B. bison is the late Holocene (extant) North American bison.

Subspecies Content:
B. b. bison is the late Holocene (extant) grassland bison of midlatitude North America.

Diagnosis:
Male horn cores triangular (isosceles) in cross section and symmetrical about the dorso-ventral axis at base; tip circular to elliptical in cross section; posterior margin straight; growth straight along arched longitudinal axis; antero-posterior plane rotated rearward from frontal plane; length along upper curve less than 250 mm; less than 5,000–4,000 BP; generally from south of boreal forest and environs of Canada (plate 13).

Female horn cores circular to elliptical in cross section at base; symmetrical about axis of greatest and least diameter; tip circular to elliptical in cross section; posterior margin straight; growth straight along straight or arched longitudinal axis; plane of greatest diameter rotated forward from vertical; lacks dorsal basal burr; length along upper curve less than 200 mm (plate 14).

Differential Diagnosis:
Male *B. a. antiquus* horn cores with antero-posterior plane parallel with frontal plane; tip cordiform to triangular in cross section; greater than 5,000–4,000 years BP. *B. a. occidentalis* horn cores frequently spiraled around arched or sinuous longitudinal axis. *B. latifrons*, *B. alaskensis*, and *B. priscus* horn cores greatly dorso-ventrally compressed; tips cordiform to triangular in cross section with prominent dorsal grooves; broadly triangular at base; all greater than 250 mm along upper

curve. Female *B. antiquus* horn cores normally more strongly arched; greater than 5,000–4,000 years BP. *B. alaskensis* and *B. priscus* have posterior margin convex or sinuous; length greater than 177 mm along upper curve. *B. latifrons* with growth spiraled around longitudinal axis; length along upper curve and basal circumference greatly exceed maxima known for *B. bison*.

Population means of *B. b. athabascae* females and males average larger than population means for *B. b. bison*, but individual specimens of either sub-species may be undifferentiable.

Biometric Summary:
Skulls, table 29.
Limb bones, table 30.
Referred Specimens:
Table 31
Description:

B. b. bison is the smallest North American bison taxon ever to live. Male horn cores are short and compact. They emanate horizontally, or occasionally subhorizontally, from the frontals at an angle of about 65° to 70° forward from the sagittal plane. The horn cores are triangular (isosceles) in cross section at the base, where the antero-posterior and dorso-ventral diameters are about equal. They taper

Table 29

Summary of *Bison bison bison* Skull Biometrics

Standard measurement	n	Range	Mean ± SE	σ	V
Males					
Spread of horn cores, tip to tip	128	510 – 778 mm	603.9 ± 3.9 mm	44.7 mm	7.4
Horn core length, upper curve, tip to burr	134	124 – 270 mm	190.7 ± 2.1 mm	24.7 mm	13.0
Straight line distance, tip to burr, dorsal horn core	132	120 – 243 mm	172.4 ± 1.9 mm	21.4 mm	12.4
Dorso-ventral diameter, horn core base	139	69 – 99 mm	81.9 ± 0.5 mm	6.4 mm	7.9
Minimum circumference, horn core base	142	199 – 324 mm	255.4 ± 1.6 mm	19.5 mm	7.6
Width of occipital at auditory openings	120	220 – 270 mm	243.9 ± 0.9 mm	9.7 mm	4.0
Width of occipital condyles	122	111 – 140 mm	126.6 ± 0.5 mm	5.7 mm	4.5
Depth, nuchal line to dorsal margin of foramen magnum	112	81 – 115 mm	98.7 ± 0.6 mm	6.2 mm	6.2
Antero-posterior diameter, horn core base	142	67 – 103 mm	83.4 ± 0.5 mm	6.3 mm	7.6
Least width of frontals, between horn cores and orbits	135	237 – 318 mm	271.1 ± 1.1 mm	12.6 mm	4.7
Greatest width of frontals at orbits	117	289 – 356 mm	324.6 ± 1.2 mm	12.9 mm	4.0
M^1–M^3, inclusive alveolar length	22	81.8 – 97.9 mm	90.6 ± 0.9 mm	4.4 mm	4.8
M^3, maximum width, anterior cusp	22	22.3 – 31.4 mm	27.7 ± 0.3 mm	1.6 mm	5.9
Distance, nuchal line to tip of premaxillae	56	500 – 583 mm	535.3 ± 2.3 mm	17.0 mm	3.2
Distance, nuchal line to nasal-frontal suture	106	214 – 279 mm	245.7 ± 1.2 mm	12.2 mm	5.0
Angle of divergence of horn cores, forward from sagittal	124	58° – 79°	67.7° ± 0.4°	4.4°	6.6
Angle between foramen magnum and occipital planes	115	118° – 159°	133.8° ± 0.7°	7.6°	5.6
Angle between foramen magnum and basioccipital planes	115	100° – 129°	110.5° ± 0.5°	5.0°	4.5
Females					
Spread of horn cores, tip to tip	28	397 – 514 mm	451.0 ± 6.6 mm	34.7 mm	7.7
Horn core length, upper curve, tip to burr	34	93 – 177 mm	124.1 ± 3.0 mm	17.5 mm	14.1
Straight line distance, tip to burr, dorsal horn core	33	92 – 161 mm	117.1 ± 2.7 mm	15.3 mm	13.1
Dorso-ventral diameter, horn core base	37	43 – 59 mm	51.2 ± 0.6 mm	3.9 mm	7.5
Minimum circumference, horn core base	37	136 – 191 mm	162.1 ± 1.9 mm	11.5 mm	7.1
Width of occipital at auditory openings	26	187 – 219 mm	201.3 ± 1.5 mm	7.7 mm	3.8
Width of occipital condyles	27	111 – 129 mm	115.7 ± 0.9 mm	4.6 mm	4.0
Depth, nuchal line to dorsal margin of foramen magnum	26	74 – 98 mm	84.6 ± 1.2 mm	5.9 mm	7.0
Antero-posterior diameter, horn core base	38	44 – 61 mm	51.5 ± 0.6 mm	3.9 mm	7.6
Least width of frontals, between horn cores and orbits	27	198 – 233 mm	216.7 ± 1.7 mm	8.6 mm	4.0
Greatest width of frontals at orbits	25	248 – 291 mm	267.5 ± 2.0 mm	10.0 mm	3.8
M^1–M^3, inclusive alveolar length	2	76.0 – 88.0 mm	82.0 ± 6.0 mm	8.4 mm	10.3
M^3, maximum width, anterior cusp	2	25.9 – 26.0 mm	26.0 ± 0.0 mm	0.1 mm	0.3
Distance, nuchal line to tip of premaxillae	19	464 – 516 mm	484.7 ± 3.2 mm	13.9 mm	2.8
Distance, nuchal line to nasal-frontal suture	24	195 – 228 mm	212.4 ± 1.6 mm	7.8 mm	3.7
Angle of divergence of horn cores, forward from sagittal	21	60° – 72°	66.0° ± 0.8°	3.7°	5.6
Angle between foramen magnum and occipital planes	24	115° – 144°	131.8° ± 1.5°	7.3°	5.5
Angle between foramen magnum and basioccipital planes	25	101° – 119°	109.7° ± 1.0°	4.9°	4.4

Table 30
Summary of *Bison bison bison* Limb Biometrics

Standard measurement	n	Range	Mean ± SE	σ	V
Males					
Humerus:					
Approximate rotational length of bone	37	295 – 345 mm	325.9 ± 1.8 mm	11.1 mm	3.4
Antero-posterior diameter of diaphysis	37	54 – 68 mm	61.5 ± 0.5 mm	3.1 mm	5.1
Transverse minimum of diaphysis	37	43 – 56 mm	50.6 ± 0.5 mm	3.1 mm	6.2
Radius:					
Approximate rotational length of bone	57	302 – 343 mm	320.5 ± 1.2 mm	9.2 mm	2.8
Antero-posterior minimum of diaphysis	59	29 – 36 mm	32.2 ± 0.2 mm	1.7 mm	5.2
Transverse minimum of diaphysis	59	48 – 61 mm	53.9 ± 0.3 mm	2.8 mm	5.2
Metacarpal:					
Total length of bone	112	192 – 227 mm	206.3 ± 0.6 mm	7.1 mm	3.4
Antero-posterior minimum of diaphysis	113	22 – 31 mm	27.2 ± 0.1 mm	1.6 mm	6.1
Transverse minimum of diaphysis	113	38 – 53 mm	45.4 ± 0.3 mm	3.4 mm	7.5
Femur:					
Approximate rotational length of bone	39	360 – 402 mm	380.7 ± 1.6 mm	10.4 mm	2.7
Antero-posterior diameter of diaphysis	43	42 – 52 mm	47.9 ± 0.3 mm	2.4 mm	5.1
Transverse minimum of diaphysis	43	43 – 52 mm	47.3 ± 0.3 mm	2.4 mm	5.1
Tibia:					
Approximate rotational length of bone	42	348 – 391 mm	371.9 ± 1.5 mm	10.0 mm	2.7
Antero-posterior minimum of diaphysis	43	30 – 38 mm	34.7 ± 0.2 mm	1.4 mm	4.2
Transverse minimum of diaphysis	43	45 – 54 mm	49.1 ± 0.3 mm	2.2 mm	4.4
Metatarsal:					
Total length of bone	118	232 – 276 mm	255.2 ± 0.7 mm	7.7 mm	3.0
Antero-posterior minimum of diaphysis	115	27 – 34 mm	30.3 ± 0.1 mm	1.3 mm	4.6
Transverse minimum of diaphysis	115	30 – 43 mm	35.3 ± 0.1 mm	1.9 mm	5.4
Females					
Humerus:					
Approximate rotational length of bone	25	259 – 301 mm	286.3 ± 1.7 mm	8.7 mm	3.0
Antero-posterior diameter of diaphysis	25	46 – 54 mm	49.8 ± 0.4 mm	2.3 mm	4.6
Transverse minimum of diaphysis	25	37 – 44 mm	40.3 ± 0.3 mm	1.8 mm	4.4
Radius:					
Approximate rotational length of bone	54	274 – 313 mm	293.0 ± 1.2 mm	9.2 mm	3.1
Antero-posterior minimum of diaphysis	56	22 – 30 mm	26.4 ± 0.2 mm	1.6 mm	6.2
Transverse minimum of diaphysis	56	38 – 50 mm	43.1 ± 0.3 mm	2.5 mm	5.8
Metacarpal:					
Total length of bone	164	182 – 219 mm	198.9 ± 0.5 mm	6.8 mm	3.4
Antero-posterior minimum of diaphysis	163	19 – 28 mm	24.0 ± 0.1 mm	1.6 mm	6.7
Transverse minimum of diaphysis	163	36 – 44 mm	36.6 ± 0.1 mm	2.3 mm	6.3
Femur:					
Approximate rotational length of bone	17	328 – 354 mm	342.2 ± 2.4 mm	10.0 mm	2.9
Antero-posterior diameter of diaphysis	18	37 – 45 mm	41.2 ± 0.5 mm	1.7 mm	5.8
Transverse minimum of diaphysis	18	38 – 45 mm	41.5 ± 0.5 mm	2.5 mm	6.0
Tibia:					
Approximate rotational length of bone	40	313 – 352 mm	335.9 ± 1.5 mm	10.0 mm	3.0
Antero-posterior minimum of diaphysis	41	28 – 34 mm	30.2 ± 0.2 mm	1.7 mm	5.8
Transverse minimum of diaphysis	41	37 – 46 mm	42.1 ± 0.3 mm	2.1 mm	5.1
Metatarsal:					
Total length of bone	177	224 – 264 mm	245.2 ± 0.6 mm	8.0 mm	3.2
Antero-posterior minimum of diaphysis	180	23 – 31 mm	27.4 ± 0.1 mm	1.5 mm	5.5
Transverse minimum of diaphysis	180	25 – 36 mm	29.3 ± 0.1 mm	1.8 mm	6.7

Table 31
Bison bison bison: Referred Specimens

Institution and identification number			Sex	Provenience
AMNH	M	331	♂	Fort McPherson, Nebraska
AMNH	M	332	♀	Fort McPherson, Nebraska
AMNH	M	5475	♂	Big Porcupine Creek, Montana
AMNH	M	5477	♂	Big Porcupine Creek, Montana
AMNH	M	5480	♂	Big Porcupine Creek, Montana
AMNH	M	16295	♂	Montana
AMNH	M	16298	♂	Montana
AMNH	M	16300	♂	Montana
AMNH	M	16301	♂	Montana
AMNH	M	16303	♂	Montana
AMNH	M	16305	♂	Montana
AMNH	M	16306	♂	Montana
AMNH	M	16309	♂	Montana
AMNH	M	16311	♂	Montana
AMNH	M	16312	♂	Montana
AMNH	M	16313	♂	Montana
AMNH	M	16314	♂	Montana
AMNH	M	16317	♂	Montana
AMNH	M	16318	♂	Montana
AMNH	M	16319	♂	Montana
AMNH	M	16320	♂	Montana
AMNH	M	16321	♂	Montana
AMNH	M	16322	♀	Montana
AMNH	M	139937	♂	Sheep Mountain, Big Horn Mountains, Wyoming
AMNH	VP	F:AM SD-2180	♀	Rocky Ford, Shannon County, South Dakota
AMNH	VP	F:AM SD-2181	♀	Rocky Ford, Shannon County, South Dakota
AMNH	VP	F:AM SD-2182	♀	Rocky Ford, Shannon County, South Dakota
AMNH	VP	F:AM SD-2183	♀	Rocky Ford, Shannon County, South Dakota
AMNH	VP	F:AM SD-2184	♀	Rocky Ford, Shannon County, South Dakota
AMNH	VP	F:AM SD-2208	♂	Rocky Ford, Shannon County, South Dakota
AMNH	VP	F:AM SD-2209	♂	Rocky Ford, Shannon County, South Dakota
AMNH	VP	F:AM SD-2210	♂	Rocky Ford, Shannon County, South Dakota
AMNH	VP	F:AM SD-2211	♂	Rocky Ford, Shannon County, South Dakota
AMNH	VP	F:AM SD-2213	♂	Rocky Ford, Shannon County, South Dakota
AMNH	VP	F:AM SD-2521	♂	East side of Pinnacles, South Dakota
AMNH	VP	F:AM SD-2524	♀	East side of Pinnacles, South Dakota
ANSP	G	12975	♀	Big Bone Lick, Boone County, Kentucky
BBL		A*	♂	Big Bone Lick, Boone County, Kentucky
BBL		B*	♂	Big Bone Lick, Boone County, Kentucky
BBL		C*	♀	Big Bone Lick, Boone County, Kentucky
BBL		D*	♀	Big Bone Lick, Boone County, Kentucky
CMNH		1221	♀	Big Bone Lick, Boone County, Kentucky
CoMNH	VP	628[a]	♂	Michies, Texas
CoMNH	VP	1362[a]	♂	Near Palmer, Nebraska
CoMNH	VP	1846	♂	Near Pagoda, Colorado
CoMNH	VP	A*	♂	9000′ elevation, Cantonment Creek, Saguache County, Colorado
DMNH	Z	16336	♂	Montana
IMNH	VP	225B	♂	Near Soda Springs, Caribou County, Idaho
IMNH	VP	6377-109	♂	Bed of Portneuf River, Power County, Idaho
IMNH	VP	16165	♂	American Falls Reservoir, Fort Hall Indian Reservation, southeast Idaho
IMNH	VP	17537	♂	Near Downey, Bannock County, Idaho
IMNH	VP	R194	♂	Gray's Lake, Idaho
MCZ	M	10-147	♂	Fort Hays, Kansas
MCZ	M	97	♂	Fort Hays, Kansas
MCZ	M	99	♂	Fort Hays, Kansas

Table 31
Bison bison bison: Referred Specimens (Continued)

Institution and identification number			Sex	Provenience
MCZ	M	105	♀	Fort Hays, Kansas
MCZ	M	1770	♂	Fort Hays, Kansas
MCZ	M	1771	♂	Fort Hays, Kansas
MCZ	VP	2047	♂	Big Bone Lick, Boone County, Kentucky
MCZ	VP	2048	♂	Big Bone Lick, Boone County, Kentucky
MCZ	VP	2050	♂	Big Bone Lick, Boone County, Kentucky
MCZ	VP	2051	♀	Big Bone Lick, Boone County, Kentucky
MCZ	VP	2057	♀	Big Bone Lick, Boone County, Kentucky
MMMN	VP	128	♂	Seddon's Corner, Manitoba
MMMN	VP	597	♂	Treesbank, Manitoba
MU		993[b]	♂	Near Wichita Falls, Wichita County, Texas
NMC	VP	17529	♂	Fort Saskatchewan, Alberta
OHS	A	Bd26	♂	Schoenbrunn Village, Ohio
PPHM	A	73-6	♀	Twilla Archeological Site, near Turkey, Hall County, Texas
PPHM	A	73-7a	♀	Twilla Archeological Site, near Turkey, Hall County, Texas
PPHM	A	73-7b	♀	Twilla Archeological Site, near Turkey, Hall County, Texas
PPHM		97-964	♂	Near Canyon, Randall County, Texas
PPHM		97-1211a	♂	Near Canyon, Randall County, Texas
PPHM		97-1211b	♂	Near Canyon, Randall County, Texas
PPHM		211-1	♂	Near Northfield, Texas
PPHM		619-14	♂	Tierra Blanca Creek, near Dawn, Texas
PPHM		630a	♂	Palo Duro Canyon, Texas
PPHM		630b	♂	Palo Duro Canyon, Texas
PPHM		762-1	♂	McFee Ranch, near Amarillo, Texas
PPHM		915	♂	Near Canyon, Randall County, Texas
PPHM		1173-1	♂	Near Canyon, Randall County, Texas
PPHM		1399-3	♂	Near Canyon, Randall County, Texas
PPHM		1482-1	♂	Canadian River, Hemphill County, Texas
PPHM		A*	♂	Near Canyon, Randall County, Texas
PPHM		B*	♂	Near Canyon, Randall County, Texas
PPHM		C*	♂	Near Canyon, Randall County, Texas
PPHM		D*	♂	Near Canyon, Randall County, Texas
PPHM		E*	♂	Near Miami, Roberts County, Texas
PPHM		F*	♂	Near Canyon, Randall County, Texas
SM		60397	♂	West Fork of Trinity River, Tarrant County, Texas
SM		61616	♂	Ash Creek (?), near Fort Worth, Tarrant County, Texas
SM		A*	♂	Near Dallas, Dallas County (?), Texas
SMNH		P195	♂	Long Creek Archeological Site, Saskatchewan
SUI		A*	♀	Dows Bog, Iowa
SUI		B*	♀	Dows Bog, Iowa
TMM		36-27	♀	Bonfire Shelter, Val Verde County, Texas
TMM		139	♂	Yellow House Canyon, near Lubbock, Texas
TMM		267-187	♀	Bonfire Shelter, Val Verde County, Texas
TMM		339-125	♀	Bonfire Shelter, Val Verde County, Texas
TMM		462-195	♀	Bonfire Shelter, Val Verde County, Texas
TMM		462-193	♀	Bonfire Shelter, Val Verde County, Texas
TMM		A*	♂	Near Beaver, Oklahoma
TTU	A	60-47-89a	♂	Lubbock Lake Archeological Site, Lubbock County, Texas
TTU	A	60-47-89b	♂	Lubbock Lake Archeological Site, Lubbock County, Texas
TTU	A	60-47-89c	♂	Lubbock Lake Archeological Site, Lubbock County, Texas
TTU	A	1917	♂	Lubbock Lake Archeological Site, Lubbock County, Texas
TTU	A	3743	♂	Lubbock Lake Archeological Site, Lubbock County, Texas
TTU	A	7440	♂	Lubbock Lake Archeological Site, Lubbock County, Texas
TTU	A	13212	♂	Lubbock Lake Archeological Site, Lubbock County, Texas
TTU	A	41LU35	♂	Canyon Lake, near Lubbock, Lubbock County, Texas
UASM	Z	198	♂	Archeological Site Y:9:11, near Yeso, De Baca County, New Mexico

Table 31

Bison bison bison: Referred Specimens (Continued)

Institution and identification number			Sex	Provenience
UCy	A	OS-825	♂	Trout Creek, Alberta
UCy	A	A*	♂	Near Hillspring, Alberta
UCy	A	B*	♂	Big Hill Springs, Alberta
UCy	A	C*	♂	Jenkin Buffalo Jump, Alberta
UCy	A	D*	♂	Big Hill Springs, Alberta
UCy	A	E*	♂	Big Hill Springs, Alberta
UCy	A	F*	♂	Big Hill Springs, Alberta
UCy	A	G*	♂	Big Hill Springs, Alberta
UCy	A	H*	♂	Big Hill Springs, Alberta
UCy	A	I*	♂	Jenkin Buffalo Jump, Alberta
UCy	A	J*	♂	Old Wives Lake, Saskatchewan
UCy	A	K*	♂	Crowsnest Pass, Alberta
UCy	A	L*	♂	Crowsnest Pass, Alberta
UCM	A	A*	♂	Near old Fort Wallace, Kansas
UCM	G	3608	♂	Near Colorado Springs, El Paso County, Colorado
UCM	G	4278	♂	Near Leadville, Lake County, Colorado
UCM	G	S29G	♂	Near Walden, Jackson County, Colorado
UCM	G	S66	♂	Mt. Audubon, Colorado
UCM	G	A*	♂	Near Ward, Colorado
UM	G	0295	♂	20 miles south of Libby, Lincoln County, Montana
UM	Z	5460	♂	Yogo Peak, Judith Basin County, Montana
UM	Z	5461	♂	Yogo Peak, Judith Basin County, Montana
UM	Z	5462	♂	Beartooth Plateau, Montana-Wyoming State Line
UM	Z	13251	♂	Beartooth Plateau, Park County, Wyoming
UNSM	M	5-6-11-26	♂	Cass County, Nebraska
UNSM	M	186-11-26	♂	Cass County, Nebraska
UNSM	VP	A*	♀	Nebraska (?)
USGS		D277	♂	Clay Basin Creek, Dagget County, Utah
USGS		A*	♂	Barger Gulch, Colorado
USGS		B*	♂	Head of Little Vasquez Creek, Grand County, Colorado
USNM	M	839	♂	Fort Union, Nebraska
USNM	M	22374	♂	Little Dry Creek, Montana
USNM	M	22664	♂	Big Dry Creek, Montana
USNM	M	120507	♂	Gardiner River, Yellowstone National Park, Montana
USNM	M	120509	♂	Gardiner River, Yellowstone National Park, Montana
USNM	M	120512	♂	Mammoth Hot Springs, Sanders County, Montana
USNM	M	122670	♂	Big Dry Creek, Montana
USNM	M	122672	♀	Big Dry Creek, Montana
USNM	M	122673	♂	Big Dry Creek, Montana
USNM	M	122676	♂	Big Dry Creek, Montana
USNM	M	122682	♂	Big Dry Creek, Montana
USNM	M	122685	♀	Big Dry Creek, Montana
USNM	M	122689	♂	Big Dry Creek, Montana
USNM	M	122690	♂	Big Dry Creek, Montana
USNM	M	122693	♂	Big Dry Creek, Montana
USNM	M	122694	♂	Big Dry Creek, Montana
USNM	M	122696	♂	Big Dry Creek, Montana
USNM	M	126620	♂	Langtry, Val Verde County, Texas
USNM	M	126621	♂	10 miles east of Langtry, Val Verde County, Texas
USNM	M	168816	♂	Big Horn Mountains, Montana
USNM	M	221089	♂	Park County, Montana
USNM	M	246529	♂	Near Izee, Grant County, Oregon
USNM	M	249844[a]	♂	Silvies River, Malheur Lake, Oregon
USNM	M	249844[b]	♀	Malheur Lake, Harney County, Oregon
USNM	M	249847	♀	Malheur Lake, Harney County, Oregon
USNM	M	249848	♀	Malheur Lake, Harney County, Oregon
USNM	M	249849	♀	Malheur Lake, Harney County, Oregon

Table 31
Bison bison bison: Referred Specimens (Continued)

Institution and identification number			Sex	Provenience
USNM	M	249895	♂	Malheur Lake, Harney County, Oregon
USNM	M	250092	♂	Malheur Lake, Harney County, Oregon
USNM	M	250145	♂	Malheur Lake, Harney County, Oregon
USNM	M	248950	♀	Malheur Lake, Harney County, Oregon
USNM	M	250089	♀	Malheur Lake, Harney County, Oregon
USNM	M	250090	♀	Malheur Lake, Harney County, Oregon
USNM	M	250091	♀	Malheur Lake, Harney County, Oregon
USNM	M	250095	♀	Malheur Lake, Harney County, Oregon
USNM	P	1718	♂	Stranger Creek, Millwood, Kansas
USNM	P	2172	♂	Deloit, Crawford County, Iowa
USNM	P	11410	juv.	Huron County, Ohio
USNM	P	A*	♂	Near Tucumcari, New Mexico
UW	A	30	♀	Glenrock, Converse County, Wyoming
UW	A	15826	♂	Glenrock, Converse County, Wyoming
UW	A	C1910	♂	Buffalo Creek, Wyoming
UW	A	C3071	♀	Buffalo Creek, Wyoming
UW	A	A*	♂	Scoggin Archeological Site, Carson County, Wyoming
UW	A	B*	♂	Union Pass, Wind River Mountains, Wyoming
UW	A	C*	♂	Big Horn Mountains, Wyoming
UW	A	D*	♂	Wind River Mountains, Wyoming
WTSU	A	A*	♂	Horn Creek, Potter County, Texas
WTSU	A	B*	♂	Timber Creek, Texas
WTSU	A	C*	♂	Red Deer Creek, Gray County, Texas
WTSU	A	D*	♂	Near Canyon, Randall County, Texas
YPM	M	01461	♂	Smokey (Hill?) River, Kansas

Sources of supplementary data: [a]Figgins, 1933; [b]Dalquest, 1959;
*No identification number was found for these specimens.

gradually toward the tip. Cores curve upward continuously from the base, but vary in the length of the arc; southern plains populations have less curved horn cores than do northern plains populations. Cores are usually rotated rearward with the tip deflected rearward, but these characters are often subtle or nonexistent in southern plains populations. A distinct dorsal groove is normally not present on the distal core. The horn tip is sharply pointed and normally occurs above the frontal plane and behind the occipital plane, although shorter horned phenotypes may have the tip below the frontal plane and/or in front of the occipital plane. The sheath tip is directed rearward and normally occurs behind the occipital plane.

The frontals are relatively robust and more consistently domed than in other taxa. The orbits are relatively low, directed more to the side than in fossil bison, and protrude strongly. Nasals are lower, broader and more distinctly triangular than in many fossil bison. The occipital is relatively flat. The nuchal line/external occipital protuberance is less prominent than in fossil bison. The occipital condyles do not protrude from the occipital plane. The ventro-lateral margins of the occipital condyles are occasionally but not normally flanged.

Female *B. b. bison* are the least robust of all North American bison. Horn cores are very small and emanate horizontally from the frontals at an angle of about 60° to 70° forward from the sagittal plane. They are circular to elliptical in cross section at the base and taper gradually toward the tip. The core varies from being nearly straight to curving distinctly upward in a continuous arc from the base. The main core axis is often rotated forward. The core tip may be directed laterally in shorter, straight cores or it may be directed rearward in longer, curving cores. The tip occurs from near to above the frontal plane, and may be either in front of or just behind the occipital plane.

The frontals are crenulated transversely, being convex near the horn cores and at the sagittal suture line, but concave between these regions. The orbits are low, directed forward, and do not protrude strongly. The occipital is relatively flat. The nuchal line/external occipital protuberance is less prominent than in earlier, larger species of bison. The occipital condyles do not protrude from the occipital plane. The ventro-lateral margins of the occipital condyles are normally not flanged.

Distribution:

B. bison is known from throughout most of the United States, western Canada, and northern Mexico (figs. 22 and 23). The highest population densities of *B. b. bison*, the smaller and southern subspecies, occurred in a north-south trending zone that extended from Alberta and Saskatchewan south to Texas and New Mexico. This principal range coincided with the central North American grassland. A larger secondary range, where population densities were lower, surrounded the primary range (fig. 23). *B. b. bison* is a grazer primarily adapted to an open grassland environment, although individual *B. b. bison* do browse if woody plant foliage and/or stems are available (pers. observ.).

Recent estimates of the first appearance of *B. bison* (as conceived herein) range from the late Wisconsin (Fuller and Bayrock, 1965; Geist, 1971*a*; Guthrie, 1970) through the early Holocene (Dalquest, 1961*b*) to the late Altithermal/Sub-Boreal (McDonald, 1976, 1978*a*; M. Wilson, 1975). Among these estimates, only the Sub-Boreal age of *B. bison* origin is supported by specifically identified radiocarbon-dated population-level samples. An early Sub-Boreal (ca. 5,000–4,000 years BP) origin of *B. bison* is recognized in this study. Table 32 lists radiocarbon ages of early *B. bison*.

B. bison probably evolved in the northern and central Great Plains and thereafter radiated both northward and southward, differentiating into *B. b.*

Table 32

Radiocarbon Ages Reported for Early *Bison bison* Localities

Locality	Radiocarbon age	Laboratory number
Del Bonita, Alberta	4,270 BP	Gx-1770
Head-Smashed-In, Alberta		
Level 10, north area	5,080 BP	RL-333
Level 16a, south area	4,050 BP	GaK-1476
Long Creek, Saskatchewan		
Upper level 8	4,635 BP*	S-52, S-53
Level 7	4,620 BP	S-50
Powers-Yonkee, Wyoming	4,450 BP	I-410
Scoggin, Wyoming	4,540 BP	RL-174

*Denotes the arithmetic average of two dates.

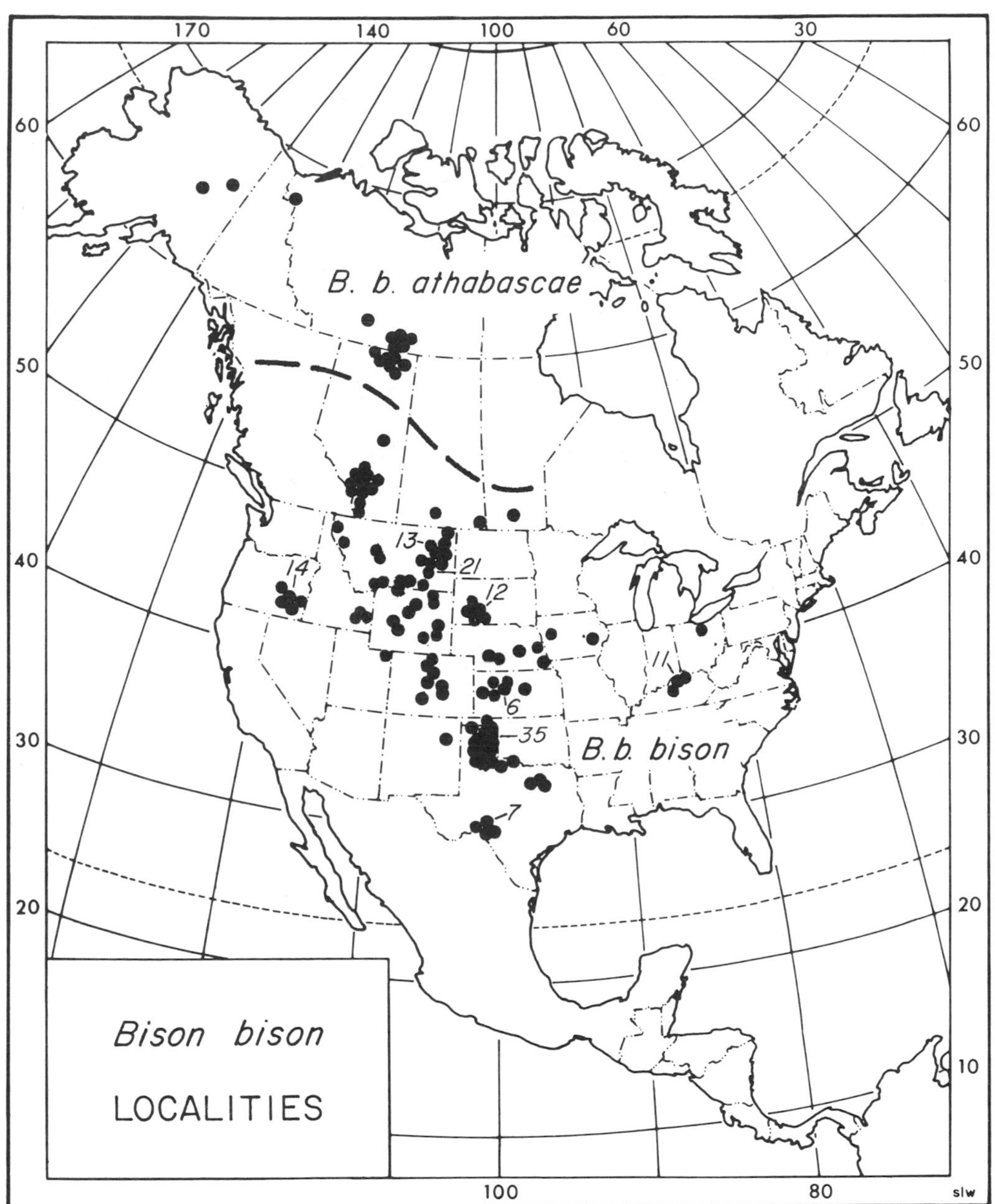

FIGURE 22 The distribution of *Bison bison* specimens used in this study. The broken line approximates the location of the maximum slope along the north–south character clines of this species; it is this co-occurrence of maximum character gradients which is used to systematically differentiate the two subspecies. Numbers indicate total specimens from clustered localities.

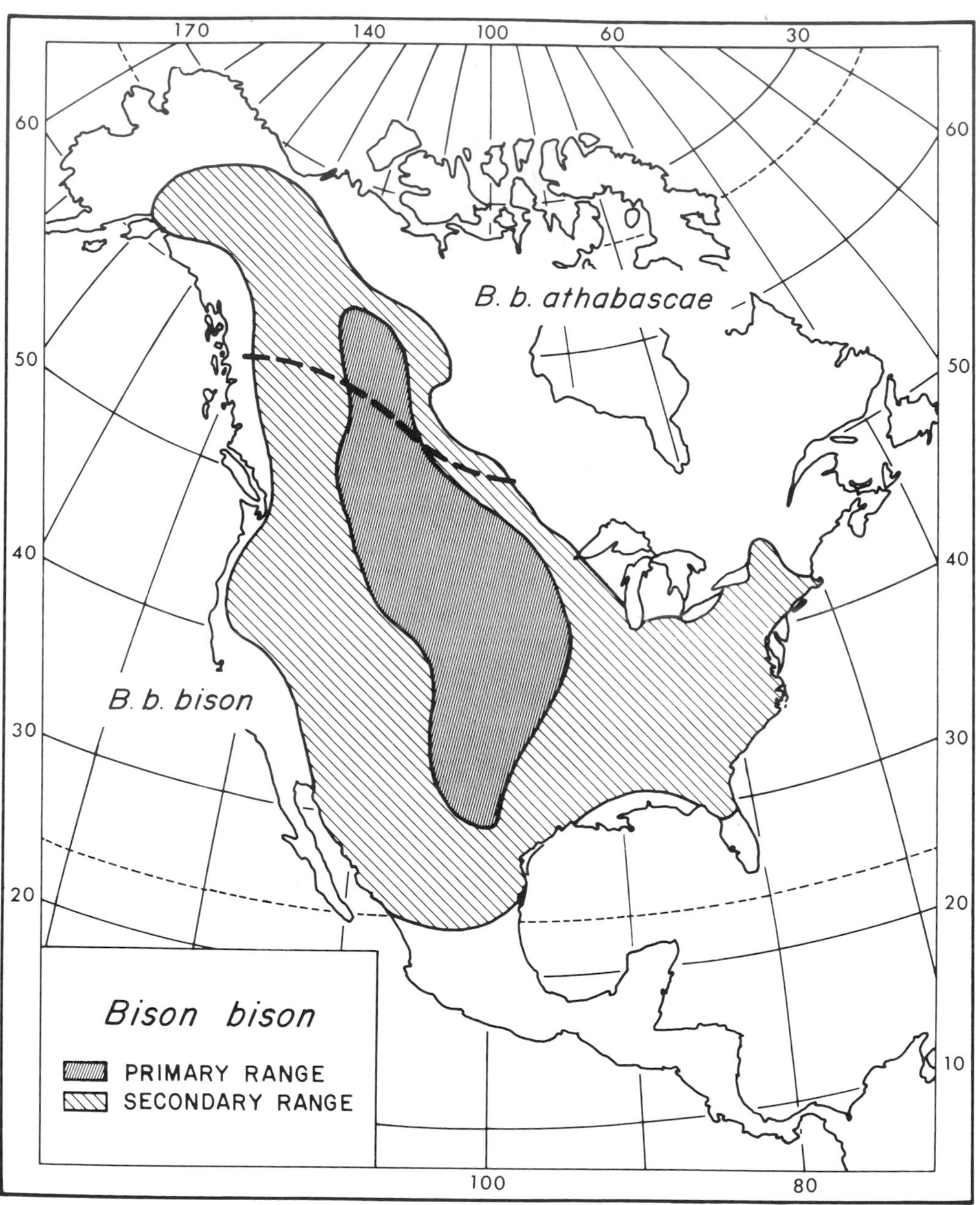

FIGURE 23 The primary and secondary ranges of *Bison bison bison* and *Bison bison athabascae*. The primary range for both subspecies aligns along a north–south axis with the central North American grassland. Although *Bison bison athabascae* did not inhabit a true grassland, populations of this taxon were probably reinforced periodically, if not regularly, by populations of *Bison bison bison* dispersing into the boreal parks and meadows from the northern central grassland.

athabascae and *B. b. bison*, respectively. *B. b. bison* probably reached maximum population numbers and densities after about 2,500 years BP, when the full development of the principal range of the taxon had been attained. Although occupation of the various sectors of the secondary range was more or less intermittent, depending on local environmental conditions (including human activities), the maximum distribution of *B. b. bison* probably occurred during the very late prehistoric period (based on the fact that early European contacts with *B. b. bison* frequently took place near the margin of the recognized maximum range of the subspecies). The temporal nature of *B. b. athabascae* distribution is poorly known, owing to the scarcity of radiocarbon-dated specimens.

Discussion:

Concept of the Subspecies: Demonstrable character clines extended north-south (and to a lesser extent east-west) through the principal range of *B. bison* (figs. 69 and 70). Hind (1860) and Allen (1876) differentiated between "plains" and "wood" bison, but applied no differentiating Linnaean name to the northern variant. Rhoads (1897) named the northern bison *B. b. athabascae* and, in doing so, created the smaller southern nominate subspecies *B. b. bison*. The ranges of the two subspecies are arbitrarily and rather loosely differentiated by a line generally separating the steppe/prairie populations and those of the boreal forest parklands (fig. 23). It is in this ecotonal zone that the *B. bison* clinal gradient is steepest (fig. 69, table 33) and here also that the most obvious and ecologically meaningful habitat change (herb dominance to woody plant dominance) occurs along the north-south axis of the primary *B. bison* range.

Synonymy: *B. sylvestris* was originally described as an extinct species of bison. The holotype (USNM P 11410) was found in a bog in Huron County, Ohio, along with *Megalonyx* bones, and was identified as the fossilized partial skull of a female animal. The assumed cooccurence of *Megalonyx* and *Bison* led Hay (1915) to consider this bison a post Wisconsin inhabitant of the Ohio forests "possibly haunting the swamps." The fragmentary holotype, however, is recent—not fossil—and represents a very young animal; it cannot be accurately sexed or specifically identified. Considering the obviously recent age of the specimen, however, and recognizing that *B. b. bison* was present throughout Ohio before and during the early historical period, *B. sylvestris* is synonymized with *B. b. bison*.

Shoemaker (1915) described *B. americanus pennsylvanicus* as being very dark, with short and crispy or curly hair, without hump, long-legged, and evenly proportioned fore and aft. The pelage was of equal length across the entire body, except that mature bulls "carried a sort of mane or crest which reached its maximum length where the hump grows on the prairie buffalo." Horns were reportedly very low on mature animals and grew upward "like horns of Ayshire cattle." This taxon was described from hearsay.

Article 1 of the *International Code of Zoological Nomenclature* (1964 revised edition) excludes names applied to hypothetical taxa such as *B. americanus pennsylvanicus*. In addition there is no material evidence to support Shoemaker's description of the taxon, even though several *B. bison* skeletons are available from the described range of the subspecies.

B. b. oregonus was described as an extinct subspecies of the modern bison from the northwestern Great Basin (Bailey, 1932). Subspecific distinctiveness was based on differences in the sizes of a syntypic series of skulls from the Malheur Lake, Oregon area, and skulls of *B. b. bison* from Texas and *B. b. athabascae* from northern Canada. Bailey found the Oregon skulls intermediate in size between those of the two taxa with which they were compared (table 33). Bailey does not, however, mention which dimensions were compared. Interestingly, Bailey chose to compare the Oregon skulls with skulls from both ends of the primary *B. bison* range. He apparently did not compare the Oregon skulls with *B. b. bison* skulls from the central portion of the species' primary range (i.e., from that part of the primary range nearest his Oregon sample). Had the Oregon skulls been compared with, for example, the Hornaday collection of *B. b. bison* skulls from Montana (both the Bailey and the Hornaday samples were at the United States National Museum when Bailey was writing), little difference would have been found.

Skinner and Kaisen (1947) synonymized *B. b. oregonus* with *B. b. athabascae.* These authors argued that bison in the montane and intermontane west were actually derived from populations of *B. b. athabascae* which moved "down the Rocky Mountain ranges and valleys where, in this case, they spread out into the now arid region around Lake Malheur from the nearby mountains. The entire Lake Malheur collection was measured and studied. It presents a population whose average size was larger than that of the plains race, precluding the chance that part of the plains race migrated across the mountains and spread out into Oregon" (p. 166). The Malheur Lake area specimens actually average larger than the central and northern plains specimens in some characters, but are smaller in other characters. Most characters in the Malheur Lake sample, however, are distinctly nearer the average for *B. b. bison*

Table 33

Mean Standard Measurements for Selected Male *Bison bison* Samples

Standard measurement	Southern Great Plains	Central Great Plains	Canadian Prairie	*B. b. athabascae*	Eastern U.S.	*B.b. oregonus*	*B.b. haningtoni*	*B.b. septemtrionalis*
Spread of horn cores, tip to tip	586.4 mm	597.1 mm	636.6 mm	681.2 mm	608.6 mm	608.8 mm	639.0 mm	617.7 mm
Horn core length, upper curve, tip to burr	173.2 mm	193.7 mm	208.6 mm	235.1 mm	184.0 mm	201.8 mm	219.0 mm	203.4 mm
Straight line distance, tip to burr, dorsal horn core	160.2 mm	172.7 mm	188.7 mm	207.0 mm	167.3 mm	178.6 mm	....	....
Dorso-ventral diameter, horn core base	80.9 mm	80.8 mm	83.1 mm	91.5 mm	78.0 mm	84.2 mm	....	....
Minimum circumference, horn core base	255.1 mm	250.0 mm	264.5 mm	289.1 mm	242.6 mm	262.2 mm	273.2 mm	246.7 mm
Width of occipital at auditory openings	241.2 mm	242.7 mm	248.7 mm	273.6 mm	233.6 mm	245.6 mm	248.7 mm	238.0 mm
Width of occipital condyles	124.6 mm	127.6 mm	124.1 mm	130.1 mm	125.6 mm	126.8 mm	124.6 mm	124.0 mm
Depth, nuchal line to dorsal margin of foramen magnum	94.2 mm	100.7 mm	99.1 mm	99.6 mm	101.0 mm	101.0 mm	....	....
Antero-posterior diameter, horn core base	83.1 mm	81.4 mm	86.9 mm	97.2 mm	79.6 mm	85.4 mm	....	....
Least width of frontals, between horn cores and orbits	271.0 mm	267.6 mm	278.2 mm	293.4 mm	274.5 mm	269.4 mm	276.0 mm	265.0 mm
Greatest width of frontals at orbits	324.0 mm	322.6 mm	327.4 mm	354.0 mm	316.5 mm	319.6 mm	327.7 mm	322.5 mm
Distance, nuchal line to tip of premaxillae	534.2 mm	533.2 mm	533.0 mm	578.6 mm	....	534.6 mm	....	....
Distance, nuchal line to nasal-frontal suture	246.8 mm	243.1 mm	238.8 mm	256.0 mm	259.5 mm	243.4 mm	....	....

Sources: Figgins, 1933; McDonald, 1978*a*.

than for *B. b. athabascae* (table 33). Spatial proximity and biometric similarity each suggest that *B. b. bison* populations from the plains were more likely the source of western bison than were southward dispersing populations of *B. b. athabascae.* Plains bison populations periodically dispersed into and across the Rocky Mountains and sporadically throughout the northern and central Great Basin. The Lake Malheur population was probably one of the small mountain and Great Basin herds that originated with this dispersal process. Reeves (1978*a*) has shown that a regular westward dispersal of plains bison into the Rocky Mountains occurred in southwestern Alberta. This same process presumably took place along the entire Rocky Mountain front when and where bison populations occurred on the adjacent plains.

B. b. haningtoni was described as a light-colored "race" of bison, larger than the plains bison, restricted to the higher mountain parks of Colorado (Figgins, 1933). The twelve skulls that Figgins used in describing this taxon could not be located when I was studying bison in the Denver Museum of Natural History collections, so no direct comparison of this series with other mountain or plains bison was possible. Figgins did, however, publish mean values for several characters of the skulls in his syntypic series (table 33). Comparison of these means with similar means from the mountain and plains samples shows that his sample means are generally larger than those for other plains or mountain samples. The *B. b. haningtoni* means are, however, nearer those of *B. b. bison* than *B. b. athabascae.*

B. b. septemtrionalis was similarly created by Figgins (1933) on the basis of small and insignificant morphological differences found in a sample of seventeen skulls presumedly from the "Valley of the North Platte River, northward to eastern Montana and westward to the foothills of the Rocky Mountains" (p. 28). These skulls also could not be located when I was studying bison at the Denver Museum of Natural History. In effect *B. b. septemtrionalis* represented intermediate size individual *B. bison* of the central and northern plains, which were larger than the small southern plains *B. bison*, but smaller than the large northern *B. b. athabascae* (table 33). When a demonstrable cline exists among contiguous populations, only two subspecies at most should be recognized, one at either end of the clinal axis (Mayr, 1969).

Distribution: *B. bison* is arbitrarily considered to date from after 5,000 BP. Admittedly, phenotypes antedating this datum may strongly resemble *B. bison*, whereas phenotypes postdating the same time may strongly resemble *B. antiquus*. Typical, not atypical, phenotypes must be the basis for taxonomic decisions, however, and there is no evidence to indicate that *B. bison* was typical before, or *B. antiquus* typical after, the 5,000–4,000 BP period. A specific date of origin for *B. bison* and termination for *B. antiquus* is, however, necessary to maintain order in the classification of these two species. The late Altithermal correlates (1) with the most apparent change in Holocene bison morphologic trends, and (2) with the improvement of environmental conditions vis-à-vis bison energy and nutrition requirements over much of the central and southern plains, a condition that created the potential for the range expansion and population increase of the new phenotype. The late Altithermal/early Sub-Boreal is, as M. Wilson (1975) has suggested, the most reasonable and logical period in which to systematically separate the late Pleistocene-early Holocene bison from the extant late Holocene bison.

IIIb. *Bison bison athabascae* Rhoads, 1897

Original Description:
Samuel N. Rhoads. 1897. "Notes on Living and Extinct Species of North American Bovidae." *Proceedings of the Academy of Natural Sciences of Philadelphia*, vol. 49, pp. 483–502.

Type Locality:
Within fifty miles southwest of Fort Resolution, Great Slave Lake, Northwest Territories, Canada.

Type Location:
Department of Zoology, Museum of Natural History, National Museum of Canada.

Type Specimen:
NMC M 299; complete skeleton and hide.

Subspecies Content:
B. b. athabascae is the late Holocene (extant) forest/woodland bison of boreal North America.

Diagnosis:
Males generally as for *B. b. bison*. Horn core growth occasionally mildly spiraled around longitudinal axis; less than 5,000–4,000 BP; generally from southern margin of boreal forest north to central Alaska (plate 15).

Females probably similar to female *B. b. bison*, with perhaps straight or concave posterior margin; growth mildly spiraled around arched or sinuous longitudinal axis; distribution as with males. Tentative diagnosis is based on presumed *B. b. bison* X *B. b. athabascae* hybrid females, with inferences made according to differences noted between these crosses and true *B. b. bison*. No true female *B. b. athabascae* skulls were available for study. Banfield and Novakowski (1960) present data for a presumedly true female *B. b. athabascae* shot in 1928, but their data are internally incompatible and therefore not used here.

Differential Diagnosis:
Males and females as for *B. b. bison*. Individual *B. b. athabascae* may not be morphologically differentiable from *B. b. bison*.

Biometric Summary:
Skulls, table 34.
Limb bones, table 35.

Referred Specimens:
Table 36

Description:

B. b. athabascae is the larger living North American bison and the largest of the world's living bison. Male horn cores are short and compact, although less so than in *B. b. bison*. Horn cores emanate horizontally or somewhat subhorizontally from the frontals at an angle of about 75° to 85° forward from the sagittal plane. The horn core is triangular (isosceles) in cross section at the base, where the antero-posterior and dorso-ventral diameters are about equal, and tapers gradually toward the tip. The core curves upward continuously from the core base, is rotated rearward, and the tips are normally deflected rearward. A distinct dorsal groove is normally not present on the distal core. The horn core tip is sharply pointed and normally occurs above the frontal plane and behind the occipital plane. Sheath tips are directed rearward and normally occur behind the occipital plane.

The frontals are moderately domed and relatively robust. The orbits are relatively low, directed more to the side than in fossil bison and protrude strongly. Nasals are low, broad, and triangular. The occipital is relatively flat. The nuchal line/external occipital protuberance is prominent, but less so than in fossil bison. The occipital condyles do not protrude from the occipital plane. The occipital condyles occasionally, but not normally, have a slight ventro-lateral flange.

Female skull characteristics of the true *B. b. athabascae* are not known, although several specimens of presumedly hybrid *B. b. athabascae* X *B. b. bison* are known. A comparison of females of these crosses and true *B. b. bison* suggests that *B. b. athabascae* females, as is to be expected, were or are very similar to *B. b. bison* females, differing mainly by being larger. Female *B. b. athabascae* horn cores especially were apparently longer and more strongly curving than are *B. b. bison* (plate 16).

Distribution:

B. b. athabascae is known from the western boreal forest and bordering ecotones of Canada north to Alaska, being generally distributed over the area incorporating northeastern British Columbia, northern Alberta, northwestern Saskatchewan, the southern and western District of Mackenzie, Yukon Territory, and the eastern half of Alaska (figs. 22 and 23). Historically, *B. b. athabascae* was most numerous in the lowlands of northeastern Alberta, northwestern Saskatchewan, and south central District of Mackenzie. This primary range of the subspecies coincides with relatively extensive parklands and is adjacent to the primary range of *B. b. bison* (fig. 23).

A small part of the primary *B. b. athabascae* range was set aside in 1922 as Wood Buffalo National Park, a national preserve for the wood bison.

B. b. athabascae is considered to date from about 5,000–4,000 BP (table 32). The arbitrarily determined time of origin coincides approximately with the radiation of *B. b. bison* across the central grassland of post Altithermal North America (see chapter 7).

Discussion:

Synonymy: No available synonyms are recognized for this taxon. Skinner and Kaisen (1947) considered both *B. b. oregonus* and *B.b. haningtoni* synonyms of *B. b. athabascae.* As discussed above, however, these two subspecies were nearer plains bison than wood bison in location and size and should, therefore, be synonymized with the former. Skinner and Kaisen's thesis that wood bison populations migrated down the Rocky Mountains, thereby giving rise to mountain and Great Basin bison populations, was rejected as unsupported, unnecessary and contrary to historic observations.

Table 34

Summary of *Bison bison athabascae* Skull Biometrics

Standard measurement	n	Range	Mean ± SE	σ	V
Males					
Spread of horn cores, tip to tip	9	542 – 848 mm	681.2 ± 30.6 mm	92.0 mm	13.5
Horn core length, upper curve, tip to burr	9	165 – 323 mm	235.1 ± 14.6 mm	43.9 mm	18.7
Straight line distance, tip to burr, dorsal horn core	9	154 – 277 mm	207.0 ± 11.4 mm	34.4 mm	16.6
Dorso-ventral diameter, horn core base	9	81 – 106 mm	91.5 ± 2.9 mm	8.7 mm	9.5
Minimum circumference, horn core base	9	254 – 322 mm	289.1 ± 7.6 mm	22.9 mm	7.9
Width of occipital at auditory openings	10	243 – 298 mm	273.6 ± 4.8 mm	15.3 mm	5.6
Width of occipital condyles	11	118 – 139 mm	130.1 ± 1.9 mm	6.4 mm	4.9
Depth, nuchal line to dorsal margin of foramen magnum	10	92 – 114 mm	99.6 ± 2.0 mm	6.6 mm	6.6
Antero-posterior diameter, horn core base	9	83 – 109 mm	97.2 ± 3.1 mm	9.5 mm	9.8
Least width of frontals, between horn cores and orbits	10	273 – 313 mm	293.4 ± 3.3 mm	10.5 mm	3.6
Greatest width of frontals at orbits	10	326 – 384 mm	354.0 ± 4.7 mm	14.8 mm	4.2
M^1–M^3, inclusive alveolar length	1	....	91.7 mm	....	....
M^3, maximum width, anterior cusp	1	....	27.9 mm	....	....
Distance, nuchal line to tip of premaxillae	7	562 – 604 mm	578.6 ± 5.7 mm	15.2 mm	2.6
Distance, nuchal line to nasal-frontal suture	9	240 – 276 mm	256.0 ± 4.4 mm	13.4 mm	5.2
Angle of divergence of horn cores, forward from sagittal	6	63° – 77°	71.0° ± 2.0°	5.0°	7.0
Angle between foramen magnum and occipital planes	9	119° – 144°	129.4° ± 2.9°	8.8°	6.8
Angle between foramen magnum and basioccipital planes	9	106 – 125°	113.8° ± 2.2°	6.6°	5.8

Note: The above sample was collected prior to 1929 and presumably contains only pure *B. b. athabascae*. I was unable to locate any female *B. b. athabascae* collected prior to 1929. Banfield and Novakowski (1960) present skull measurements from a female reportedly shot in 1928, but their data are internally incompatible and have not been included in this study.

Table 35
Summary of *Bison bison athabascae* Limb Biometrics

Standard measurement	n	Range	Mean ± SE	σ	V
Males					
Humerus:					
Approximate rotational length of bone	2	350 – 359 mm	354.5 ± 4.4 mm	6.3 mm	1.8
Antero-posterior diameter of diaphysis	2	67 – 68 mm	67.5 ± 0.4 mm	0.7 mm	1.0
Transverse minimum of diaphysis	2	58 – 59 mm	58.5 ± 0.4 mm	0.7 mm	1.2
Radius:					
Approximate rotational length of bone	3	328 – 356 mm	341.3 ± 8.0 mm	14.0 mm	4.1
Antero-posterior minimum of diaphysis	3	34 – 36 mm	35.0 ± 0.5 mm	1.0 mm	2.8
Transverse minimum of diaphysis	3	56 – 63 mm	59.7 ± 2.0 mm	3.5 mm	5.8
Metacarpal:					
Total length of bone	3	224 – 232 mm	228.7 ± 2.3 mm	4.1 mm	1.8
Antero-posterior minimum of diaphysis	3	29 – 31 mm	30.3 ± 0.6 mm	1.1 mm	3.8
Transverse minimum of diaphysis	3	50 – 54 mm	52.0 ± 1.1 mm	2.0 mm	3.8
Femur:					
Approximate rotational length of bone	3	408 – 429 mm	421.3 ± 6.6 mm	11.5 mm	2.7
Antero-posterior diameter of diaphysis	3	51 – 59 mm	54.0 ± 2.4 mm	4.3 mm	8.0
Transverse minimum of diaphysis	3	51 – 52 mm	51.7 ± 0.2 mm	0.5 mm	1.1
Tibia:					
Approximate rotational length of bone	3	400 – 424 mm	414.3 ± 7.2 mm	12.6 mm	3.0
Antero-posterior minimum of diaphysis	3	36 – 39 mm	38.0 ± 0.9 mm	1.7 mm	4.5
Transverse minimum of diaphysis	3	52 – 55 mm	53.3 ± 0.8 mm	1.5 mm	2.8
Metatarsal:					
Total length of bone	3	270 – 281 mm	275.7 ± 3.1 mm	5.5 mm	2.0
Antero-posterior minimum of diaphysis	3	32 – 34 mm	33.0 ± 0.5 mm	1.0 mm	3.0
Transverse minimum of diaphysis	3	38 – 41 mm	39.0 ± 0.9 mm	1.7 mm	4.4
Females					
Humerus:					
Approximate rotational length of bone	1	....	304.0 mm	....	...
Antero-posterior diameter of diaphysis	1	....	50.0 mm	....	...
Transverse minimum of diaphysis	1	....	40.0 mm	....	...
Radius:					
Approximate rotational length of bone	1	....	313.0 mm	....	...
Antero-posterior minimum of diaphysis	1	....	28.0 mm	....	...
Transverse minimum of diaphysis	1	....	46.0 mm	....	...
Metacarpal:					
Total length of bone	1	....	220.0 mm	....	...
Antero-posterior minimum of diaphysis	1	....	26.0 mm	....	...
Transverse minimum of diaphysis	1	....	39.0 mm	....	...
Femur:					
Approximate rotational length of bone	1	....	362.0 mm	....	...
Antero-posterior diameter of diaphysis	1	....	44.0 mm	....	...
Transverse minimum of diaphysis	1	....	42.0 mm	....	...
Tibia:					
Approximate rotational length of bone	1	....	361.0 mm	....	...
Antero-posterior minimum of diaphysis	1	....	32.0 mm	....	...
Transverse minimum of diaphysis	1	....	43.0 mm	....	...
Metatarsal:					
Total length of bone	1	....	264.0 mm	....	...
Antero-posterior minimum of diaphysis	1	....	29.0 mm	....	...
Transverse minimum of diaphysis	1	....	31.0 mm	....	...

Note: The above sample was collected prior to 1929 and presumably contains only pure *B. b. athabascae.*

Autochthonous Eurasian Species

Two species of bison which appear to have originated in Eurasia occurred in North America. These species—*B. priscus* and *B. alaskensis*—entered North America through Beringia and differentially dispersed southward into midlatitude and tropical North America.

The recognition of two Eurasian autochthons in North America is based primarily on the study of specimens with North American provenience, and secondly on more limited data for specimens with Eurasian provenience. The data base is admittedly imbalanced and, insofar as the Eurasian record is concerned, inadequate to formulate a final opinion on the classification of middle and late Pleistocene Eurasian bison. Indeed, the available data do not unquestionably indicate the presence of two "good" species of pre-Holocene Eurasian bison, but neither do these data support the recognition of only one species (fig. 24). Considering the available data, including patterns of autochthonous North American bison biometrics, adaptations, and distributions, the decision was made to recognize two Eurasian autochthons. The reasoning is as follows:

1. Recognizing that quantifiable differences exist between larger forest/woodland and smaller savanna/grassland North American autochthons, as exemplified by pre-Holocene *B. latifrons*/*B. antiquus* and late Holocene *B. b. athabascae*/*B. b. bison*, it is reasonable to assume that a similar difference existed in pre-Holocene Eurasian autochthons.

Table 36
Bison bison athabascae: Referred Specimens

Institution and identification number		Sex	Provenience
AMNH	M 18163	♂	Near Great Slave Lake, Northwest Territories
AMNH	M 73615	♂	Near Lake Luke, Wood Buffalo National Park (1926)
AMNH	VP F:AMA479-4783	♂	Yukon River, near Circle City, Alaska
MCZ	M 24017	♂	Near Great Slave Lake (1925)
NMC	M 299	♂	Within 50 miles southwest of Fort Resolution, Northwest Territories (1892)
NMC	M 625	♂	Near Lake Athabasca, Alberta
PMA	D 235	♂	Wood Buffalo National Park (1925)
USNM	M 172689	♂	Northern Alberta (1910)
USNM	M 177630	♂	Near Fort Providence, Northwest Territories (1910)
USNM	M 177631	♂	West of Fort Smith, Northwest Territories (ca. 1909)
USNM	M 177632	♂	Near Fort Smith, Northwest Territories (1909)
USNM	M 223292	♂	30 miles upstream from mouth of Tanana River, Alaska
USNM	P 16861	♂	Near Fort McPherson, Peel River, Northwest Territories
UAI	D 233[a]	♂	Wood Buffalo National Park (1925)
UAI	D 234[a]	♂	Wood Buffalo National Park (1925)
UAI	626	♂	Wood Buffalo National Park (1922)

Source of supplementary data:
[a]Bayrock and Hillerud, 1964.
Note: The above specimens were collected prior to 1929 and all are presumed to be pure *B. b. athabascae*. The actual or approximate date of collection, when known, is given in parentheses.

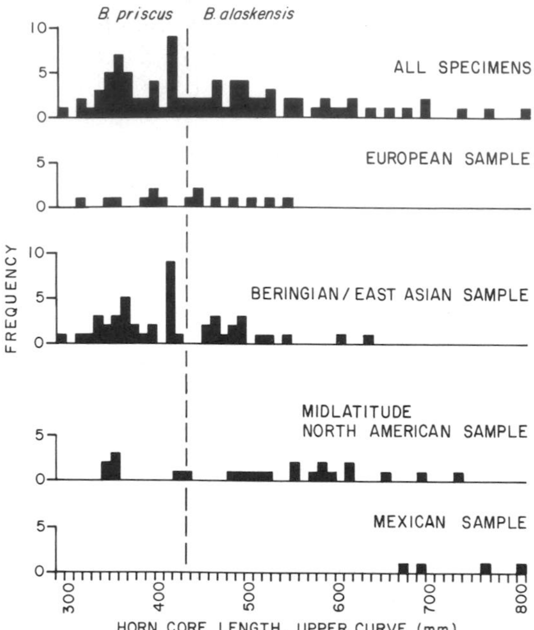

FIGURE 24 Horn core lengths in selected regional groupings of male Eurasian autochthons. The top histogram includes all specimens for which data were available, whereas the lower four histograms divide that sample into regional samples. Both character range and maximum specimen size increase from Europe to North America, suggesting the operation of selective factors favoring larger specimens in Beringia and North America (see also figs. 25, 57).

2. Larger and smaller horned bison identified as Eurasian autochthons are differentially represented in North America (fig. 24). If these phenotypes were not differentially adapted to specific environments, there should be no strong spatial bias in their dispersal history, but there is such a bias. Larger horned Eurasian autochthons are much more strongly represented in central and southern North America than are shorter horned Eurasian autochthons. This differential pattern of dispersal success is interpreted as a reflection of different adaptations by the larger and smaller phenotypes, the larger phenotypes probably being adapted to a forest/woodland environment and the smaller phenotypes probably being adapted to a savanna/grassland environment.
3. The frequency distribution of the horn core length along the upper curve for all Eurasian autochthons (fig. 24) forms less of a normal curve than is expected for a sample of this size (figs. 30–44). The range of horn core lengths for all Eurasian autochthons is also much greater than is expected for a variate of this size (fig. 25). When the total sample of Eurasian autochthons is divided into two samples, separated midway between the two obvious near-normal clusters of cases in the Beringian sample, more expected patterns of distribution and

Table 37

Horn Core Length Variation Among Selected Regional Samples of Male *Bison priscus* and *Bison alaskensis*

	Europe		East Asia & Beringia		Midlatitude & tropical North America		
	B. priscus	*B. alaskensis*	*B. priscus*	*B. alaskensis*	*B. priscus*	*B. alaskensis*	**Combined sample**
n	8	7	31	16	6	21	89
Maximum	424 mm	530 mm	416 mm	622 mm	413 mm	795 mm	795 mm
Minimum	307 mm	436 mm	285 mm	441 mm	331 mm	426 mm	285 mm
Range	117 mm	94 mm	131 mm	181 mm	82 mm	369 mm	510 mm
Mean	367.3 mm	477.1 mm	364.4 mm	499.1 mm	351.2 mm	591.4 mm	450.4 mm
SE	13.0 mm	14.2 mm	6.4 mm	13.4 mm	12.5 mm	21.6 mm	11.9 mm
σ	36.7 mm	37.7 mm	36.1 mm	53.9 mm	30.6 mm	98.8 mm	112.1 mm
V	10.0	7.9	9.9	10.8	8.7	16.7	24.9

Sources: McDonald 1978*a*; Skinner and Kaisen, 1947; Vialli, 1954.
Note: The horn core length is standard measurement 3.

variation occur, however (figs. 24 and 25, table 37). The conspicuous column of cases in the 400–409 mm class of the Asian/Beringian sample (fig. 24) is interpreted as probable *B. priscus* X *B. alaskensis* hybrids. Hybridization is not uncommon among mammalian populations (see chapter 6), and character equilibrium is not uncommon among parapatric or sympatric species which hybridize (E. Wilson, 1975).

I. *Bison priscus* (Bojanus, 1827)

Original Description:
L. H. Bojanus. 1827. "De Uro Nostrate Eiusque Sceleto Commentatio." *Nova Acta Leopoldina*, 13 Bd. (Leipzig, etc.: J. A. Barth, etc.). (Name used).
Max Hilzheimer. 1918. "Dritter Beitrag zur Kenntnis der Bisonten." *Arch. Naturgeschichte*, vol. 84, div. A., no. 6, pp. 41–87. (Selection of lectotype from Bojanus's syntypic series).
Giulia Sacchi Vialli. 1954. "I Bisonti Fossili delle Alluvioni Quaternarie Pavesi," *Atti dell' Istituto Geologico della Universita di Pavia*, vol. 5, pp. 1–27. (Identification of lectotype from Pavia series of skulls).

Type Locality:
Po River Valley, Lombardy, Italy.

Type Location:
Questionable. Museum of the Institute of Geology, University of Pavia, Pavia, Italy (?); Museum of the Institute of Geology, University of Torino, Torino, Italy (?).

Type Specimen:
Questionable; nearly complete skull.

Principal Synonymy:
Bison crassicornis Richardson, 1852–1854
Bison willistoni Martin, 1924
Bison antiquus barbouri Schultz and Frankforter, 1946
Bison geisti Skinner and Kaisen, 1947
Bison preoccidentalis Skinner and Kaisen, 1947

Species Content:
B. priscus is the late-early, middle, and

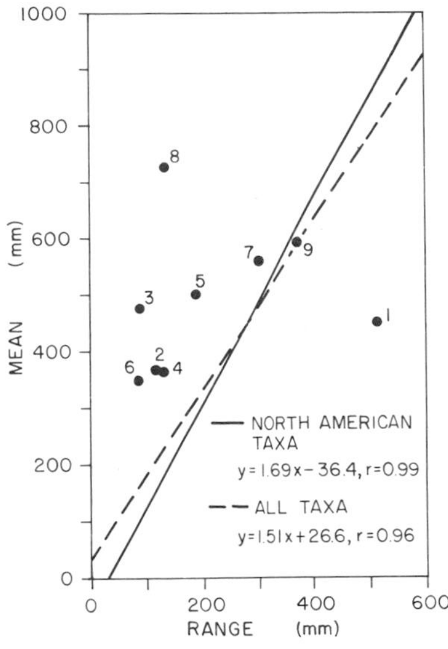

FIGURE 25 Observed and expected horn core length variation in selected regional samples of male Eurasian autochthons. The regression lines are based upon the ranges and means observed among all North American autochthons, and North American autochthons with Eurasian autochthons added. Points identify samples plotted in figure 24, as follows: 1 = all specimens; 2 = European *Bison priscus*; 3 = European *Bison alaskensis*; 4 = East Asian/Beringian *Bison priscus*; 5 = East Asian/Beringian *Bison alaskensis*; 6 = midlatitude North American *Bison priscus*; 7 = midlatitude North American *Bison alaskensis*; 8 = Mexican *Bison alaskensis*; and 9 = midlatitude North American and Mexican *Bison alaskensis* combined. Based on all horn core length data gathered in this study, the combined sample represents too great a range, relative to the mean, for a natural population or normal taxon. When this sample is divided into two taxa, and they into regional samples, most show an unexpectedly narrow range of variation relative to their means. The European sample of both taxa is inadequately represented, so the observed variation should not be considered typical of the region until and unless it is further substantiated. If the unusually well-represented *Bison priscus* specimens in the 400-410 mm class from East Asia/Beringia (fig. 24) were removed (they could represent *Bison priscus* X *Bison alaskensis* hybrids), point 4 would move nearer the regression lines. Points 7 and 9 most nearly approximate expected patterns of character variation.

late Pleistocene savanna/steppe bison of Eurasia, including glacial Beringia. Late Pleistocene dispersal extended the range into midlatitude and tropical North America.

Diagnosis:

Male horn cores broadly triangular (scalene) in cross section and asymmetrical about dorso-ventral axis at base; tip cordiform to triangular in cross section; posterior margins sinuous; growth spiraled around sinuous longitudinal axis; antero-posterior plane usually, but sometimes not, rotated forward from frontal plane; length along upper curve greater than 250 mm, less than 425 mm.

Female horn core circular to elliptical in cross section at base; symmetrical about greatest and least diameter at base; tip circular to elliptical in cross section; posterior margin sinuous or convex; growth straight along arched longitudinal axis; plane of greatest diameter usually rotated forward from vertical plane; length along upper curve greater than 150 mm, less than 250 mm.

Differential Diagnosis:

Male *B. bison, B. antiquus* and *B. latifrons* horn cores symmetrical about dorso-ventral axis at base; posterior margins straight; antero-posterior plane parallel with or rotated rearward from frontal plane; growth straight or mildly spiraled around arched longitudinal axis (except in *B. a. occidentalis*, where spiraled growth is stronger). *B. bison* and *B. a. occidentalis* have circular or elliptical cross section at tip. *B. alaskensis* length on upper curve greater than 425 mm. Female *B. bison* and *B. antiquus* posterior margin straight; antero-posterior plane usually rotated less far forward than in *B. priscus. B. latifrons* posterior margin usually straight; length along upper curve greater than 500 mm. *B. alaskensis* length along upper curve greater than 250 mm.

Biometric Summary:

Skulls, table 38.

Referred Specimens:

Table 39.

Description:

B. priscus was the smaller species of late Pleistocene Eurasian bison. The horn cores of male *B. priscus* emanate horizontally or subhorizontally from the frontals at an angle of about 65° to 75° forward from the sagittal plane. The horn cores are broadly triangular (scalene) in cross section at the base with the greatest dorso-ventral diameter situated toward the front of the core. The horn cores either maintain or increase their antero-posterior diameter in the proximal region, and thereafter taper gradually toward the tip. The tip of the core is normally well above the frontal plane and usually behind the occipital plane, although tips are often situated in front of the occipital plane because the cores are rotated forward. A distinct dorsal groove is frequently present along the distal 10–20 percent of the core. The distal core is normally deflected to the rear and occasionally recurved, although more subtly and less frequently recurved than in *B. alaskensis*. The sheath tip is normally deflected to the rear and terminates behind the occipital plane (fig. 17, plate 17).

The frontals vary from relatively flat to relatively domed. The orbits are high, directed forward and protrude strongly. The nasals are low, broad, and rectangular. The occipital is moderately concave. The nuchal line/external occipital protuberance is moderately developed. The occipital condyles do not usually protrude strongly from the occiput. The ventro-lateral margin of the occipital condyle is occasionally flanged.

Female horn cores are slender and they emanate horizontally from the frontals at about the same angle forward from the sagittal plane as do male horn cores (65° to 75°). The horn cores are circular to elliptical in cross section at the base. They either maintain or increase their greatest diameter in the proximal section of the core. Thereafter the core tapers gradually toward the tip. The core has a gradual upward and outward, sometimes forward, curvature. The tip occurs tangential to or above the frontal plane on and in front of the occipital plane (plate 18).

Frontals are relatively flat. Orbits are high, directed forward, and do not protrude strongly. Nasals are narrow, high, and triangular. The occipital is relatively flat. The nuchal line/external occipital protuberance is not prominent. The occipital condyles do not protrude from the occipital plane, and they are not flanged.

Distribution:

B. priscus is known from throughout mid- and high latitude Eurasia and Beringia, but from only a few localities in midlatitude and tropical North America (fig. 26). The distribution pattern for this species suggests that it evolved in late-early Pleistocene midlatitude Eurasia and that it was probably a grazer/browser primarily adapted to savanna or steppe environments. *B. priscus*

Table 38

Summary of *Bison priscus* Skull Biometrics

Standard measurement	n	Range	Mean ± SE	σ	V
Males					
Spread of horn cores, tip to tip	23	751 – 1064 mm	887.8 ± 20.3 mm	97.4 mm	11.0
Horn core length, upper curve, tip to burr	36	285 – 416 mm	362.7 ± 5.9 mm	35.3 mm	9.7
Straight line distance, tip to burr, dorsal horn core	36	268 – 379 mm	318.4 ± 5.3 mm	31.5 mm	9.9
Dorso-ventral diameter, horn core base	52	84 – 116 mm	98.3 ± 1.1 mm	8.3 mm	7.0
Minimum circumference, horn core base	52	293 – 387 mm	330.1 ± 2.9 mm	21.0 mm	6.4
Width of occipital at auditory openings	21	248 – 310 mm	276.3 ± 3.2 mm	14.6 mm	5.3
Width of occipital condyles	25	127 – 165 mm	141.1 ± 1.4 mm	6.9 mm	4.9
Depth, nuchal line to dorsal margin of foramen magnum	26	91 – 119 mm	102.9 ± 1.6 mm	8.3 mm	8.1
Antero-posterior diameter, horn core base	52	98 – 130 mm	114.9 ± 1.1 mm	8.0 mm	7.0
Least width of frontals, between horn cores and orbits	22	269 – 336 mm	293.7 ± 14.4 mm	20.5 mm	7.0
Greatest width of frontals at orbits	14	313 – 415 mm	350.4 ± 6.8 mm	25.6 mm	7.3
M^1–M^3, inclusive alveolar length	0	....	....	....	....
M^3, maximum width, anterior cusp	0	....	....	....	....
Distance, nuchal line to tip of premaxillae	1	....	608.0 mm	....	....
Distance, nuchal line to nasal-frontal suture	5	254 – 286 mm	272.4 ± 5.5 mm	12.3 mm	4.5
Angle of divergence of horn cores, forward from sagittal	21	63° – 79°	70.8° ± 0.9°	4.0°	5.7
Angle between foramen magnum and occipital planes	25	119° – 156°	133.2° ± 1.8°	8.9°	6.7
Angle between foramen magnum and basioccipital planes	26	107° – 133°	117.2° ± 1.1°	5.8°	5.0
Females					
Spread of horn cores, tip to tip	8	505 – 673 mm	598.5 ± 21.1 mm	59.6 mm	10.0
Horn core length, upper curve, tip to burr	8	177 – 245 mm	208.9 ± 9.4 mm	26.7 mm	12.8
Straight line distance, tip to burr, dorsal horn core	8	169 – 227 mm	197.8 ± 7.3 mm	20.5 mm	10.4
Dorso-ventral diameter, horn core base	10	58 – 70 mm	63.4 ± 1.4 mm	4.5 mm	7.1
Minimum circumference, horn core base	9	179 – 225 mm	199.3 ± 4.4 mm	13.1 mm	6.6
Width of occipital at auditory openings	3	203 – 232 mm	219.0 ± 8.5 mm	14.7 mm	6.7
Width of occipital condyles	4	114 – 130 mm	121.8 ± 4.0 mm	7.9 mm	6.5
Depth, nuchal line to dorsal margin of foramen magnum	3	89 – 94 mm	91.0 ± 1.5 mm	2.6 mm	2.9
Antero-posterior diameter, horn core base	10	57 – 74 mm	64.0 ± 1.5 mm	4.8 mm	7.5
Least width of frontals, between horn cores and orbits	10	211 – 246 mm	228.5 ± 3.4 mm	10.8 mm	4.7
Greatest width of frontals at orbits	8	257 – 297 mm	279.6 ± 5.3 mm	15.1 mm	5.4
M^1–M^3, inclusive alveolar length	0	....	....	....	....
M^3, maximum width, anterior cusp	0	....	....	....	....
Distance, nuchal line to tip of premaxillae	0	....	....	....	....
Distance, nuchal line to nasal-frontal suture	4	224 – 243 mm	234.5 ± 3.9 mm	7.8 mm	3.4
Angle of divergence of horn cores, forward from sagittal	10	60° – 78°	71.2° ± 2.0°	6.3°	8.8
Angle between foramen magnum and occipital planes	4	129° – 142°	134.8° ± 2.7°	5.4°	4.0
Angle between foramen magnum and basioccipital planes	4	101° – 115°	110.3° ± 3.1°	6.3°	5.7

Table 39

Bison priscus: Referred Specimens

Institution and identification number		Sex	Provenience
AMNH	VP 17733	♂	Combs, England
AMNH	VP F:AMA235-6640	♀	Goldstream, Alaska
AMNH	VP F:AMA251-8071	♀	Engineer Creek, Alaska
AMNH	VP F:AMA369-1369	♀	Ester Creek, Alaska
AMNH	VP F:AMA424-1259	♂	Head of Ester Creek, Alaska
AMNH	VP F:AMA432-4024	♂	Cripple Creek, Alaska
AMNH	VP F:AMA500-2025	♂	Cripple Creek Sump, Alaska
AMNH	VP F:AMA500-3613	♂	Cripple Creek Sump, Alaska
AMNH	VP F:AMA501-1007	♀	Fairbanks Creek, Alaska
AMNH	VP F:AMA501-2016	♂	Cripple Creek Sump, Alaska
AMNH	VP F:AMA508-2002	♂	Cripple Creek, Alaska
AMNH	VP F:AMA570-1015	♀	Myers Fork, 40 Mile region, Alaska
AMNH	VP F:AMA580-2006	♂	Fairbanks Creek, Alaska
AMNH	VP F:AMA580-2017	♂	Fairbanks Creek, Alaska
AMNH	VP F:AMA580-2038	♀	Fairbanks Creek, Alaska
AMNH	VP F:AMA605-1006	♂	Fairbanks Creek, Alaska
AMNH	VP F:AMA606-1078	♀	Fairbanks Creek, Alaska
AMNH	VP F:AMA625-3114	♂	Dome Creek, Alaska
AMNH	VP F:AMA625-4188	♂	Gold Hill, Alaska
AMNH	VP F:AMA645-6075	♂	Cripple Creek Sump, Alaska
AMNH	VP F:AMA684-1588	♂	Gold Hill, Alaska
AMNH	VP F:AMA685-1583	♂	Gold Hill, Alaska
AMNH	VP F:AMA685-3249	♂	Dome Creek, Alaska
AMNH	VP F:AMA698-2025	♂	Fairbanks Creek, Alaska
AMNH	VP F:AMA698-2032	♂	Fairbanks Creek, Alaska
AMNH	VP F:AMAFAM 30547	♂	Cleary Creek, Alaska
AMNH	VP F:AMAFAM 30552	♂	Cleary Creek, Alaska
AMNH	VP F:AMAFAM 30581	♂	Cleary Creek, Alaska
AMNH	VP F:AMAFAM 30609	♂	Cleary Creek, Alaska
AMNH	VP F:AMAFAM 30623	♂	Cleary Creek, Alaska
AMNH	VP F:AMAFAM 30635	♂	Goldstream Banks at Fox, Alaska
AMNH	VP F:AMAFAM 46885	♂	Upper Cleary Creek, Alaska
AMNH	VP F:AMAFAM 46893	♂	Cripple Creek, Alaska
AMNH	VP F:AMAFAM 46906	♂	Engineer Creek, Alaska
FMNH	P 25480	♂	Point Barrow, Alaska
INAH(DP)	516	♂	San Vicente Chicoloapan, Mexico
IMNH	VP 2000	♂	American Falls Reservoir, southeast Idaho
IMNH	VP 6377-808A	♂	American Falls Reservoir, southeast Idaho
KU	VP 8347	♂	Cripple Creek, Alaska
NMC	VP 8144	♂	Bonanza Creek, Dawson District, Yukon Territory
NMC	VP 8146	♂	Dawson District, Yukon Territory
NMC	VP 9923	♂	Hunker Creek, Yukon Territory
NMC	VP 11351	♂	Cripple Hill, Yukon Territory
NMC	VP 11352	♂	Cripple Hill, Yukon Territory
NMC	VP 11782A	♂	Dominion Creek, Yukon Territory
NMC	VP 11782B	♂	Dominion Creek, Yukon Territory
NMC	VP 13507	♂	Gold Run Creek, Yukon Territory
NMC	VP 13508	♂	Gold Run Creek, Yukon Territory
NMC	VP 13559	♂	Gold Run Creek, Yukon Territory
NMC	VP 13560	♂	Gold Run Creek, Yukon Territory
NMC	VP 17330	♂	Quartz Creek, Yukon Territory
SMP	272-74-z	♀	Cherokee County, Iowa
TMM	1123	♂	Alaska
UMo	VP 270	♂	Missouri (?)
UNSM	VP 3-54	♂	Near Platte River, Dodge County, Nebraska
UNSM	VP 2107-53	♀	Nebraska (?)
UNSM	VP 30950	♂	Red Willow County, Nebraska
UNSM	VP 36534	♂	Nebraska (?)
UNSM	VP 55174	♂	Clovis area, Curry County (?), New Mexico
UNSM	VP A251-7923	♀	Goldstream, Alaska
USNM	P 25601	♂	Near mouth of Lost Chicken Creek, east central Alaska
YPM	VP 34620	♂	Southwold, Suffolk, England

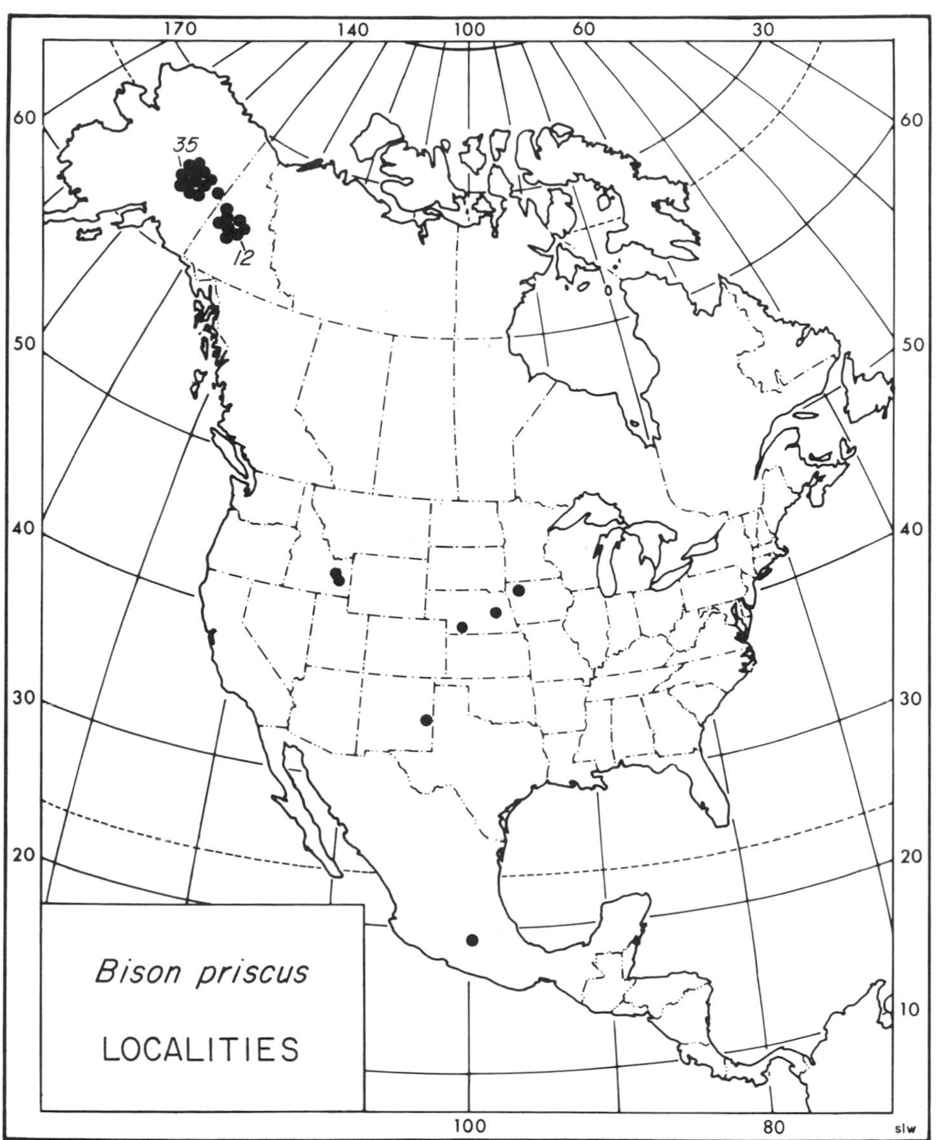

FIGURE 26 The distribution of *Bison priscus* specimens used in this study. *Bison priscus* was a Eurasian autochthon that periodically dispersed into midlatitude and tropical North America during glacial periods. The taxon was apparently well represented in eastern Beringia, ecologically an extension of the Palearctic, but was less well represented elsewhere in North America. Numbers indicate total specimens from clustered localities.

probably reached its maximum distribution and population densities during Wisconsin (and possibly Illinoian) glacial stades.

B. priscus first appears in the Nihowan and other contemporary faunas (= middle and late Villafranchian) of northeastern China (Aigner, 1972; Huang and Chang, 1966; Teilhard de Chardin and Piveteau, 1930). Early middle Pleistocene records of *B. priscus* (and unidentified bison) are from midlatitude Eurasia, whereas late middle Pleistocene records show a shift to higher latitudes, especially in Siberia and Beringia (Aigner, 1972; Flerow, 1971; Kurten, 1968; Péwé, 1975; Péwé and Hopkins, 1967). *B. priscus* was widespread throughout Europe, Siberia, and glacial Beringia during the late Pleistocene, but became extinct over most of its range during the late Würm/late Wisconsin.

The sporadic occurrence of *B. priscus* in midlatitude and tropical North America precludes a satisfactory determination of the geologic age of this species in North America. Limited evidence indicates that *B. priscus* was present in North America during the mid-Wisconsin and late Wisconsin, but in small numbers only. McDonald and Anderson (1975) described a *B. priscus* (IMNH VP 2000) from the Michaud Gravels, American Falls Reservoir, Idaho, which dates from about 31,000 to 21,000 BP. MCZ VP 4425, presumed to be a *B. priscus* X *B. antiquus* hybrid, from Harvard, Massachusetts, has been radiocarbon dated at about 21,200 BP. Harington and Clulow (1973) report *B. priscus* from Gold Run Creek, Yukon Territory, dated at about 20,000 BP. SMP 272-74-z is from gravels near Des Moines, Iowa, assumed (but not proven) to be late Wisconsin in age. *B. priscus* probably disappeared from North America during the late Wisconsin or very early Holocene.

Discussion:

Identity of the lectotype: Bojanus (1827) created the name *Urus priscus* to accommodate the large horned extinct bison. The group of five specimens which Bojanus studied included skulls from the United States, Germany, Siberia, and Italy. Three of these skulls—the United States, Siberian, and one German skull—had previously received Linnaean names (see below). The second German skull was lost sometime after Bojanus studied it. Hilzheimer (1918) determined that the only known and previously unnamed specimen of Bojanus's series—the skull from Italy—should be regarded as the lectotype for *B. priscus* (fig. 17). Although all three of the known Eurasian skulls were similar, because the names given to the remaining German skull (*B. cesaris*) and Siberian skull (*B. pallasii*) have not been kept in use, these names are herein informally synonymized with *B. priscus*. The U.S. skull in the syntypic series was the holotype for *B. latifrons*, a species distinctly different from *B. priscus*.

Vialli (1954) reviewed the fossil *Bison* from the Pavia, Italy, area and concluded that "Skull No. 1," located in the Geology Institute Museum, University of Pavia, was the *B. priscus* lectotype illustrated by Cuvier (1825) (fig. 17). Cuvier had stated that the specimen he illustrated was in the cabinet of the University of Pavia, but Cuvier's illustration much more strongly resembles Vialli's "Skull No. 2," now located in the Geology Institute Museum, University of Torino. Comparison of Cuvier's illustration with skulls number 1 and 2 of Vialli (1954) [plate 19] reveals that the horn core, fronto-parieto-occipital, nasal, and maxillary morphology of Cuvier's figure more nearly resemble skull number 2 (Torino) than skull number 1 (Pavia), whereas the orbits of Cuvier's figure are nearer skull number 1. Cuvier may never have seen the specimen(s) before publishing the description and illustration, which may be either poorly drawn, or a composite of characters of more than one skull. The matter is of great importance because the horn core length of skull number 1 (Pavia) falls within the *B. alaskensis* range, whereas the same character of skull number 2 (Torino) falls within the *B. priscus* range. Considering the greater resemblance of Cuvier's figure to skull number 2 than skull number 1, skull number 2 is recognized here as the probable lectotype of *B. priscus*.

Synonymy: *B. crassicornis* has long been one of the most perplexing "species" of bison. Richardson (1852–1854) described this species from the Eschscholtz Bay region, Alaska; the lectotype (BM 1A), which I did not examine, consists of the frontals, occipital, and the proximal horn cores. Richardson established the distinctiveness of *B. crassicornis* on characters of the occipital and the apparent "heaviness" of the horn cores. The occipital of bison is not specifically diagnostic, and an insufficient amount of horn core remains to positively identify the specimen.

Judging from Richardson's illustrations, the basal horn cores appear to be rotated forward, which suggests affinity with the late Pleistocene European bison. No reliable estimate of the relative flatness/roundness of the horn core base cross section or of the posterior profile can be made. The posterior profile appears, however, to lack the proximal, postbasal posterior bulge that gives the *B. priscus-B. alaskensis* group its distinctive posterior profile. The occipital condyles are flanged along the ventro-lateral edge, an occasional but not diagnostic characteristic of *B. priscus, B. alaskensis,* and *B. latifrons*. The forward rotation of the horn core, the occipital condyle flange, and the relatively small horn circumference (ca. 320 mm) preclude referring this specimen to *B. latifrons, B. alaskensis, B. antiquus,* or *B. bison*. The holotype of *B. crassicornis* is, therefore, referred to *B. priscus* because (1) it lacks sufficient diagnostic characters to justify specific distinctiveness; (2) it conforms to the range and complex of characters expected of *B. priscus*; and (3) it was found in a place and circumstance which suggests affinity with *B. priscus*. The concept of *B. crassicornis* which has become widespread recognizes this species as a highly variable taxon occurring throughout the Beringian region, but with some populations occurring southward as far as the northern United States. *B. crassicornis* has traditionally been viewed as (1) a taxon phenotypically and phyletically intermediate between earlier larger horned bison and later short horned bison; or (2) a synonym of *B. priscus* which (combined with *B. alaskensis*) creates the image of a highly variable *B. priscus* (Fuller and Bayrock, 1965; Gromova, 1935; Guthrie, 1970; Wilson, 1969, 1974*a*, 1974*b*).

A high proportion of the Beringian skulls do show characters of both late Pleistocene Eurasian and late Pleistocene North American species, yet no discrete canalization of diagnostic characters exists in this sample. The presence of diagnostic characters of both Eurasian and North American species in many of the Beringian skulls, evaluated in light of the dynamic environmental history of Pleistocene Beringia, has led to the conclusion that this highly variable group of bison constitutes a set of hybrid individuals. This concept does not require that all intermediate specimens be first-generation hybrids; the frequency and magnitude of specific traits represented in individuals probably varied with changing directions and rate of gene flow. Hypothetically, Eurasian bison traits would have been more common after glacial periods as North American populations reentered the region, and North American traits would have been more common after interglacial periods as Eurasian populations reentered the region. Complicating the pattern of variation were the varied and changing local selection regimes and the periodic existence of small, inbreeding populations.

B. willistoni was described from a left horn core and a partial maxillary (KU VP 390) found along the Arkansas River near Garden City, Finney County, Kansas (Martin, 1924). Martin stated, "In the following measurements, notice should be particularly directed to those of the transverse and vertical diameters of the horn core, which indicate that in *B. willistoni* the core is more flattened vertically along its whole length than is the case in *B. occidentalis*" (pp. 274–275). Martin also considered the length of the horn core to be significantly longer in this specimen than in *B. kansensis, B. occidentalis*, or *B. bison*.

B. willistoni differs from all recognized species—it does not have a diagnostic horn core complex. Rather, it shares traits of both North American and Eurasian bison and is, accordingly, considered unique and a hybrid. The holotype horn core is broadly triangular (scalene) at the base and has a sinuous posterior profile, suggesting an affinity with Eurasian bison. The antero-posterior plane is apparently parallel with the frontals, the longitudinal axis is arched, and the distal horn core resembles that of *B. latifrons* more so than that of *B. priscus*, suggesting affinity with North American bison. This specimen is considered a *B. priscus* X *B. latifrons* hybrid, with stronger resemblance to *B. priscus*.

B. antiquus barbouri was considered taxonomically distinct on the basis that its holotype (UNSM VP 30310) was larger and more massive than bison from later deposits, and smaller than middle Pleistocene bison (Schultz and Frankforter, 1946). This description contains no distinguishing criteria—it simply refers to intermediate size fitted into both the retrogressive horn core phylogenesis and the river terrace chronology models espoused by Schultz and others (Schultz and Frankforter, 1946; Schultz and Hillerud, 1977; Schultz, Martin, and Tanner, 1972; Schultz and Stout, 1948). The truly diagnostic criteria—flat horn cores rotated forward, sinuous posterior profile, dorso-ventral compression, shape of basal cross section, etc.—were not recognized or considered by Schultz and Frankforter. In fact, Schultz (oral comm., July 30, 1975) attributes these characters to crushing by overburden! The horns have in fact been broken and restored, but the dorso-ventral flattening is clearly a developmental character. In addition to dorso-ventral compression the horn cores of UNSM VP 30310 exhibit postbasal widening, a (scalene) triangular basal cross section, and rapid thinning at the tip; these are characters typical of *B. priscus*. The cores have a straight axis and growth pattern, lacking distal twist, distal deflection, or forward rotation, however; these are characters more typical of *B. a. antiquus*. UNSM VP 30310 shares characters typical of two taxa and is therefore considered a hybrid individual. *B. priscus* characters are dominant; *B. a. barbouri* is thus synonymized with *B. priscus*.

B. preoccidentalis was named and described by Skinner and Kaisen (1947) from a small group of generally similar skulls withdrawn from a large suite of late Pleistocene Alaska skulls. The characters used to specifically differentiate *B. preoccidentalis* amount to no more than individual variation within the expected range of variation for *B. priscus* or hybrids of *B. priscus* and North American autochthons.

B. geisti was differentiated from *B. pallasii* (= *B. priscus*) on the basis of minor and nondiagnostic differences in horn core length and angle of posterior deflection (Skinner and Kaisen, 1947). The authors noted the similarity of *B. geisti* to *B. alaskensis*, but did not consider it related to *B. priscus*. The *B. geisti* holotype (AMNH VP F:AM46893; plate 17), however, possesses all the diagnostic characters of *B. priscus*.

II. *Bison alaskensis* Rhoads, 1897

Original Description:
Samuel N. Rhoads. 1897. "Notes on Living and Extinct Species of North American Bovidae." *Proceedings of the Academy of Natural Sciences of Philadelphia*, vol. 49, pp. 483–502.

Type Locality:
Tundra back of Point Barrow, Alaska.

Type Location:
Department of Paleontology, Field Museum of Natural History, Chicago, Illinois.

Type Specimen:
FMNH P 25226; frontals, occipital, basioccipital, and horn cores.

Principal Synonymy:
Bison chaneyi Cook, 1928
Bison aguascalentensis Mooser and Dalquest, 1975

Species Content:
B. alaskensis was the late Pleistocene forest/woodland bison of Eurasia, including glacial Beringia. Periodic dispersal southward expanded the range through midlatitude and into tropical North America.

Diagnosis:
Male horn cores broadly triangular (scalene) in cross section and asymmetrical about dorso-ventral axis at base; tip cordiform to triangular in cross section; posterior margin sinuous; growth spiraled around sinuous longitudinal axis; anteroposterior plane usually rotated forward from frontal plane; length along upper curve greater than 425 mm.

Female horn cores circular to elliptical in cross section and symmetrical about greatest and least diameters at base; tip circular to elliptical in cross section; posterior margin sinuous or convex; growth spiraled or straight around arched or sinuous axis; plane of greatest diameter rotated forward from vertical; length along upper curve greater than 250 mm.

Differential Diagnosis:
B. bison, B. latifrons, B. antiquus as with *B. priscus* for both sexes. Male *B. priscus* length along upper curve less than 425 mm. Female *B. priscus* length along upper curve less than 250 mm.

Biometric Summary:
Skulls, table 40.
Referred Specimens:
Table 41.
Description:

B. alaskensis is the largest species of late Pleistocene Eurasian bison. The horn cores of male *B. alaskensis* are long and relatively slender. They emanate horizontally or subhorizontally from the frontals at an angle of about 65° to 75° forward from the sagittal plane. The horn cores are broadly triangular (scalene) in cross section at the base, with the greatest dorso-ventral diameter situated toward the front of the core. The horn core either maintains a constant antero-posterior diameter, or increases this diameter, in the proximal region, and thereafter tapers gradually toward the tip. A distinct dorsal groove is usually, but not always, present along the distal 10–15 percent of the core. The tip of the core normally occurs above the frontal plane and may be either behind or in front of the occipital plane. The tip often occurs in front of the occipital plane because the core is rotated forward. In these cases the medial core normally passes through the occipital plane. The distal core is normally deflected to the rear and is often recurved. The sheath tip is deflected to the rear and ends behind the occipital plane (plate 20).

Table 40

Summary of *Bison alaskensis* Skull Biometrics

Standard measurement	n	Range	Mean ± SE	σ	V
Males					
Spread of horn cores, tip to tip	25	800 – 1540 mm	1150.5 ± 30.7 mm	153.6 mm	13.3
Horn core length, upper curve, tip to burr	36	426 – 795 mm	552.9 ± 15.8 mm	94.6 mm	17.1
Straight line distance, tip to burr, dorsal horn core	35	294 – 667 mm	458.7 ± 13.5 mm	79.6 mm	17.4
Dorso-ventral diameter, horn core base	49	100 – 153 mm	117.0 ± 1.6 mm	11.4 mm	9.7
Minimum circumference, horn core base	44	341 – 493 mm	397.4 ± 6.1 mm	40.5 mm	10.2
Width of occipital at auditory openings	28	268 – 340 mm	302.2 ± 3.6 mm	19.2 mm	6.3
Width of occipital condyles	30	133 – 175 mm	152.2 ± 1.7 mm	9.5 mm	6.2
Depth, nuchal line to dorsal margin of foramen magnum	28	97 – 134 mm	111.2 ± 2.0 mm	10.3 mm	9.3
Antero-posterior diameter, horn core base	48	122 – 171 mm	141.8 ± 2.1 mm	14.3 mm	10.1
Least width of frontals, between horn cores and orbits	23	281 – 394 mm	332.1 ± 5.3 mm	25.5 mm	7.7
Greatest width of frontals at orbits	18	356 – 440 mm	395.8 ± 5.5 mm	23.2 mm	5.9
M^1–M^3, inclusive alveolar length	0	....	....	....	....
M^3, maximum width, anterior cusp	0	....	....	....	....
Distance, nuchal line to tip of premaxillae	3	602 – 676 mm	650.3 ± 24.1 mm	41.8 mm	6.4
Distance, nuchal line to nasal-frontal suture	9	265 – 338 mm	298.4 ± 7.3 mm	21.8 mm	7.3
Angle of divergence of horn cores, forward from sagittal	23	64° – 85°	73.7° ± 1.2°	5.8°	7.9
Angle between foramen magnum and occipital planes	24	118° – 148°	129.6° ± 1.7°	8.4°	6.5
Angle between foramen magnum and basioccipital planes	24	105° – 136°	120.5° ± 1.6°	7.8°	6.5
Females					
Spread of horn cores, tip to tip	3	703 – 937 mm	808.3 ± 68.6 mm	118.7 mm	14.7
Horn core length, upper curve, tip to burr	6	288 – 388 mm	319.2 ± 14.9 mm	36.6 mm	11.5
Straight line distance, tip to burr, dorsal horn core	6	253 – 348 mm	291.0 ± 13.6 mm	33.3 mm	11.5
Dorso-ventral diameter, horn core base	5	73 – 91 mm	81.2 ± 3.1 mm	6.9 mm	8.6
Minimum circumference, horn core base	6	210 – 290 mm	252.8 ± 12.5 mm	30.6 mm	12.1
Width of occipital at auditory openings	1	....	260.0 mm	....	....
Width of occipital condyles	2	123 – 134 mm	128.5 ± 5.4 mm	7.7 mm	6.0
Depth, nuchal line to dorsal margin of foramen magnum	1	....	100.0 mm	....	....
Antero-posterior diameter, horn core base	5	75 – 96 mm	84.6 ± 4.6 mm	10.2 mm	12.1
Least width of frontals, between horn cores and orbits	3	245 – 279 mm	260.0 ± 10.0 mm	17.3 mm	6.7
Greatest width of frontals at orbits	2	308 – 320 mm	314.0 ± 6.0 mm	8.5 mm	2.7
M^1–M^3, inclusive alveolar length	0	....	....	....	....
M^3, maximum width, anterior cusp	0	....	....	....	....
Distance, nuchal line to tip of premaxillae	1	....	575.0 mm	....	....
Distance, nuchal line to nasal-frontal suture	2	245 – 255 mm	250.1 ± 5.0 mm	7.1 mm	2.8
Angle of divergence of horn cores, forward from sagittal	1	....	64.0°	....	....
Angle between foramen magnum and occipital planes	1	....	112.0°	....	....
Angle between foramen magnum and basioccipital planes	1	....	134.0°	....	....

Table 41
Bison alaskensis: Referred Specimens

Institution and identification number			Sex	Provenience
AMNH	VP	F:AMA421-1216	♂	Dawson Cut, Engineer Creek, Alaska
AMNH	VP	F:AMA426-4025	♂	Cripple Creek, Alaska
AMNH	VP	F:AMA502-2007	♂	Cripple Creek, Alaska
AMNH	VP	F:AMA514-2006	♂	Cripple Creek, Alaska
AMNH	VP	F:AMA625-3113	♂	Dome Creek, Alaska
AMNH	VP	F:AMA625-6549	♂	Engineer Creek, Alaska
AMNH	VP	F:AMA625-6550A	♂	Engineer Creek, Alaska
AMNH	VP	F:AMA----1308	♂	Gertrude Creek, Livengood District, Alaska
AMNH	VP	F:AMAFAM 46901	♂	Dawson Cut, Engineer Creek, Alaska
AMNH	VP	F:AMAFAM 46903	♂	Fairbanks Creek, Alaska
AMNH	VP	F:AMAFAM 46909	♀	Ester Creek, Alaska
AMNH	VP	F:AMAFAM 46939	♂	Cripple Creek, Alaska
AMNH	VP	F:AMAFAM 54041	♂	Fairbanks area, Alaska
CoMNH	VP	1147	♂	Vernon, Wilbarger County, Texas
FC		658[a]	♂	Near Aguascalientes, Aguascalientes
FMNH	P	25226	♂	Near Point Barrow, Alaska
IG		49-33[b]	♂	Tajo del Desague Viejo, Tequixquiac, Mexico
IMNH	VP	548	♂	McCammon Pit (?), Bannock County, Idaho
IMNH	VP	1676	♂	American Falls Reservoir, southeast Idaho
IMNH	VP	1902	♀	American Falls Reservoir, southeast Idaho
IMNH	VP	2448	♂	American Falls Reservoir, southeast Idaho
IMNH	VP	2597	♂	Near Acequia, Minidoka County, Idaho
IMNH	VP	6377-133	♂	Chicken Ramsey Pit, Idaho
IMNH	VP	16424	♂	American Falls Reservoir, southeast Idaho
IMNH	VP	17384	♂	Near Acequia, Minidoka County, Idaho
IMNH	VP	17769	♂	McCammon Pit, Bannock County, Idaho
IMNH	VP	17876	♂	Kloepher Company Pit, Minidoka County, Idaho
IMNH	VP	A*	♂	Idaho (?)
KU	VP	4634	♂	Near Newton, Harvey County, Kansas
KU	VP	4927	♂	Near Fredonia, Wilson County, Kansas
KU	VP	7387	♂	Fall River, Wilson County, Kansas
LACM	VP	20834	♂	Hollywood, Los Angeles County, California
MCZ	VP	9113	♂	Texas
MNHN		15[b]	♂	Tequixquiac, Mexico
MRG		A*	♂	Lago de Chapala, Jalisco
MU		538[c]	♂	Near Electra, Wilbarger County, Texas
NMC	VP	7392	♂	Gold Run Creek, Klondike District, Yukon Territory
NMC	VP	11348	♂	Cripple Hill, Yukon Territory
NMC	VP	11646	♂	Gold Run Creek, Yukon Territory
NMC	VP	13506	♂	Gold Run Creek, Yukon Territory
NMC	VP	23349	♂	Old Crow River, Yukon Territory
NMC	VP	24202	♀	Old Crow River, Yukon Territory
NMC	VP	26055	♂	Apex Gravel Pit, Edmonton, Alberta
NMC	VP	31000	♂	Near Dawson, Yukon Territory
PMA	P	71.7.1	♂	A & R Gravel Pit, near Watino, Alberta
SM		61364[d]	♀	Moore Pit, near Dallas, Texas
SM		A*	♂	Oklahoma
SMNH		AC9961	♂	Echo Lake, Saskatchewan
SU	G	A*	♂	American Falls Reservoir, southeast Idaho
TMM		31041-30	♂	Near Swenson, Stonewall County, Texas
TMM		A*	♂	Near Beeville, Bee County, Texas

Table 41

Bison alaskensis: Referred Specimens (Continued

Institution and identification number			Sex	Provenience
UCMP		33025	♂	Tequixquiac, Mexico
UCMP		39048	♂	Cool Quarry, near Cool, El Dorado County, California
UCMP		58484	♂	California Sand and Gravel Company Pit, Alameda County, California
UMMP		31752	♂	Wichita County, Texas
UMMP		60836	♂	Near Hollywood Corner, Cleveland County, Oklahoma
UNSM	VP	5443	♂	Hamilton County, Nebraska
UNSM	VP	30356	♂	Saunders County, Nebraska
UNSM	VP	30358	♂	Saunders County, Nebraska
UNSM	VP	31027	♀	Cuming County, Nebraska
UNSM	VP	55170	♂	Stanton County, Nebraska
UNSM	VP	AFAM 30580	♂	Gilmore, Alaska
UNSM	VP	AFAM 46972	♀	Goldstream, Alaska
USGS		D901-P327B	♂	Near Acequia, Minidoka County, Idaho
USGS		D901-A*	♂	Near Acequia, Minidoka County, Idaho
USNM	P	13692	♂	American Falls Reservoir, southeast Idaho
USNM	P	13693	♂	American Falls Reservoir, southeast Idaho
USNM	P	25913	♂	Near Parker, Douglas County, Colorado

Sources of supplementary data:
[a]Mooser and Dalquest, 1975;
[b]Hibbard and Villa R., 1950;
[c]Dalquest, 1957;
[d]Slaughter, 1966.
*No identification number was found for these specimens.

The frontals vary from relatively flat to extremely domed. The orbits are high and directed relatively forward. Nasals are broad, low, and triangular. The occipital is more concave than in any other species of North American bison except *B. latifrons*, a result of both the prominent nuchal line/external occipital protuberance and the exaggerated protrusion of the occipital condyles. The ventrolateral margin of the occipital condyles is frequently flanged.

Female horn cores are long and relatively slender. They emanate horizontally from the frontals at about the same angle forward from the sagittal plane as do male horn cores. The horn cores are circular to elliptical in cross section at the base. They either maintain or increase their greatest diameter in the proximal section of the core. Thereafter the core gradually tapers toward the tip. The core curves gradually upward and outward and, occasionally, forward. The tip occurs above the frontal plane and forward of the occipital plane (plate 21).

The frontals are relatively flat. The orbits are high, directed forward, and do not protrude strongly. Nasals are narrow, high, and triangular. The occipital, more strongly featured than in any species except *B. latifrons*, has a relatively prominent nuchal line/external occipital protuberance and moderately protruding occipital condyles. The occipital condyles are normally not flanged and their long axes are arranged in a nearly straight line.

Distribution:

B. alaskensis is known from throughout middle and higher latitude Eurasia, Beringia, and much of midlatitude and subtropical North America (fig. 27). The distribution pattern for this species suggests that it evolved in late-middle or late Pleistocene Eurasia and

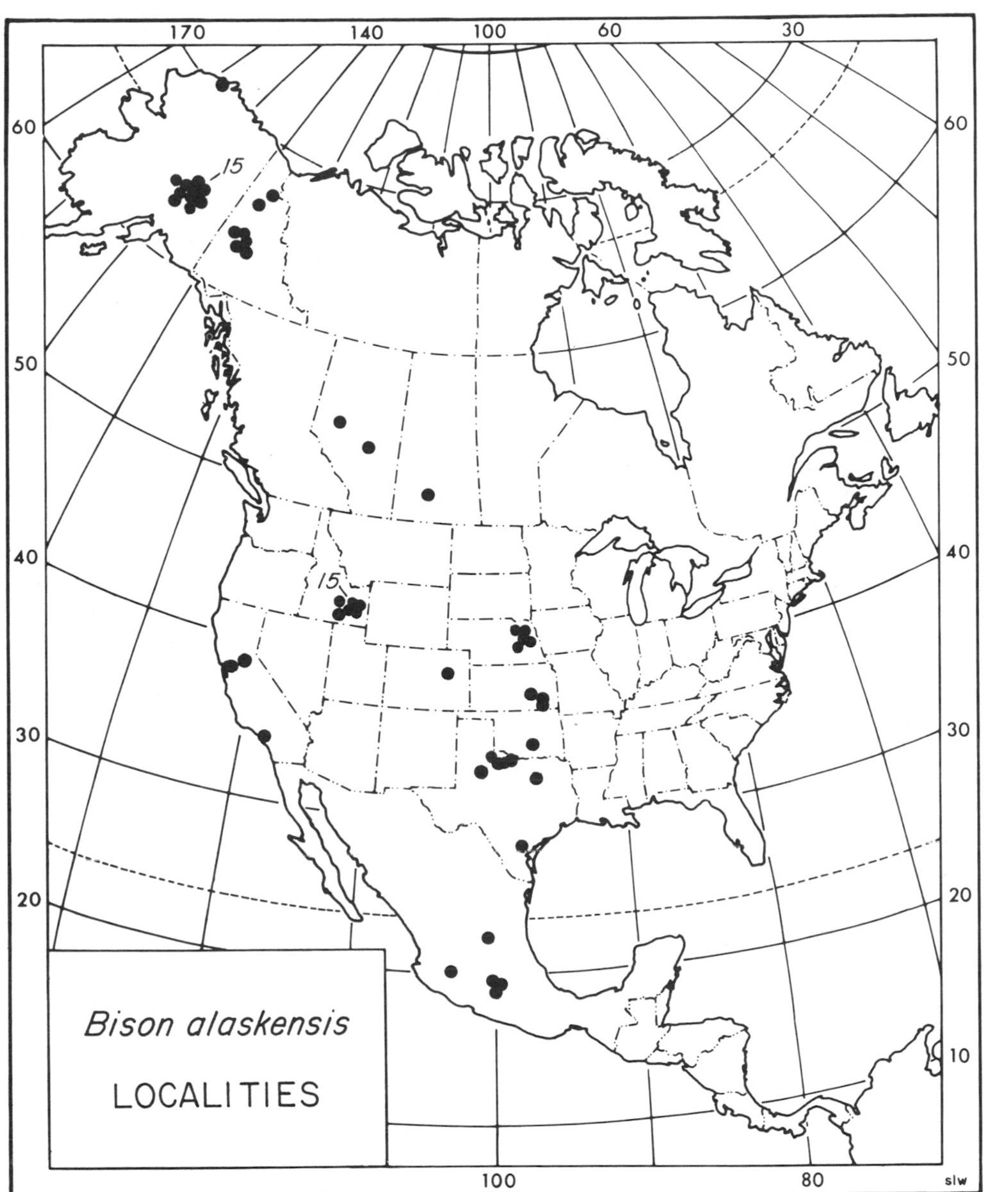

FIGURE 27 The distribution of *Bison alaskensis* specimens used in this study. *Bison alaskensis* was a Eurasian autochthon that periodically dispersed into midlatitude and tropical North America during glacial periods. *Bison alaskensis* was well represented in eastern Beringia, ecologically an extension of the Palearctic, but it was also well represented elsewhere in North America. *Bison alaskensis* was apparently more capable of competing in midlatitude and tropical North America than was *Bison priscus*. Numbers indicate total specimens from clustered localities.

that it was a browser/grazer primarily adapted to forest openings or woodlands. *B. alaskensis* probably reached its maximum distribution and population size during the Sangamon interglacial or Wisconsin interstades when forest/woodland habitat was most widespread throughout its range.

B. alaskensis is usually undifferentiated from *B. priscus* by European writers, so it is difficult to assess the antiquity of this larger Eurasian bison. *B. alaskensis* is probably somewhat more recent than *B. priscus*. There is, for example, no evidence of *B. alaskensis* from the early Pleistocene, but there are inconclusive indications that this species may have occurred in the late-middle or early-late Pleistocene of both Eurasia and North America (Hopkins, 1951; Hopkins, Bonnichsen, and Fortsch, 1969; Mooser and Dalquest, 1975; Péwé and Hopkins, 1967; Schultz and Hillerud, 1977). Most specimens are known from the late Pleistocene of both Eurasia and North America. The species apparently became extinct during the late Wisconsin.

The earliest records of *B. alaskensis* in North America place this species in the Illinoian. Radiocarbon dates indicate that *B. alaskensis* was present during the middle Wisconsin (table 42). Fuller and Bayrock (1965) described an isolated horn core from Alberta, found in late Wisconsin or early Holocene deposits, which may be a *B. alaskensis* X *B. antiquus* hybrid. This specimen is worn and appears to have been redeposited and is, therefore, probably older than the age indicated by Fuller and Bayrock. Based on available evidence, *B. alaskensis* is considered to have been present in mid-latitude and subtropical North America from the Illinoian to the middle Wisconsin and probably into the late Wisconsin in Beringia.

Discussion:

Synonymy: *B. chaneyi* was differentiated on the basis of the holotype's "long, heavy, strongly curved" horn cores, the tips of which were recurved inward and the base strongly flattened "fore and aft" [sic] (Cook,

Table 42

Geologic Ages Reported for *Bison alaskensis* Specimens

Locality	Reported age	Dating method	Source
Watino, Alberta	Pre-Wisconsin	Geostratigraphy	Storer, 1971
Gold Run, Yukon Territory	>39,000 – 22,000 BP	Radiocarbon	I-5405*, I-3570*
Aguascalientes, Aguascalientes	Illinoian (?)	Biostratigraphy	Mooser and Dalquest, 1975
Lago de Chapala, Jalisco	Late Pleistocene	Biostratigraphy	Downs, 1958
	30,000 BP ±	Radiocarbon	F. R. Solorzano, oral comm., July 14, 1976
Tequixquiac, Mexico	Late Sangamon–early Wisconsin	Geostratigraphy	Hibbard, 1955b
Parker, Colorado	Early Wisconsin	Geostratigraphy	Lewis, 1970; Scott and Lindvall, 1970
American Falls Reservoir, Idaho	Illinoian	Geostratigraphy	Hopkins, 1951; Hopkins, Bonnichsen, and Fortsch, 1969
	Pre-Illinoian–early Illinoian	Geostratigraphy	Schultz and Hillerud, 1977
Wilson County, Kansas	31,000 BP	Radiocarbon	M-997*
Oklahoma	Sangamon	Geostratigraphy	B. Slaughter, written comm., May 19, 1978
Electra, Texas	Late Illinoian–Sangamon	Biostratigraphy	Dalquest, 1959
Moore Pit, Texas	Mid-Wisconsin	Geostratigraphy	Slaughter, 1966

*Denotes radiocarbon date laboratory number.

1928). Cook perceived the posterior "bulge" near the base of the horn cores (which is responsible for much of the sinuous posterior profile that characterizes Eurasian Pleistocene bison) and stated that this character suggested "Asiatic relationships." *B. chaneyi* possesses characters diagnostic of *B. alaskensis* and falls within the range of individual variation to be expected for a taxon the size of *B. alaskensis.*

B. aguascalentensis is the most recently named and described of North American bison (Mooser and Dalquest, 1975). The only apparent characters that differentiated the holotype of this species from other similar species were the angle of horn core emergence from the frontals (used to differentiate the holotype from *B. chaneyi*) and horn core length (used to differentiate the holotype from *B. alleni*). The magnitude of these differences is insufficient to justify specific distinctiveness.

NORTH AMERICAN BISON PHYLOGENESIS

Bison are considered to have originated during the upper Pliocene in eastern Asia. *Leptobos* is currently regarded as the most probable direct antecedent of *Bison* (Pilgrim, 1947; Sinclair, 1977), although Sahni and Khan (1968) regard *Probison dehmi* from the Upper Siwalik Tatrot deposits as the most proximate antecedent. The earliest recognized bison species is *B. sivalensis,* probably from the Upper Siwalik Pinjor beds, near Pinjor, India, and from the Nihowan beds about 150 km west of Peking, China (figs. 14 and 15). Wilson (1974*a*) has suggested the synonymy of *Probison* and *Bison,* but because of marked differences in the skull of each, they should be kept as separate genera.

B. sivalensis is here considered directly ancestral to both the *B. priscus-B. alaskensis* (Eurasian) lineage and the *B. latifrons-B. antiquus* (North American) lineage. *B. sivalensis* was apparently widespread throughout much of eastern Asia during the early Pleistocene. During the early middle Pleistocene primitive *B. priscus* appeared in much of midlatitude Eurasia, then shifted northward into higher latitude Eurasia. *B. alaskensis* also appeared during the middle Pleistocene in Eurasia, as did *B. latifrons* and perhaps even *B. antiquus* in North America. Characters shared by *B. sivalensis* and the Eurasian lineage include the antero-posterior spiral growth around the longitudinal axis of the core and the posterior deflection of the horn core tip. Characters shared by *B. sivalensis* and the North American lineage include the triangular (isosceles) shape of the horn core in cross section, the straight posterior profile, the arched longitudinal axis, and, in *B. latifrons,* the antero-posterior spiral growth. These characters in *B. latifrons* and *B. antiquus* are qualitatively similar to *B. sivalensis,* suggesting a closer morphological affinity between *B. sivalensis* and the later North American autochthons than between *B. sivalensis* and the later Eurasian autochthons. *B. sivalensis* is, however, morphologically more like both the Eurasian and the North American lineages than are these two lineages to each other (fig. 28).

B. sivalensis probably entered North America no later than the Illinoian, and possibly during one or more earlier glaciation(s). If *B. sivalensis* was a relatively rare subdominant in North

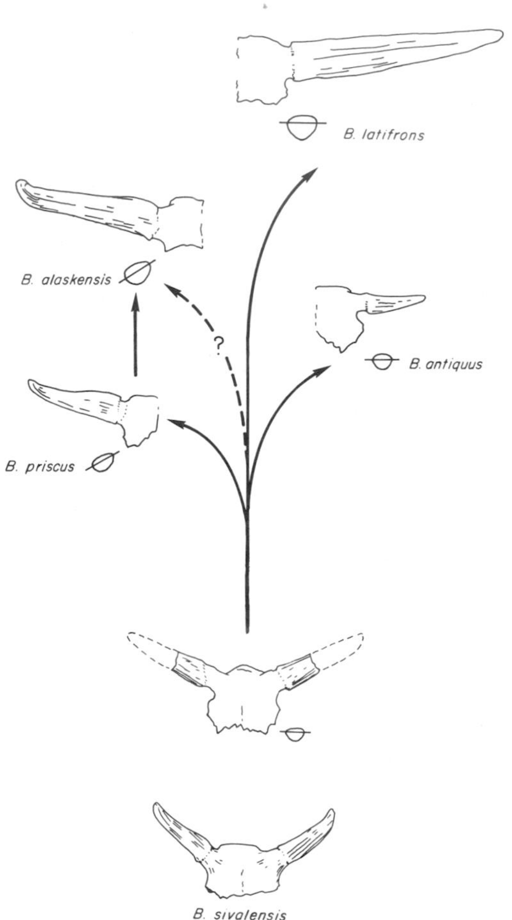

FIGURE 28 Character relations between *Bison sivalensis* and subsequent species. *Bison sivalensis* probably gave rise to both the Eurasian and North American lineages of bison. Eurasian bison retained the compressed basal horn cores, the sinuous posterior profile, the distal twist, and the sinuous axis. North American bison retained the isosceles triangle cross section shape, horn cores not rotated relative to the frontals, and the straight growth apparent in the *Bison sivalensis* holotype. *Bison priscus* probably appeared before *Bison alaskensis,* based on fossil evidence, but both could have appeared synchronously as, apparently, did *Bison latifrons* and *Bison antiquus.*

America, a negative fossil record would not be unexpected; there is no North American fossil record for *B. sivalensis,* but neither is there a Eurasian fossil record for *B. latifrons* or *B. antiquus.* The earliest record of bison in Alaska comes from sediments of pre-Illinoian age (Péwé, 1975; Péwé and Hopkins, 1967), but these cannot yet be specifically identified. The earliest acceptably dated bison remains from central North America, some of which cannot be specifically identified, are assigned to the Illinoian (Green, 1962; Hibbard, 1955*a*; Hopkins, 1951; Hopkins, Bonnichsen, and Fortsch, 1969). The earliest specifically identifiable species of North American bison—*B. latifrons*—also is assigned to the Illinoian and later time. Available evidence leads to the conclusion that populations of *B. sivalensis* dispersed into North America sufficiently early to allow the speciation of *B. latifrons,* and possibly of *B. antiquus,* during or soon after the Illinoian glaciation.

B. latifrons and *B. antiquus* both existed as distinctly different species by the late Sangamon-early Wisconsin, and both persisted until *B. latifrons* became extinct during or immediately after the middle Wisconsin interstadial. *B. antiquus* remained the only autochthonous North American species until after the middle Holocene. During the late Wisconsin-early Holocene, however, *B. antiquus* phyletically bifurcated into subspecies, the conservative but less abundant *B. a. antiquus* and the highly variable and more abundant *B. a. occidentalis.* Following range shifts and population increases these two subspecies fused during the late early and middle Holocene. By about 5,000–4,000

BP, *B. antiquus* evolved into *B. bison*, whose two subspecies, *B. b. bison* and *B. b. athabascae*, occupied southerly grasslands and northerly parklands, respectively, at either end of the species' primary range (fig. 23).

In Eurasia *B. priscus* evolved from *B. sivalensis* by the late early Pleistocene. *B. alaskensis* probably evolved from *B. priscus* during the middle Pleistocene. These species were probably contemporaries, but *B. priscus* probably inhabited savanna/steppe environments whereas *B. alaskensis* probably inhabited forest/woodland environments. Both species are known from North America, and either could have entered North America at any time that favorable environmental conditions existed. One need not have preceded the other; probably, each dispersed into North America on more than one occasion. Both species apparently became extinct over most of their range during the late Wisconsin.

B. schoetensacki, which very strongly resembles *B. antiquus*, appeared in Eurasia during the very late Wisconsin. *B. schoetensacki* may have evolved from *B. priscus* (which is unlikely considering the degree of morphological difference between the two similarly sized species) or from eastward dispersing populations of *B. a. antiquus* which reached Beringia during the late Wisconsin. (Kurten [1968] has proposed a Nearctic origin for late Wisconsin and Holocene Eurasian bison.) *B. schoetensacki* apparently evolved into *B. bonasus* during or after the very late Wisconsin-early Holocene, paralleling the evolution of *B. antiquus* into *B. bison* in North America, except that the process probably took place earlier in Eurasia than in North America. The Eurasian record is too incomplete to allow a final conclusion to be reached on this matter, but broadly similar selection regimes in both continents, undergoing similar reorganization, have operated on late Pleistocene and Holocene bison and produced broadly similar results. Bohlken (1967) recognizes two historic subspecies of *B. bonasus*, *B. b. bonasus*, which inhabited the forests of central Europe, and *B. b. caucasicus*, which inhabited the highland steppes of the Caucasus region. There were, however, small bison—similar in size to *B. bonasus* and *B. bison*—in Siberia during the very late Wisconsin and at least part of the Holocene.

Figure 29 summarizes the phyletic relationship among *Bison* recognized in this classification and provides a chronology for the speciation sequence.

HORN CORE COMPLEX CHARACTER CHART

Elements of the horn core complex of characters for both sexes of all taxa recognized in this classification are summarized in table 43. This chart should permit reliable assignment of a specimen to a taxon and sex when the diagnostic characters of the horn core are normally developed.

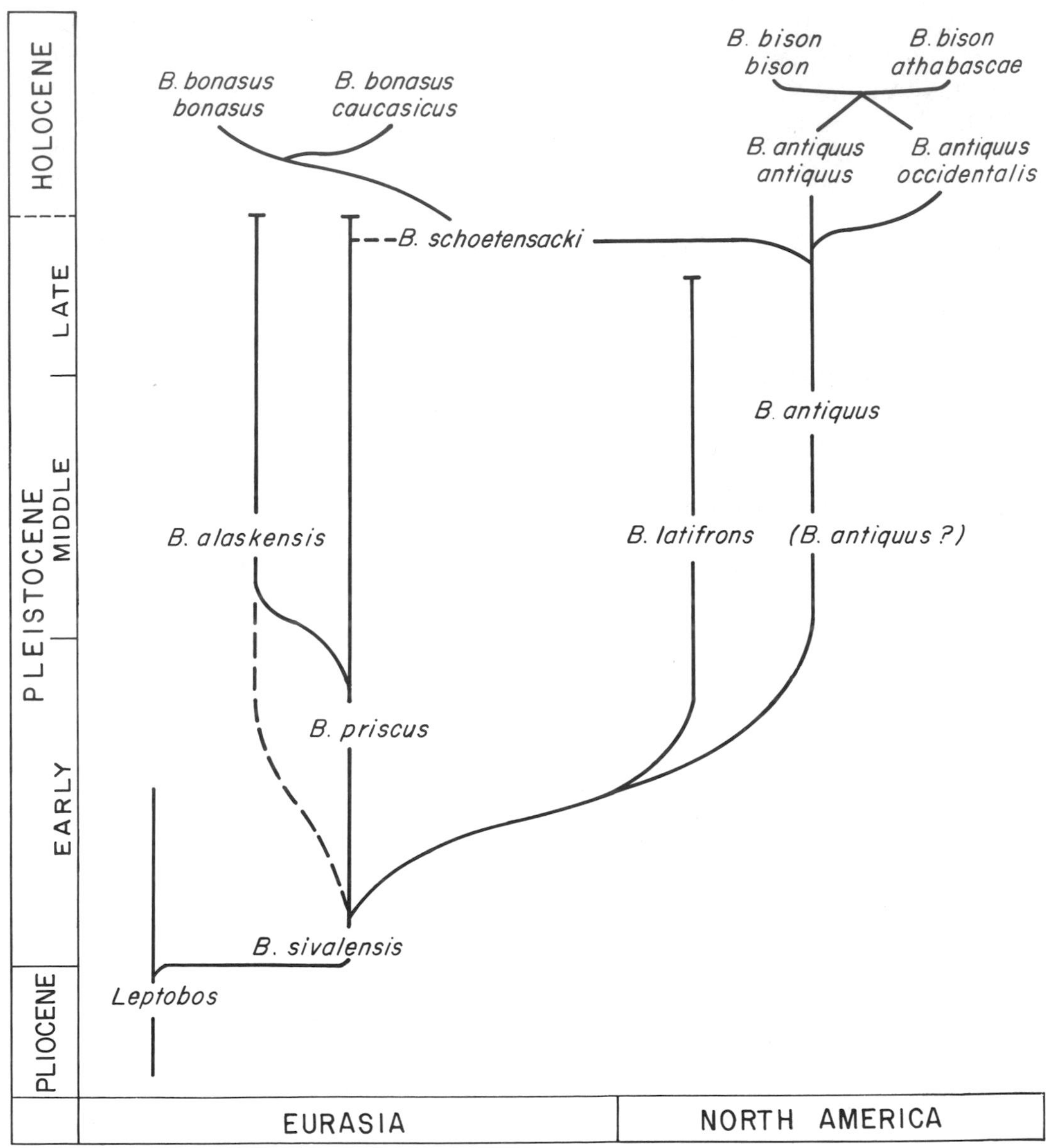

FIGURE 29 A model of *Bison* phylogenesis. Solid lines indicate phyletic routes preferred here; broken lines indicate possible alternate phyletic routes.

Table 43

Horn Core Character Chart

Character	*B. latifrons* ♂	*B. latifrons* ♀	*B. a. antiquus* ♂	*B. a. antiquus* ♀	*B. a. occidentalis* ♂	*B. a. occidentalis* ♀	*B. b. bison* ♂	*B. b. bison* ♀	*B. b. athabascae* ♂	*B. b. athabascae* ♀	*B. priscus* ♂	*B. priscus* ♀	*B. alaskensis* ♂	*B. alaskensis* ♀
Cross section shape at base:														
isosceles triangle	x		x		x		x		x					
scalene triangle											x		x	
circular/ellipitical		x		x		x		x		x		x		x
Cross section shape at tip:														
cordiform/triangle	x		x								x		x	
circular/elliptical		x		x	x	x	x	x	x	x		x		x
Posterior margin of core:														
convex												x		x
straight	x	x	x	x	x	x	x	x	x	x				
concave					x		x		x	x				
sinuous		x									x	x	x	x
Growth along axis:														
straight			x	x	x	x	x	x	x	x		x		x
mildly spiraled	x	x			x				x	x				
spiraled					x						x		x	x
Shape of longitudinal axis:														
straight								x						
arched	x	x	x	x	x	x	x	x	x	x		x		x
sinuous					x					x	x		x	x
Symmetry of cross section at base:														
symmetrical	x	x	x	x	x	x	x	x	x	x		x		x
asymmetrical											x		x	x
Rotation relative to frontals:														
forward		x		x		x		x		x	x	x	x	x
not rotated	x		x											
rearward					x		x		x					

[a]No true *B. b. athabascae* female skulls were available to me for study. The characters indicated here are therefore tentative, and are based upon inferences made from what are presumed to be *B. b. bison* x *B. b. athabascae* hybrid females.

CHAPTER 3

MORPHOLOGICAL VARIATION AND VARIABILITY

Genotypic and phenotypic variation are qualities of sexually reproducing populations and the source material for the evolution of species. This chapter presents information on selected facets of phenotypic variation and variability observed among bison skulls and limb bones, including patterns of continuous variation within each sex of each taxon, sexual dimorphism, allometry, variability of characters, and choroclines, in order to illustrate generalized patterns of variation.

This chapter presents data on all specifically identified skulls that yielded biometric information and selected postcranial samples. The taxonomic identity of postcranial samples has been determined for historic period collections (Hornaday, Richardson and Elliot, Rowan, and so on) and numerous archeologic samples with associated horn cores and good provenience information, as well as a few individual fossil bison found with skull and limb bones articulated. Many postcranial samples have not, however, been positively identified. These samples are of two types: those which probably contain bones predominantly, if not entirely, of one taxon (Rancho La Brea, Big Bone Lick, Dows Bog, Minnesota bogs); and those that predominantly contain bones representing two or more taxa (American Falls region, Fairbanks area, Dawson area).

The *B. b. bison* limb sample used in this chapter consists of the Hornaday, Richardson and Elliot, Buffalo Creek, Dows Bog, Glenrock, Rocky Ford, Bonfire (bed 3), Sitter Ranch, Lubbock Lake (Archaic and late Prehistoric levels), Southern Plains Archaic, and Big Bone Lick collections. The *B. b. athabascae* limb sample has been assembled from several individual specimens. *B. a. antiquus* limbs are from Rancho La Brea, Blackwater Draw, and Big Bone Lick. *B. a. occidentalis* limbs are from Lindenmeier, Jones-Miller, Bonfire (bed 2), Lubbock Lake (Clovis and Folsom levels), Folsom, Olsen-Chubbock, Des Moines, Casper, Agate Basin, Finley, Jurgens, Hudson-Meng, Plainview, James Allen, Wasden and Hawken archeological sites, and Minnesota bogs and Interstate Park Bog sites.

Few *B. latifrons* limb bones have been positively identified (tables 44 and 45). Many skulls and isolated limb bones have been found around American Falls Reservoir in southeast Idaho. A majority of the identifiable skulls found here (25 of 34; cf. tables 19, 23, 39, and 41) belongs to *B. latifrons*. The majority of limb bones probably also belongs to *B. latifrons*, although limbs of all other species (especially *B. alaskensis*) are also probably present. The

Table 44

Measurements of *Bison latifrons* Skulls with Articulated Limb Bones

Standard measurement	IMNH VP 27[a]	IMNH VP 26594	IMNH VP 26602[a]	UF VP 7559[b]
Spread of horn cores, tip to tip	1886 mm	....	1830 mm	(1342 mm)
Horn core length, upper curve, tip to burr	870 mm	519 mm	930 mm	(795 mm)
Straight line distance, tip to burr, dorsal horn core	....	506 mm	890 mm	....
Dorso-ventral diameter, horn core base	....	107 mm	175 mm	145 mm
Minimum circumference, horn core base	....	316 mm	560 mm	(414 mm)
Width of occipital at auditory opening	330 mm	....	....	....
Width of occipital condyles	....	....	155 mm	....
Antero-posterior diameter, horn core base	160 mm	97 mm	175 mm	(126 mm)
Least width of frontals, between horn cores and orbits	330 mm	....	....	368 mm
Greatest width of frontals at orbits	....	....	....	430 mm
$M^1 - M^3$, inclusive alveolar length	....	....	....	....
M^3, maximum width, anterior cusp	....	....	....	....
Distance, nuchal line to tip of premaxillae	....	....	....	....
Distance, nuchal line to nasal-frontal suture	340 mm	....	....	....

Sources:
[a]John A. White, field notes, August 9, 1975;
[b]Robertson, 1974.
Note: Figures in parentheses are estimates. UF VP 7559 is not fully mature.

Table 45
Measurements of *Bison latifrons* Limb Bones Articulated with Skull Material

Standard measurement	IMNH VP 27[a]	IMNH VP 26594	IMNH VP 26602[a]	UF VP 7559[b]
Humerus:				
Approximate rotational length of bone	394 mm	....	....	379 mm
Antero-posterior diameter of diaphysis	86 mm	....	....	73 mm
Transverse minimum of diaphysis	68 mm	....	....	60 mm
Radius:				
Approximate rotational length of bone	384 mm	(366 mm)	....	362 mm
Antero-posterior minimum of diaphysis	46 mm	37 mm	....	41 mm
Transverse minimum of diaphysis	72 mm	64 mm	....	65 mm
Metacarpal:				
Total length of bone	249 mm	....	....	241 mm
Antero-posterior minimum of diaphysis	37 mm	....	....	35 mm
Transverse minimum of diaphysis	69 mm	....	....	58 mm
Femur: (no articulated specimens available)				
Tibia:				
Approximate rotational length of bone	464 mm	....	....	459 mm
Antero-posterior minimum of diaphysis	54 mm	....	....	46 mm
Transverse minimum of diaphysis	69 mm	....	....	63 mm
Metatarsal:				
Total length of bone	302 mm	....	310 mm	286 mm
Antero-posterior minimum of diaphysis	40 mm	....	42 mm	37 mm
Transverse minimum of diaphysis	54 mm	....	52 mm	48 mm

Sources:
[a]John A. White, field notes, August 9, 1975;
[b]Robertson, 1974.
Note: The figure in parentheses is an estimate. Robertson (1974) provides measurements for other postcranial bones of UF VP 7559 (which is not fully mature).

American Falls Reservoir area sample (which does not include specimens from Minidoka County) represents the closest approximation to *B. latifrons* limb size and variation possible at this time. True *B. latifrons* limb dimensions would possibly average slightly larger than the mean dimensions reported in the American Falls area sample.

B. priscus and *B. alaskensis* limb bones have not been positively identified from North America, but limb bones of both species (and hybrid individuals) are certainly present in the various Beringian samples. Here limb bones from the sexable Old Crow River and Lost Chicken Creek collections are grouped into a single sample. This sample is probably representative of *B. priscus* limb size. The Fairbanks and Yukon areas samples could not be adequately sexed because there was no bimodal clustering of cases within these samples. Both of these samples probably contain undifferentiable limb bones of *B. priscus*, *B. alaskensis*, and hybrid bison, but the samples do provide some indication of the overall range of limb sizes which occurred in Beringia and can therefore be used to approximate the upper and lower limits of limb bone variation among the various species known to have inhabited the region.

CONTINUOUS VARIATION

Continuous variation among phenotypes is characteristic of mammalian populations. Usually, phenotypic characters are normally or near normally distributed within populations and, usually, skull characters are more variable than postcranial characters (Olsen, 1964; Yablokov, 1974). I will discuss the nature of case distribution within samples, the relationship between sample

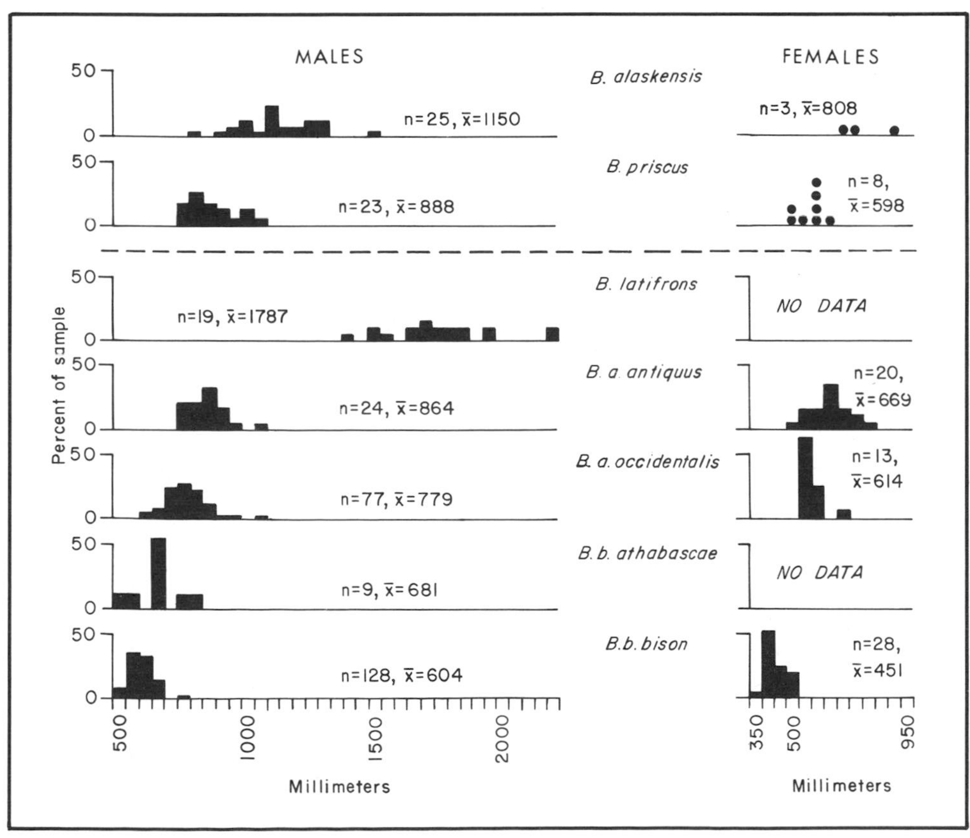

FIGURE 30 Frequency distribution: spread of horn cores, tip to tip.

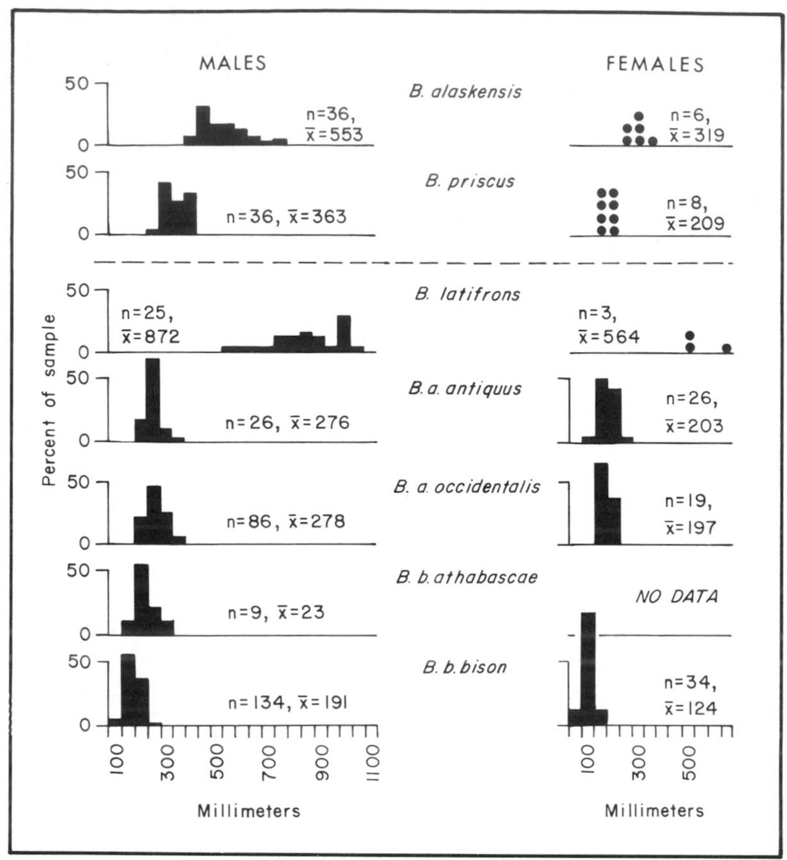

FIGURE 31 Frequency distribution: horn core length, upper curve, tip to burr.

means and absolute ranges, and the direction and magnitude of skew. Patterns of variation observed in *B. b. bison*, both sexes of which are represented by large sample sizes, are useful as reference standards. Patterns of variation in other taxa need not, however, be identical to those observed for *B. b. bison*, and the small sample size for some characters needs to be kept in mind when evaluating variation.

Frequency distributions of linear skull variates are illustrated in figures 30 to 44. Most samples are normally or near normally distributed, although variations in the degree and direction of kurtosis and skew are apparent. Generally, the larger the character, the greater is the observed range of variation (fig. 57). The greater the

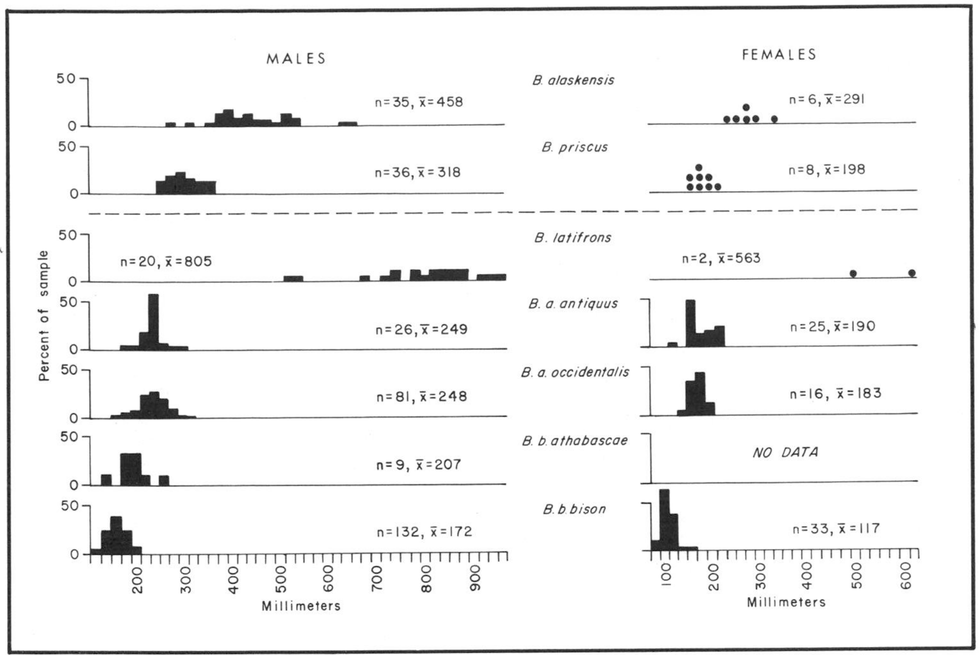

FIGURE 32 Frequency distribution: horn core length, straight line, tip to burr.

range, the lower is the observed kurtosis. Both positive and negative skewing are evident in figures 30 to 44.

Frequency distributions of limb lengths are illustrated in figures 45 to 50. Limb bone lengths are usually normally or near normally distributed and show less kurtotic variation and skew than do skull variates. As was true for skulls, the larger the character the greater the observed range (fig. 58). *B. a. occidentalis*, however, is conspicuously more variable, skewed, and kurtotic than the preceding generalizations suggest should be the case. This pattern of distribution is probably due in great part to the evolutionary role of this taxon as the adaptive transition from *B. a. antiquus* to *B. bison*.

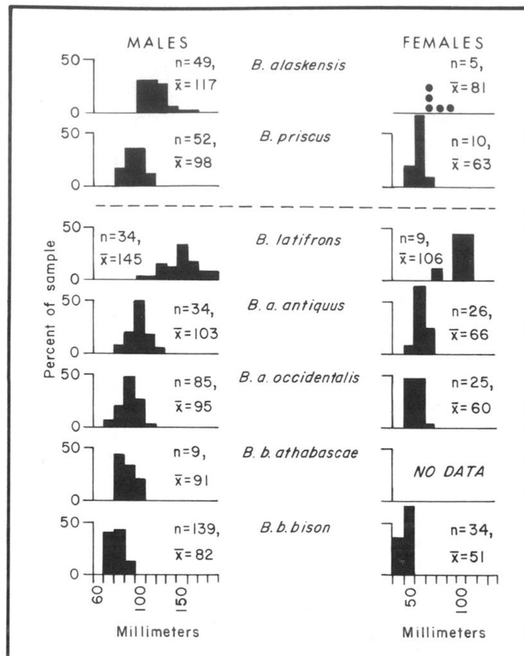

FIGURE 33 Frequency distribution: dorso-ventral diameter, horn core base.

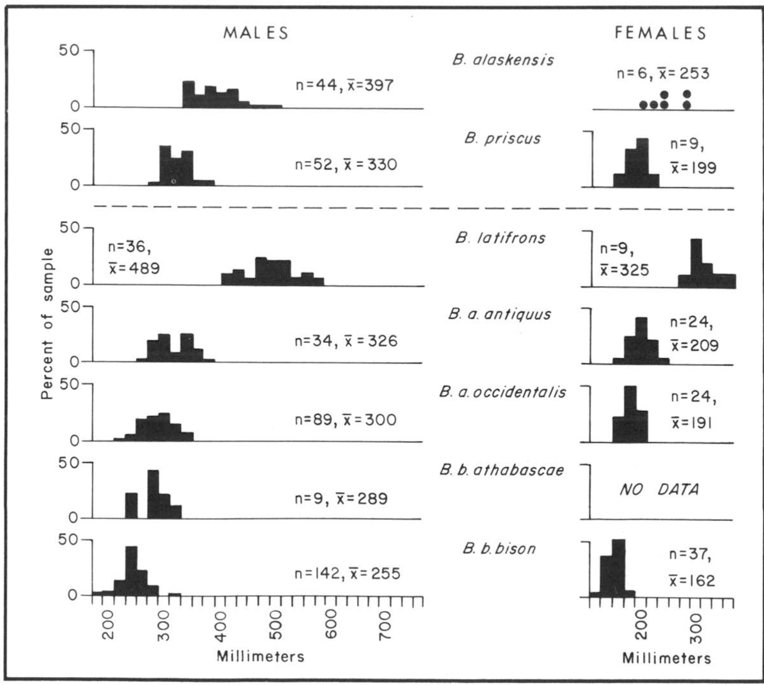

FIGURE 34 Frequency distribution: minimum circumference, horn core base.

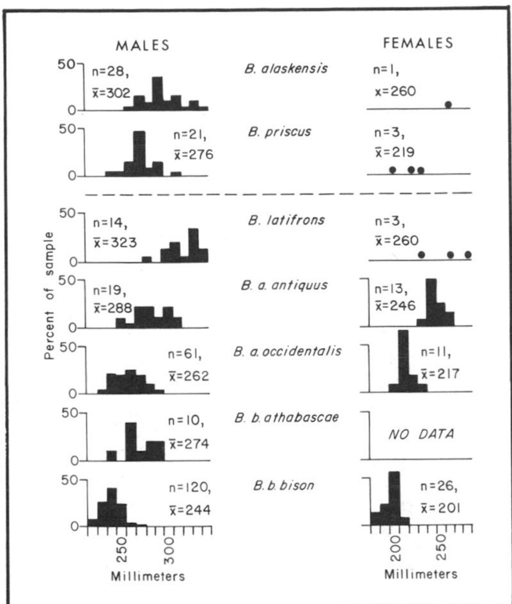

FIGURE 35 Frequency distribution: width of occipital at auditory openings.

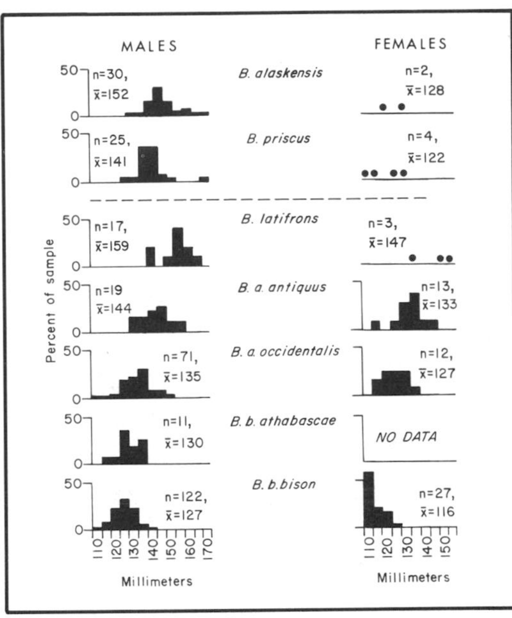

FIGURE 36 Frequency distribution: width of occipital condyles.

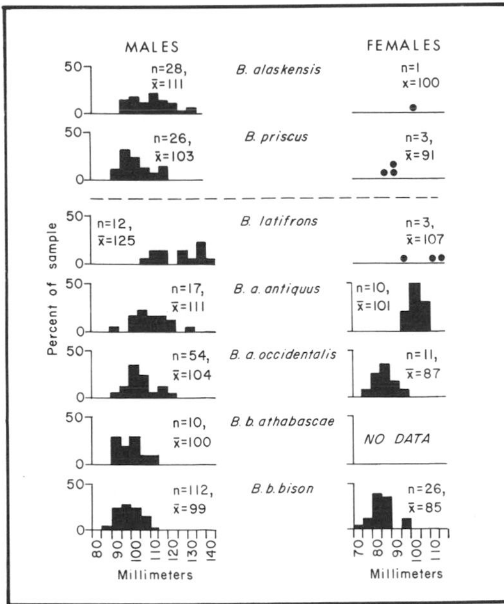

FIGURE 37 Frequency distribution: depth, nuchal line to foramen magnum.

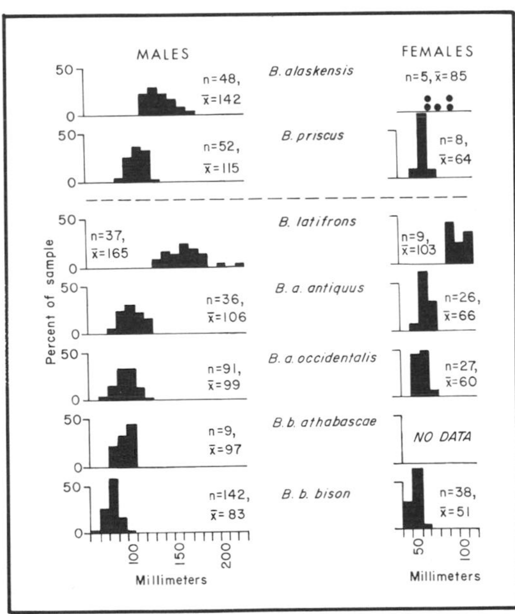

FIGURE 38 Frequency distribution: antero-posterior diameter, horn core base.

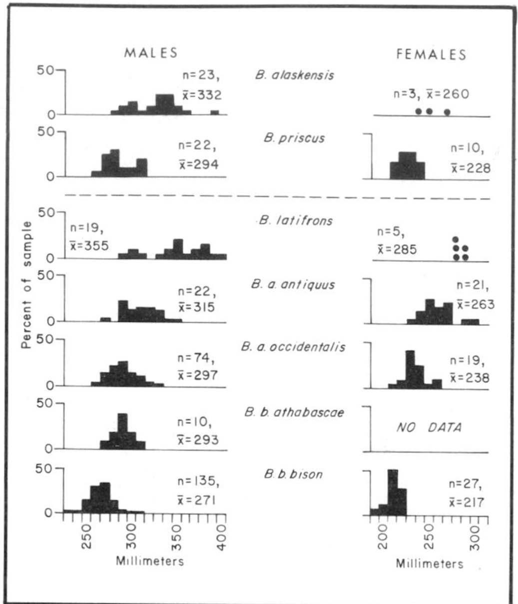

FIGURE 39 Frequency distribution: least width of frontals.

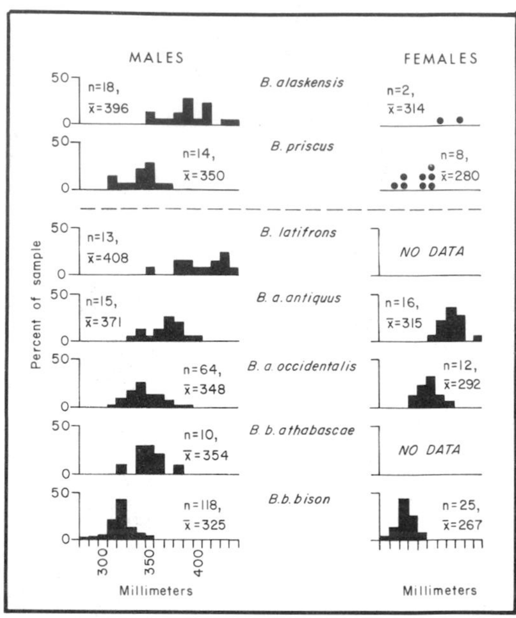

FIGURE 40 Frequency distribution: greatest width of frontals.

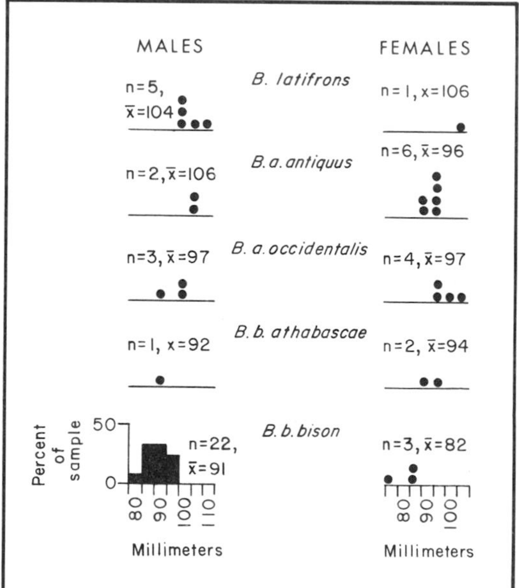

FIGURE 41 Frequency distribution: superior molar alveolar length.

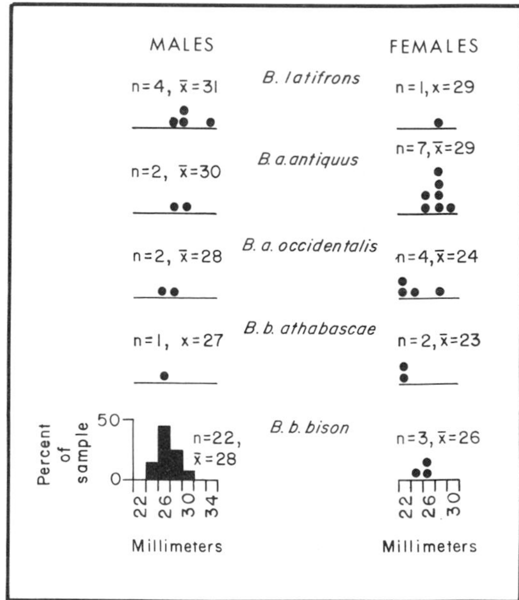

FIGURE 42 Frequency distribution: anterior M^3 cusp width.

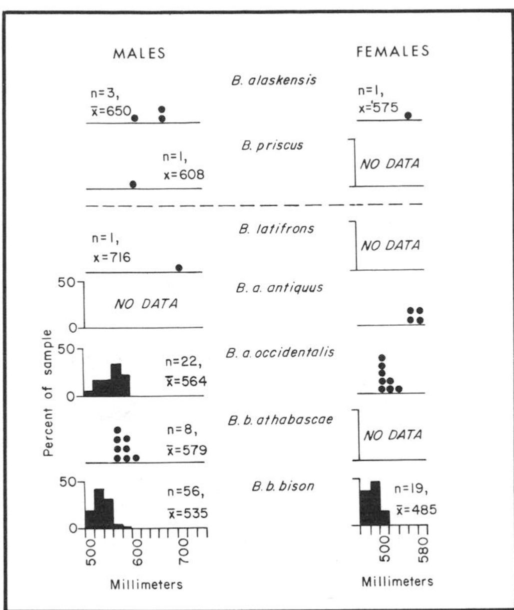

FIGURE 43 Frequency distribution: distance, nuchal line to tip of premaxillae.

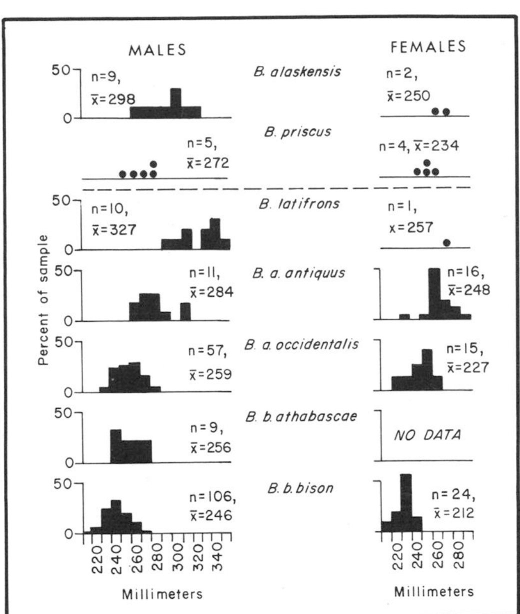

FIGURE 44 Frequency distribution: distance, nuchal line to nasal-frontal suture.

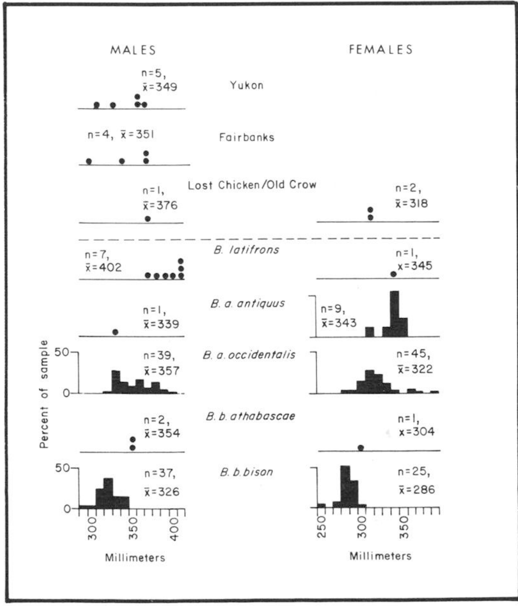

FIGURE 45 Frequency distribution: humerus length.

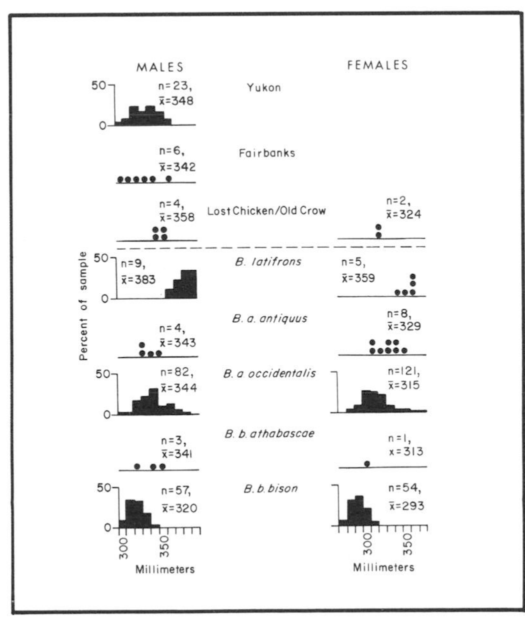

FIGURE 46 Frequency distribution: radius length.

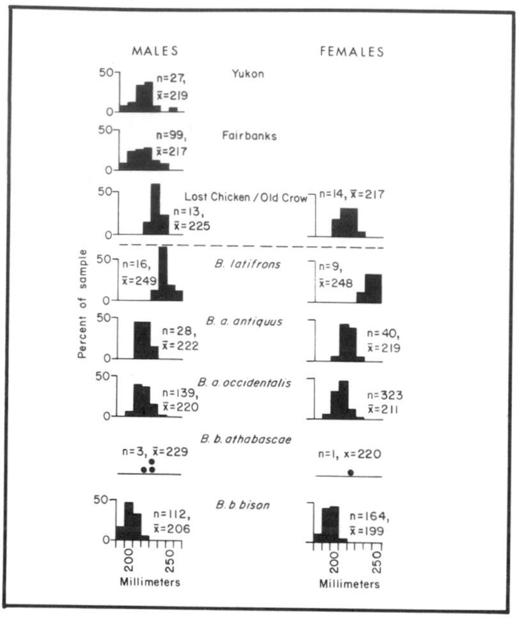

FIGURE 47 Frequency distribution: metacarpal length.

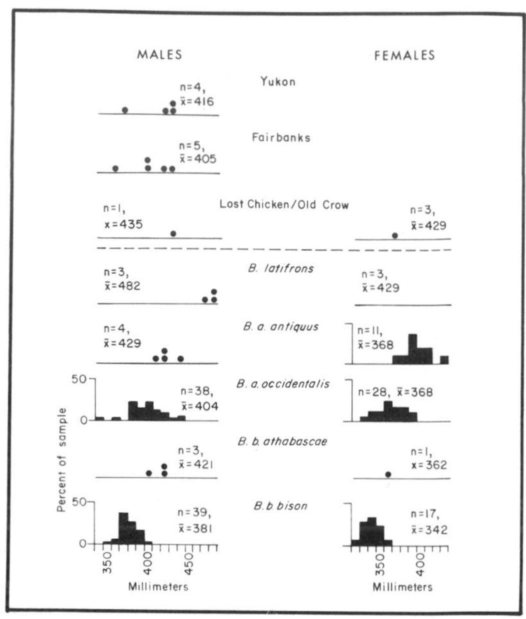

FIGURE 48 Frequency distribution: femur length.

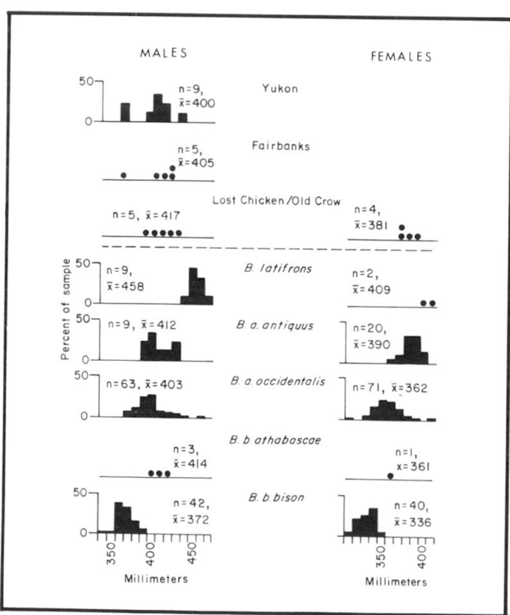

FIGURE 49 Frequency distribution: tibia length.

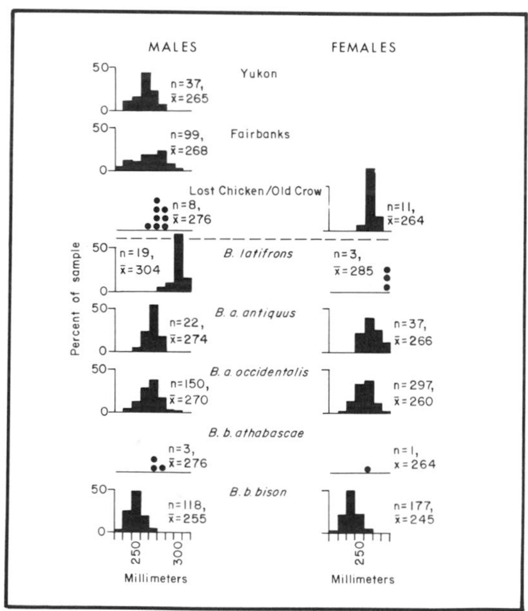

FIGURE 50 Frequency distribution: metatarsal length.

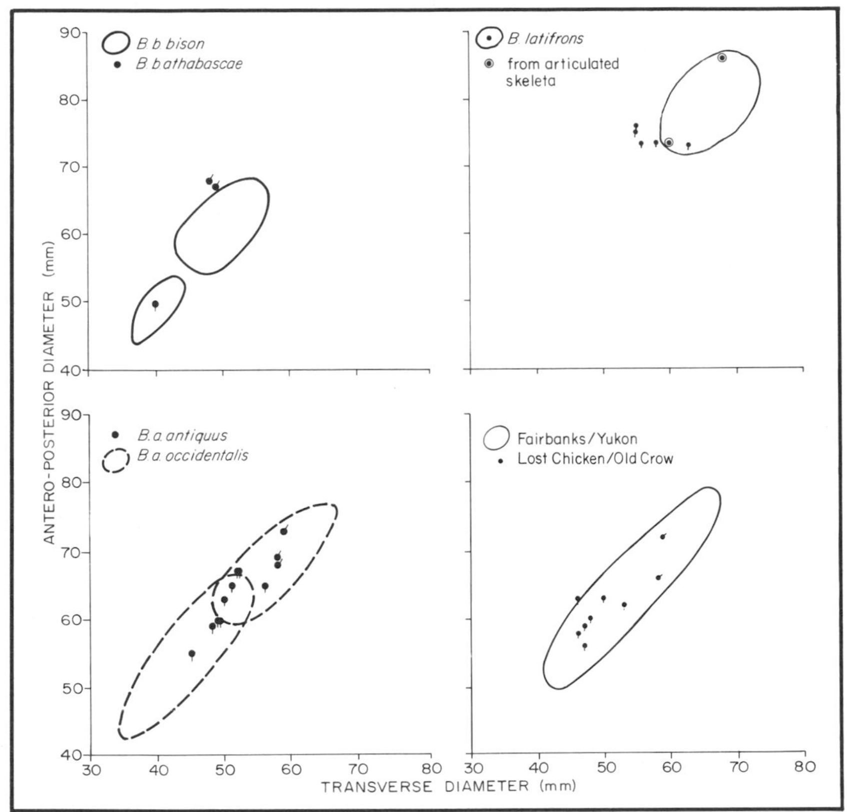

FIGURE 51 Bivariate distribution: humerus diaphysis.

Absolute variation among limb diaphysial variates is relatively small. Variation among samples of diaphysial variates is summarized here in the form of scatterplots (figs. 51–56). The large number of cases for some samples necessitated enclosing areas of case distribution rather than actually plotting points. So that the illustrations were comparable, points were plotted only where the sample size was very small ($n < 9$). Generally, a positive relationship exists between the anteroposterior and transverse diameters of the diaphysis in all six limb bones in both sexes.

Continuous variation among linear variates of bison skulls and limb bones may be summarized as follows: (1) Distribution of cases within samples is usually normal or near-normal; (2) the range of variation is greater for larger characters and less for smaller characters; (3) kurtosis tends to be greater for smaller characters and less for larger characters; (4) skewing of samples is common, but the direction of skew is unpredictably variable; (5) generally, the skull is more variable than are the limb bones; and (6) males and females exhibit similar patterns of continuous variation.

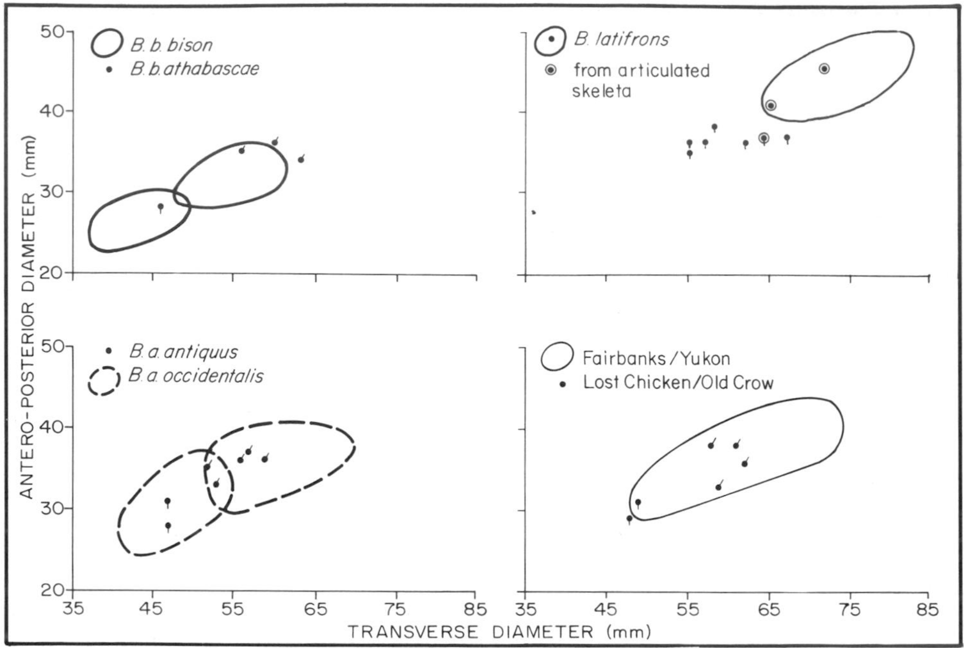

FIGURE 52 Bivariate distribution: radius diaphysis.

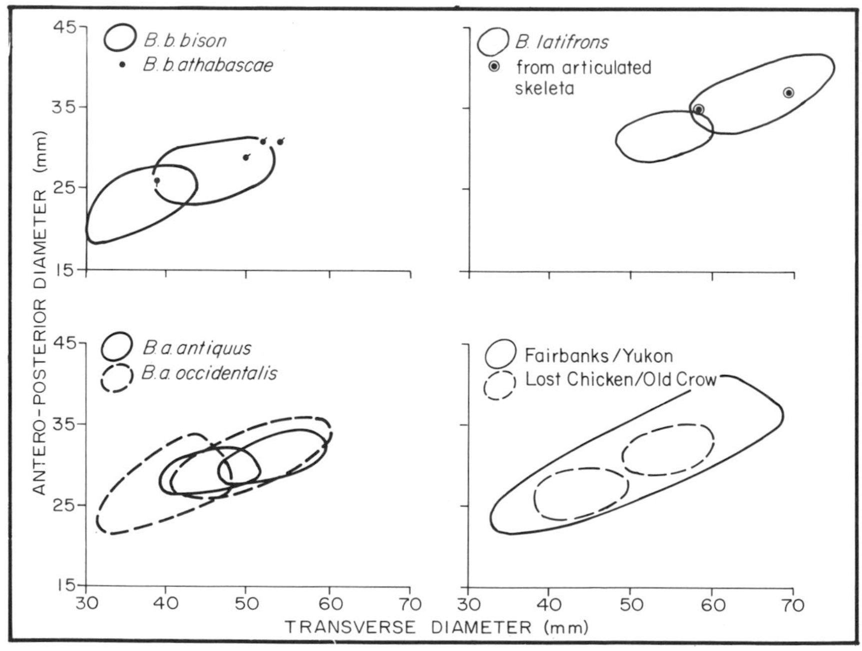

FIGURE 53 Bivariate distribution: metacarpal diaphysis.

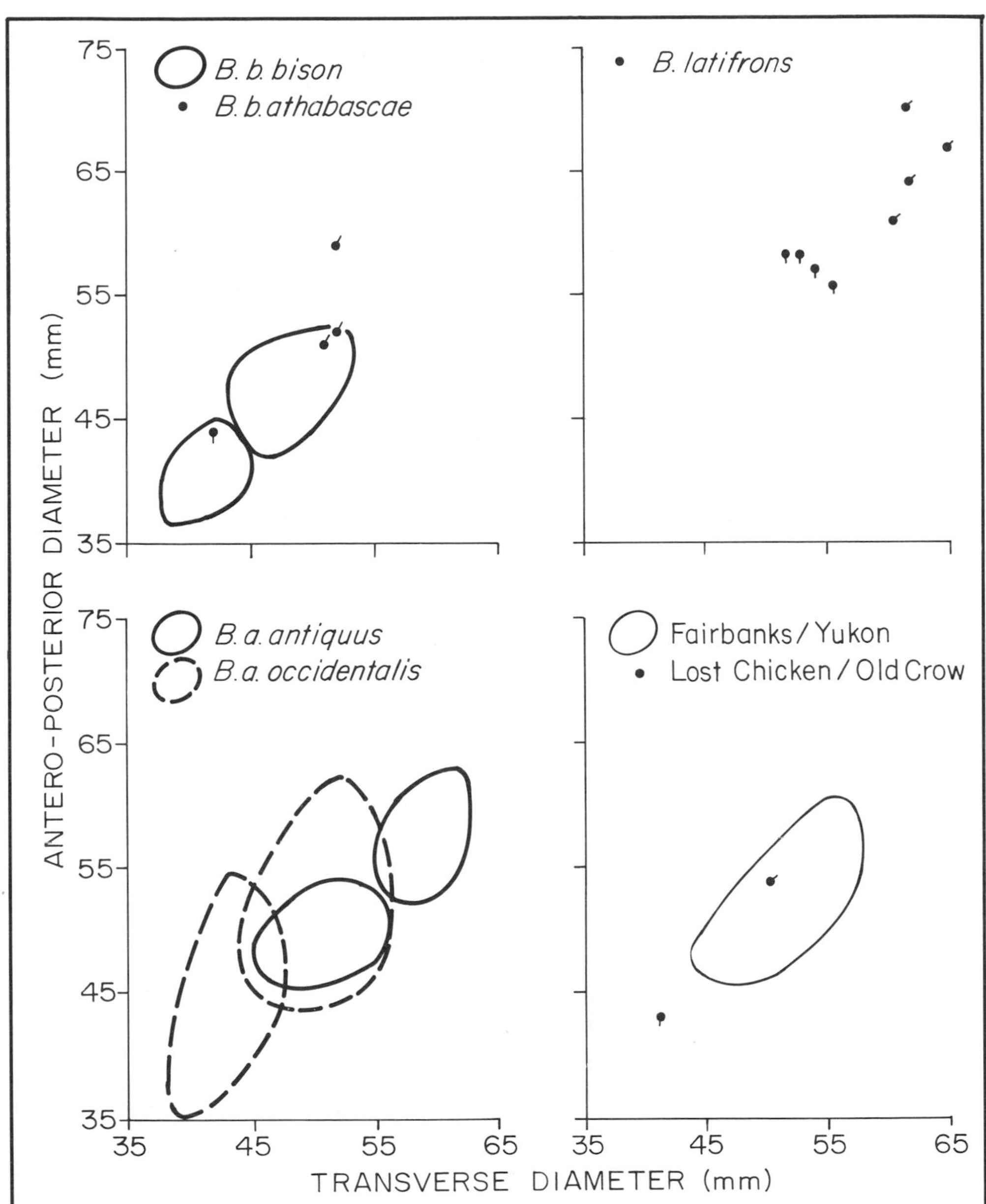

FIGURE 54 Bivariate distribution: femur diaphysis.

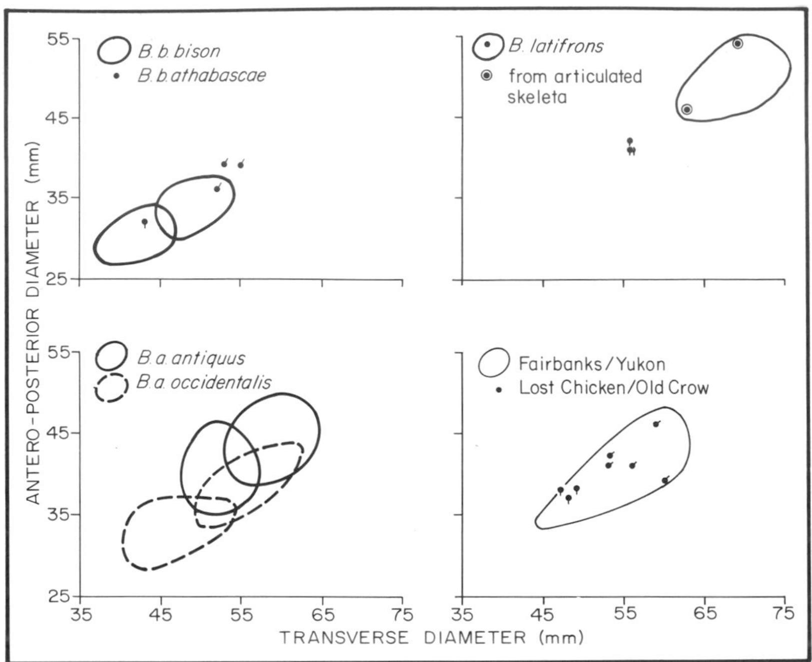

FIGURE 55 Bivariate distribution: tibia diaphysis.

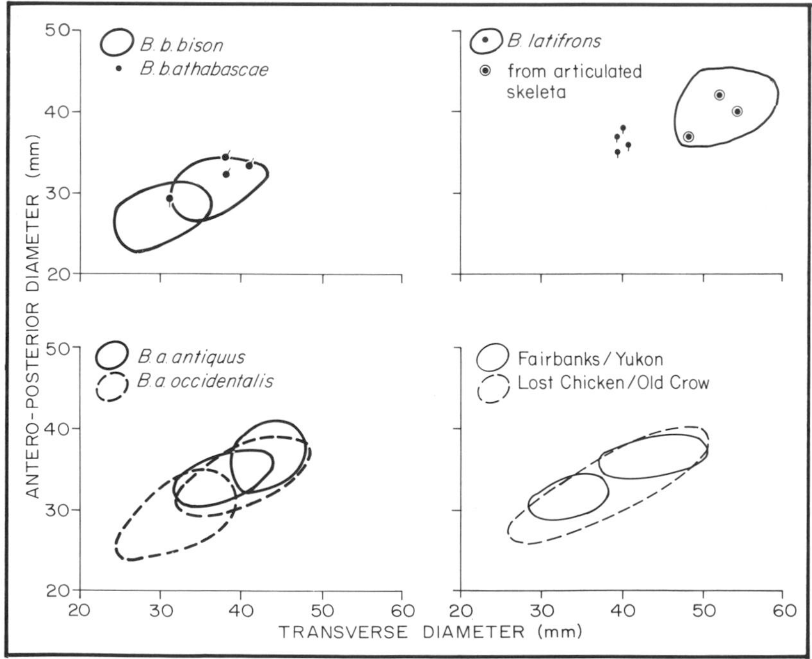

FIGURE 56 Bivariate distribution: metatarsal diaphysis.

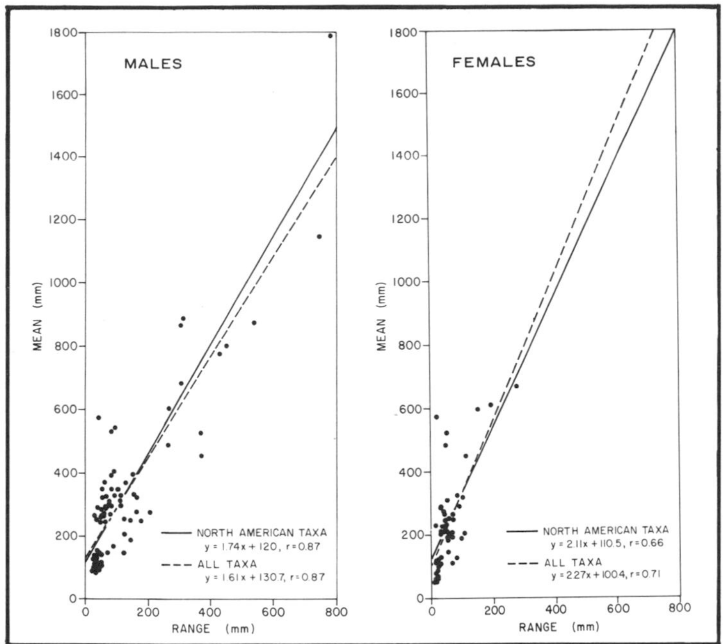

FIGURE 57 Means versus ranges for skull characters.

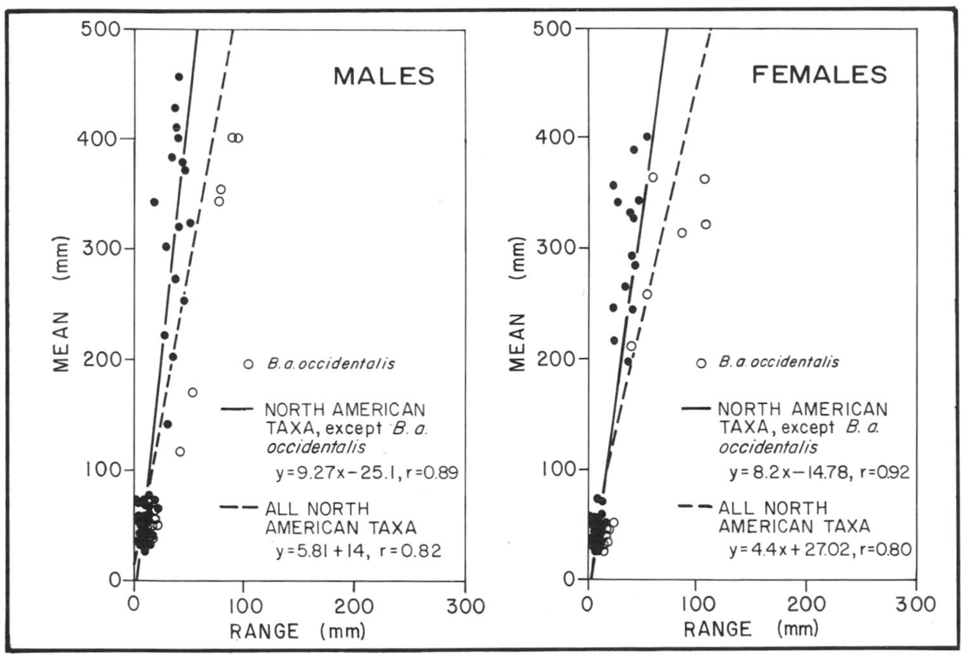

FIGURE 58 Means versus ranges for limb characters.

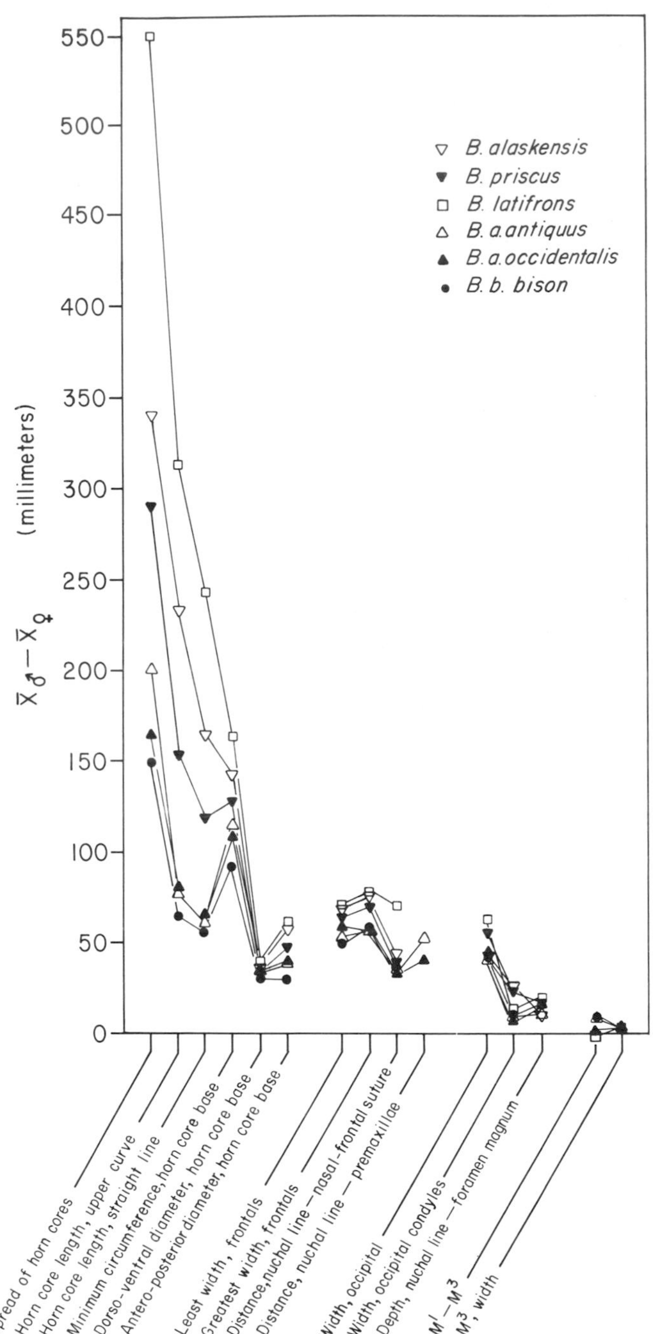

FIGURE 59 Absolute sexual dimorphism among skull characters.

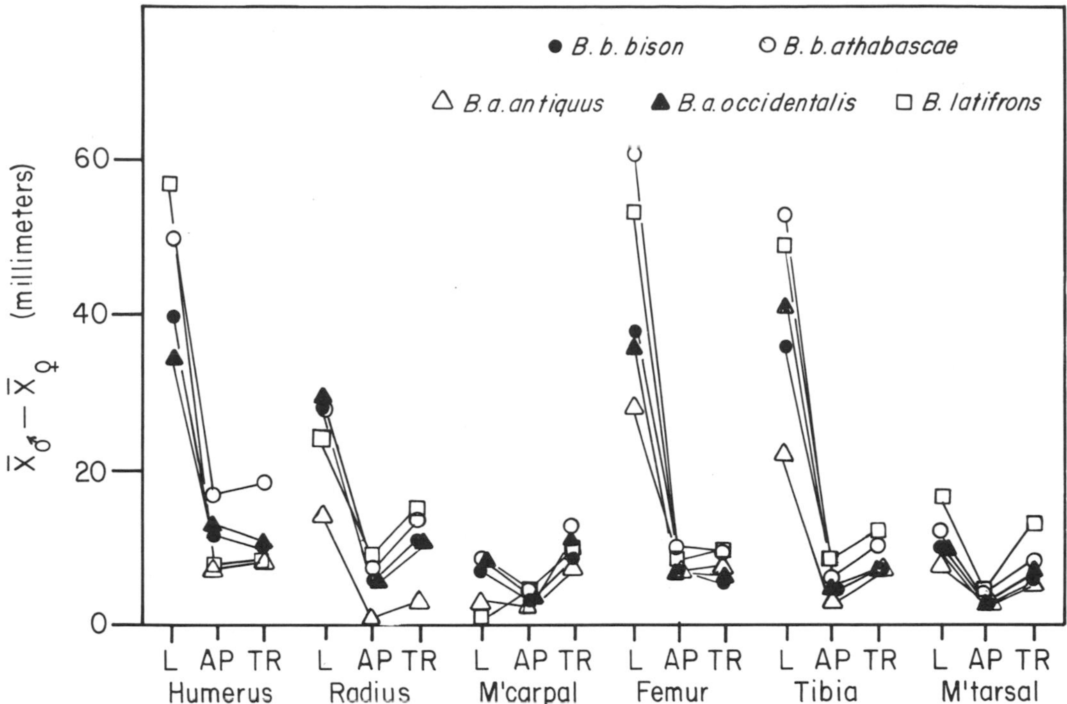

FIGURE 60 Absolute sexual dimorphism among limb characters.

SEXUAL DIMORPHISM

Sexual dimorphism is a usual characteristic of sexually reproducing organisms. The magnitude of both absolute and relative sexual dimorphism in characters of bison differs among characters within a taxon and among the various taxa. Absolute dimorphism in linear skull characters is generally greatest among horn core characters, less among facial and frontal characters, and least among occipital characters (fig. 59). Absolute dimorphism in limb bones is greatest for length and less for the two diaphysial diameters (fig. 60). In both limbs the most proximal bone shows greatest dimorphism and the most distal bone shows least dimorphism (figs. 45–56 and 60). Generally, in both skulls and limb bones the larger the character the greater the absolute dimorphism.

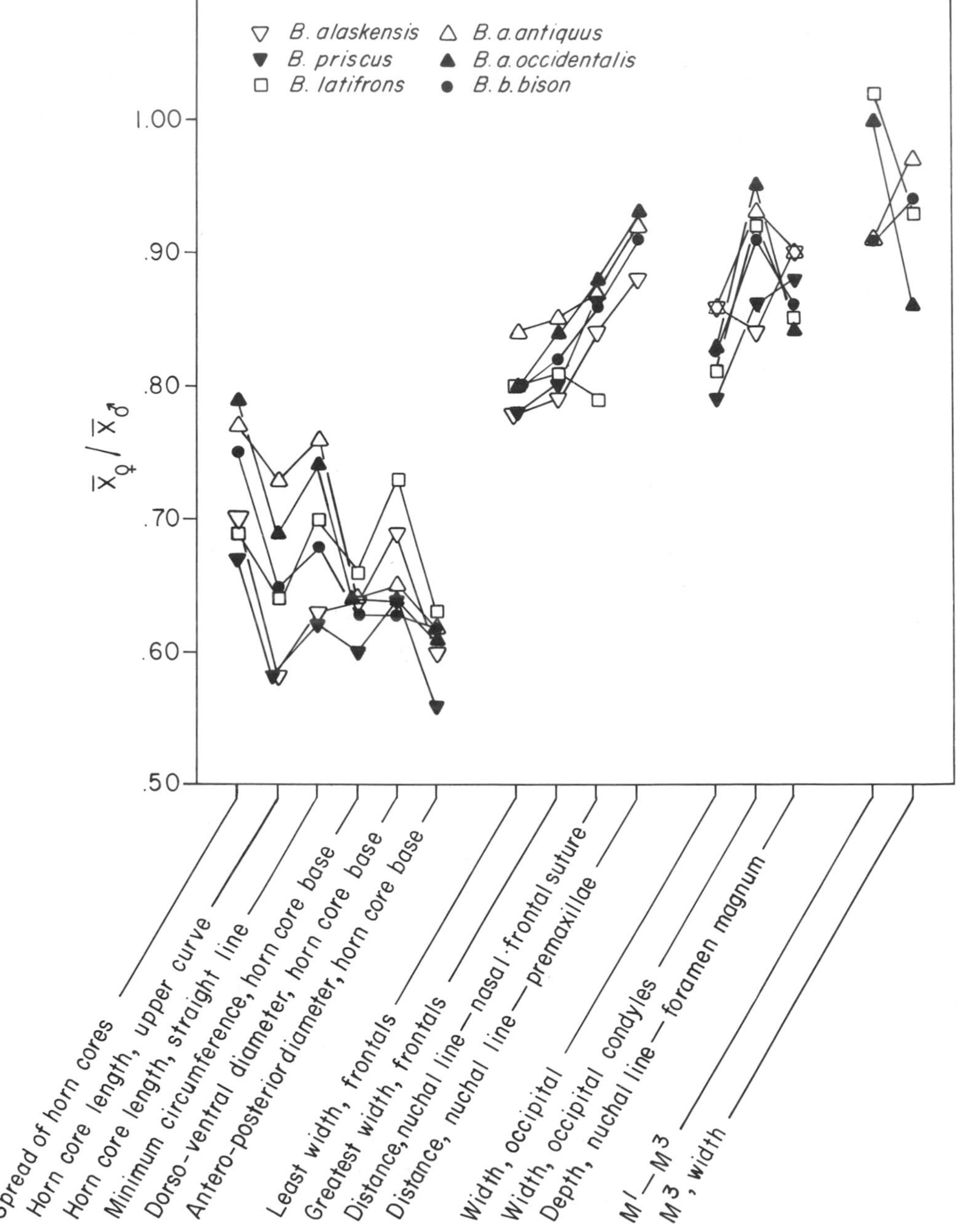

FIGURE 61 Relative sexual dimorphism among skull characters. The vertical axis expresses the female mean for each taxon as a multiple of the male mean.

Pl. 1 The Mexican bull, "*Taurus Mexicanus*," as pictured in Dr. Francisco Hernandez's natural history of Mexico (1651). Linnaeus used Hernandez's description as the type description for *Bos bison*, now *Bison bison*. *Bison bison* is the type species for the genus.

Pl. 2 The holotype of *Bison latifrons* (ANSP G 12993) was found in a small creek about twelve miles north of Big Bone Lick, Kentucky. This was the first fossil bison specimen from North America to be described and figured.

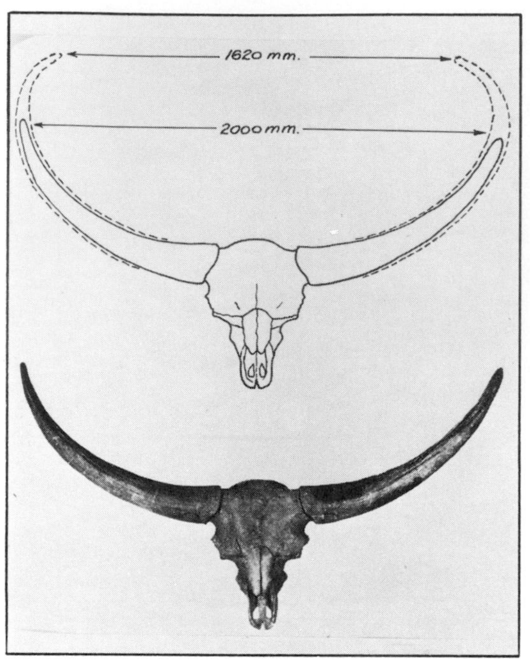

Pl. 3 Few complete skulls of *Bison latifrons* are known. One of the finest (UCMP 4067) was discovered at MacArthur, Shasta County, California in 1933. The only known likeness of a *Bison latifrons* horn sheath occurred as a sand cast in association with this skull. This cast provided important information on the shape of the horn sheath and placement of the horn tips relative to other skull characters.

Pl. 4 Male *Bison latifrons* horn cores (AMNH VP 26828) from the Bradenton Canal, Manatee County, Florida, show typical horn core, frontal, and occipital characters.

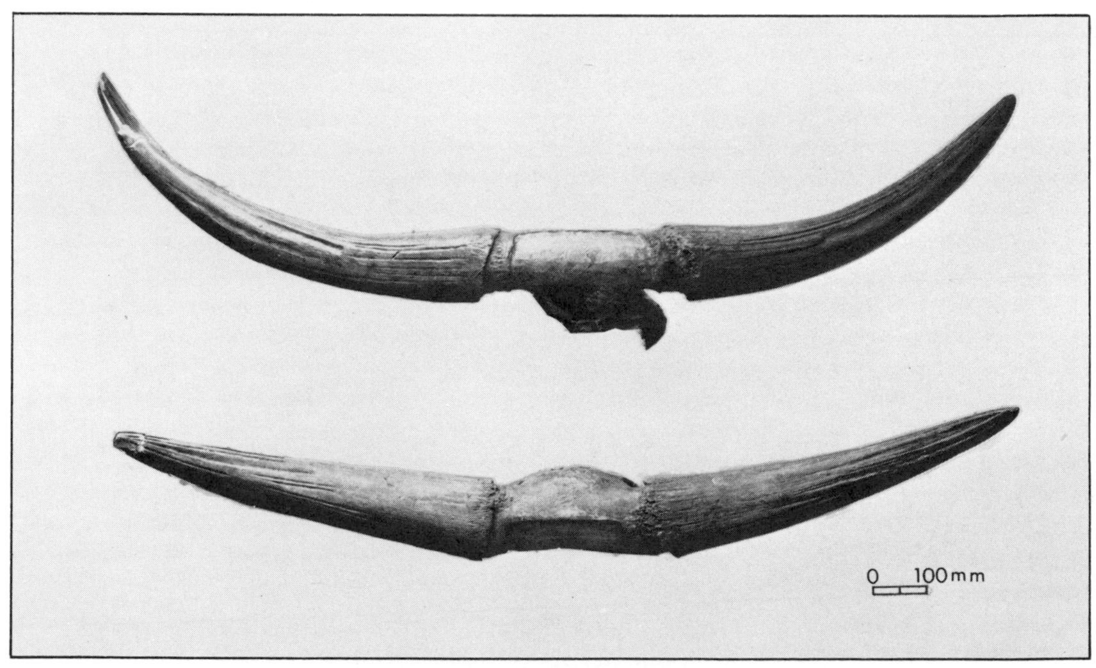

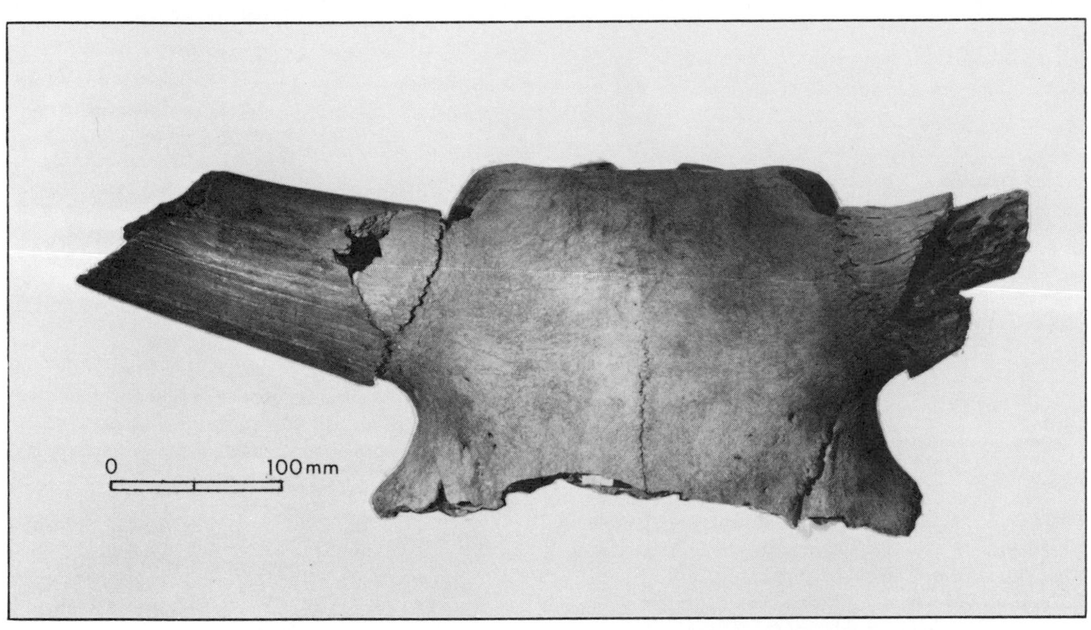

Pl. 5 Female *Bison latifrons* have more robust skulls and well defined burrs on the dorsal horn core than do females of other species. These characters show clearly on this skull (IMNH VP 17536) from the American Falls Reservoir in southeast Idaho.

Pl. 6 This life size restoration of a *Bison latifrons* head was based on the MacArthur, California, skull (UCMP 4067) and associated horn sheath cast. *Bison bison* was used as a guide for this restoration, a circumstance which probably resulted in a model with more body hair and rearward rotation of the horns than was actually characteristic of *Bison latifrons.*

Pl. 7 This cast of the *Bison alleni* holotype (YPM VP 10910) shows typical *Bison latifrons* characters—straight posterior margin of horn cores, mildly spiraled growth around an arched axis, and an isosceles triangle shaped basal cross section. These characters are not typical of *Bison alaskensis* or *Bison priscus*, specimens of which have been frequently identified as *Bison alleni.*

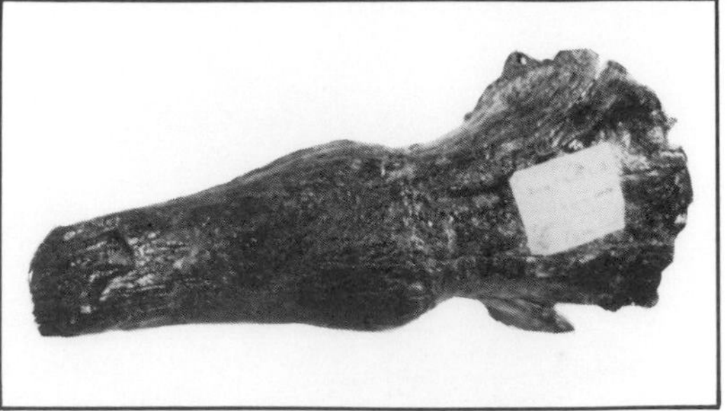

Pl. 8 The *Bison antiquus antiquus* holotype (ANSP G 12990) is from Big Bone Lick, Kentucky. This specimen is actually atypical of the taxon. It possibly represents a *Bison antiquus* X *Bison priscus* hybrid, but *Bison antiquus* characters are dominant.

Pl. 9 This male *Bison antiquus antiquus* (TMM 30967-423) from the Tedford Farm, San Patricio County, Texas, is more typical of the taxon than is the holotype.

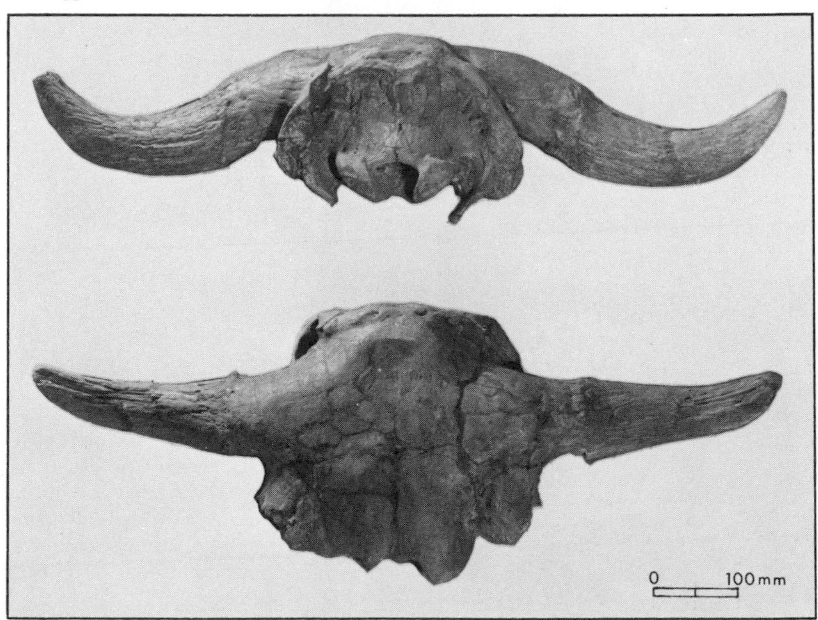

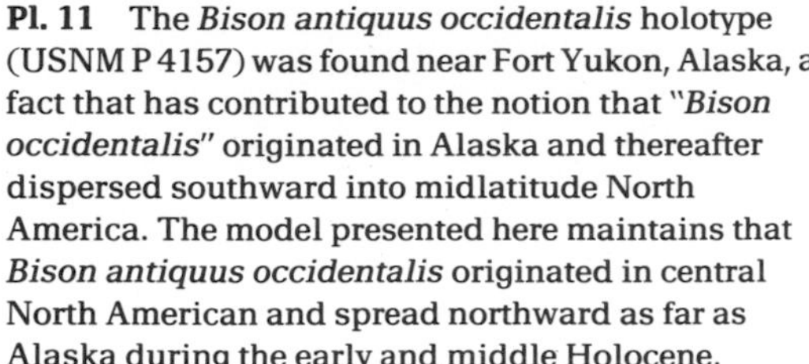

Pl. 10 The holotype of *Bos scaphoceras* (= *Bison antiquus antiquus*; UP—no catalog number) from northern Nicaragua represents the southernmost specifically identifiable specimen of North American bison. Lucas (1899) thought this was the horn core of a sheep (*Ovis*), but it is actually an abnormal bison horn core not unlike other bison horn cores from tropical North America.

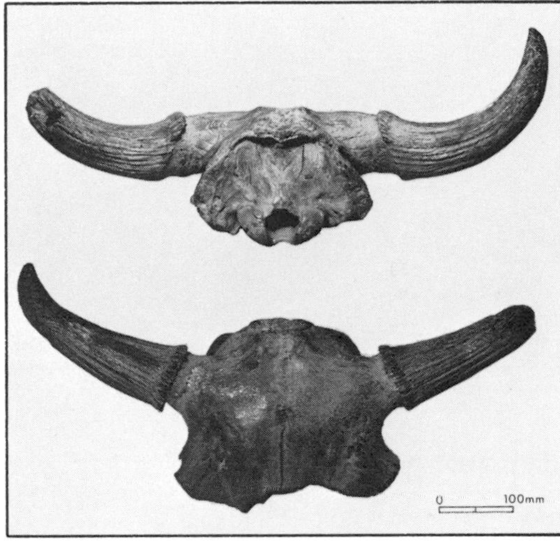

Pl. 11 The *Bison antiquus occidentalis* holotype (USNM P 4157) was found near Fort Yukon, Alaska, a fact that has contributed to the notion that "*Bison occidentalis*" originated in Alaska and thereafter dispersed southward into midlatitude North America. The model presented here maintains that *Bison antiquus occidentalis* originated in central North American and spread northward as far as Alaska during the early and middle Holocene.

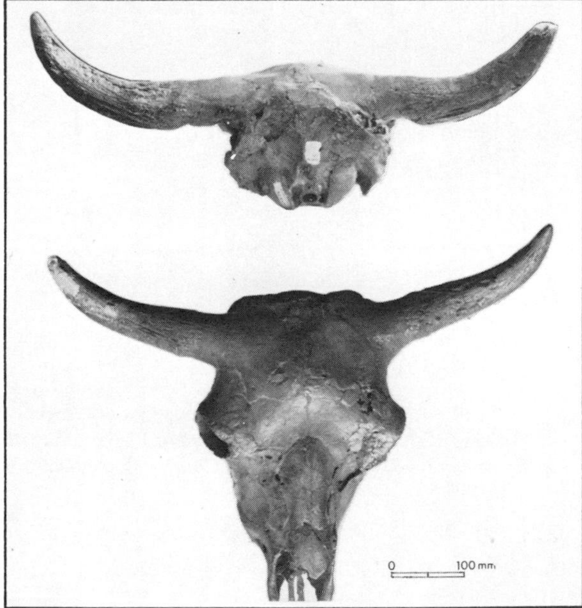

Pl. 12 A female *Bison antiquus occidentalis* (UNSM VP 30809) from the Scottsbluff archeological site, Nebraska.

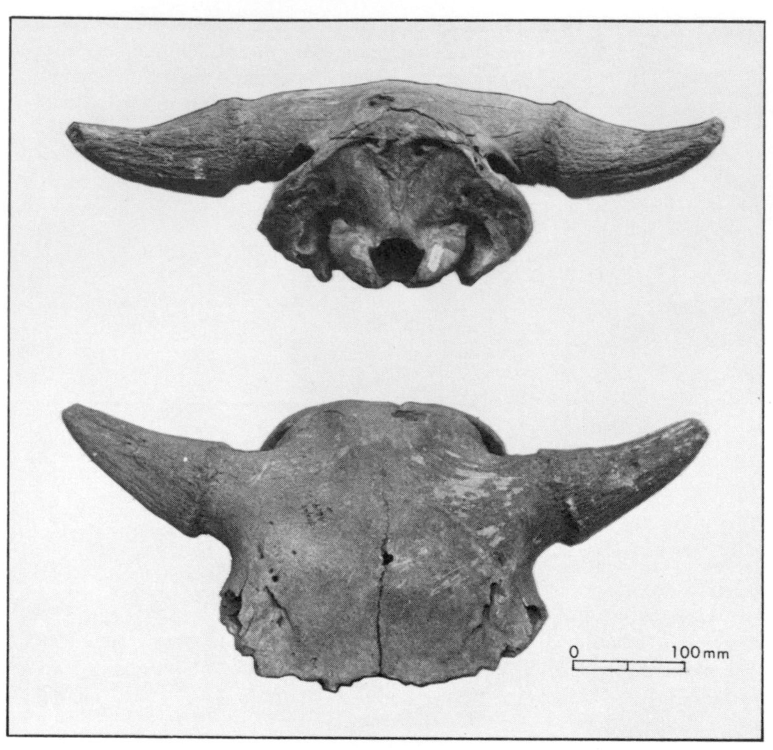

Pl. 13 *Bison bison bison* from the southern Great Plains were characterized by somewhat smaller body size and distinctly smaller horn cores than northern populations. This specimen (PPHM 619–14) is from Tierra Blanca Creek in the Texas panhandle, and shows horn core characters which might now be lost as a result of the commercial hunting during the 1870s which destroyed the southern plains population.

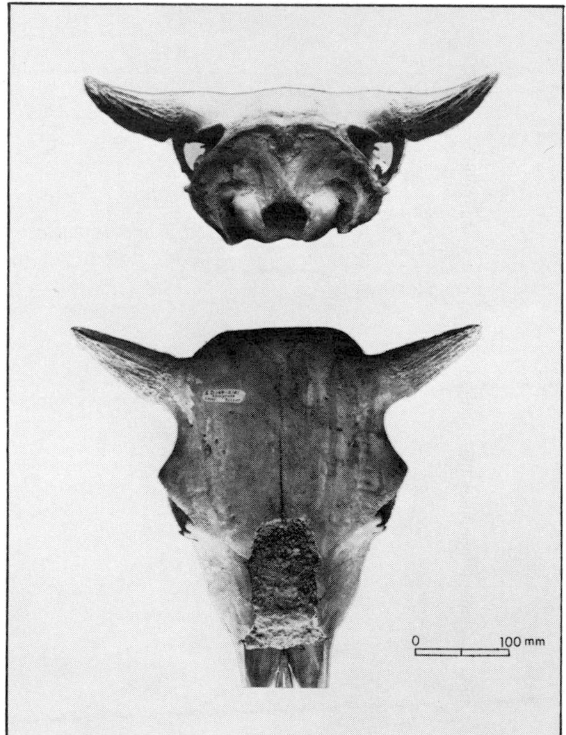

Pl. 14 A female *Bison bison bison* skull (AMNH VP F:AMSD109–2181) from Rocky Ford, a crossing of the White River, in Shannon County, South Dakota.

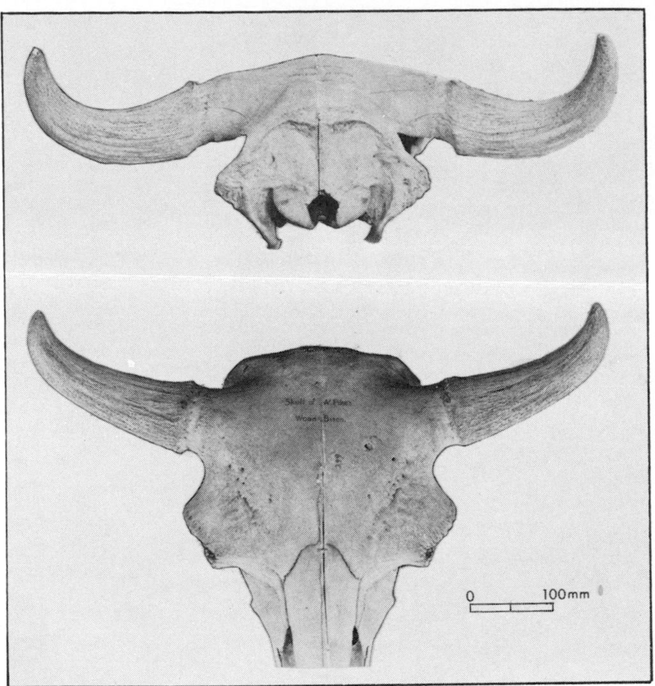

Pl. 15 The *Bison bison athabascae* holotype (NMC M 625) was shot in 1892 by an Indian about 50 miles southwest of Fort Resolution, Northwest Territories.

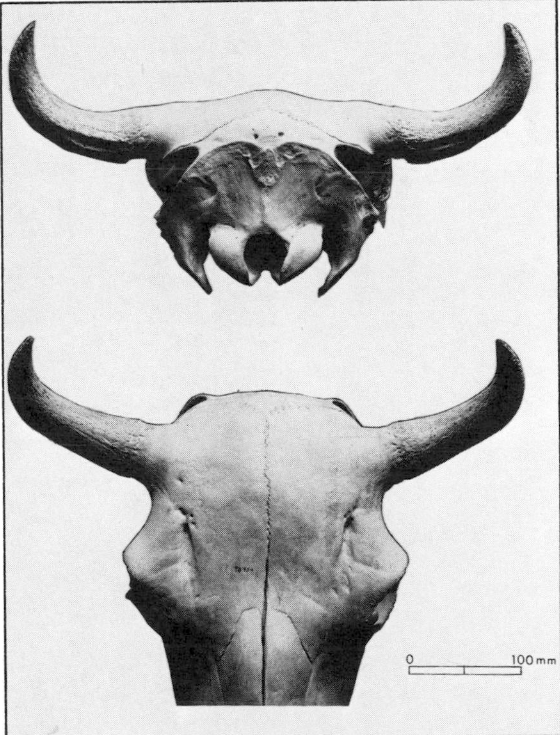

Pl. 16 More than 6,000 plains bison were released into Wood Buffalo National Park between 1925 and 1928, and most bison in the park are now considered plains X woods bison hybrids. This skull (AMNH M 98954) of an assumed hybrid female probably shows characters near those of true *Bison bison athabascae* females.

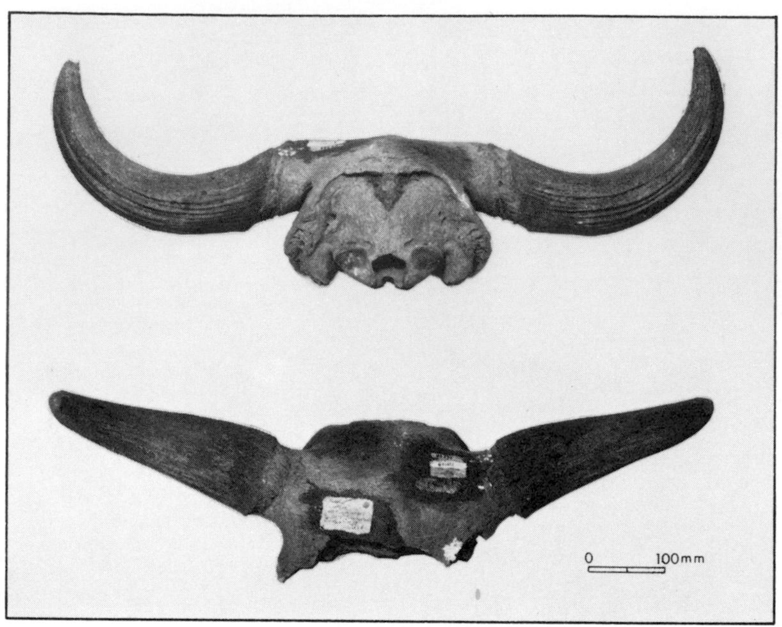

Pl. 17 The holotype of *Bison geisti* (AMNH VP F:AM46893), found near Fairbanks, Alaska, clearly shows characters typical of *Bison priscus*, with which it is here synonymized. *Bison priscus* is well represented in eastern Beringia, but specimens also occur in midlatitude and tropical North America.

Pl. 18 This female *Bison priscus* (SMP 272-74-z) from Cherokee County, Iowa, is the most southerly female specimen of this taxon studied. The forward rotation and sinuous posterior profile of the horn cores differentiate this specimen from *Bison antiquus* females.

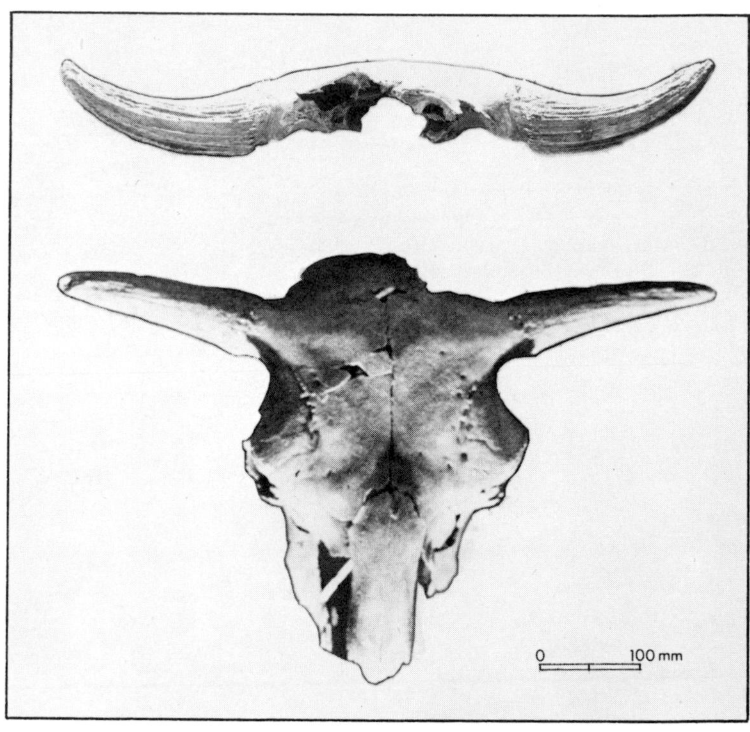

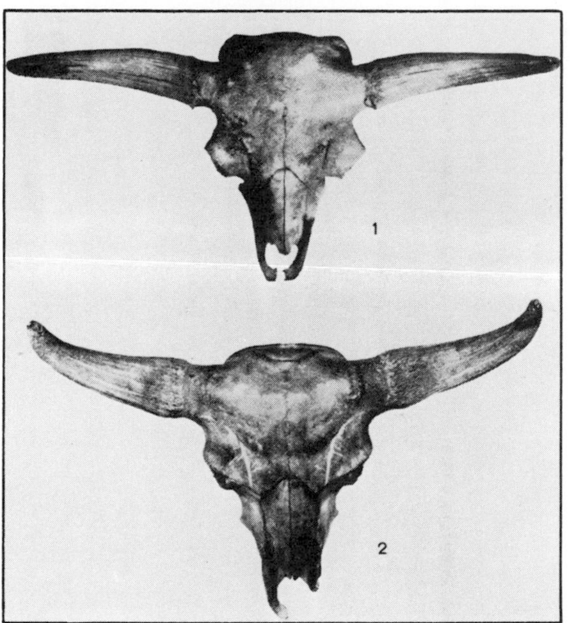

Pl. 19 Locating and identifying the *Bison priscus* lectotype has been a long and complicated process, and it may not yet be over. Most bison taxonomists accept Cuvier's 1825 figure as the type illustration for the taxon. Cuvier indicated that the figured specimen was in the University of Pavia. Vialli attempted to identify the figured specimen from among the Pavia bison and selected the top specimen shown above. The bottom skull, however, in the University of Turin, more strongly resembles Cuvier's figure and is here considered more likely the lectotype if, in fact, Cuvier's figure was based on a single specimen.

Pl. 20 The *Bison alaskensis* holotype (FMNH P 25226) was found near Point Barrow, Alaska. This specimen is typical of larger Eurasian autochthons.

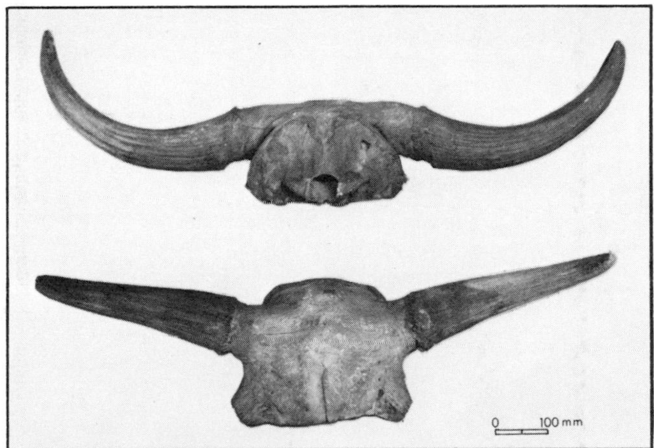

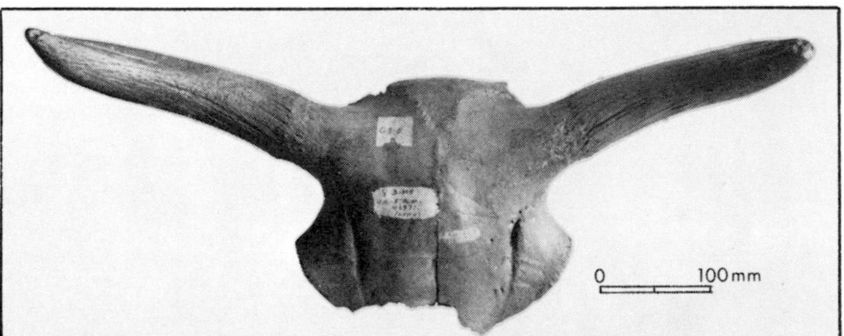

Pl. 21 This female *Bison alaskensis* (UNSM VP F:AM46972) from Goldstream, Alaska, shows horn core characters typical of Eurasian bison—sinuous posterior margin, sinuous axis with obviously spiraled growth, and forward rotation of the antero-posterior axis.

Pl. 22 Escalated fighting between two male *Bison bison bison* during the summer rut. These bulls have just collided head-to-head, culminating a mutual charge of several yards toward each other. The wool-covered frontals were the surfaces of impact in both cases. The individual on the right was reeled as a result of the collision. (Photo by S. L. Woodward; used with permission.)

Pl. 23 Head wrestling following the collision in plate 22. Pushing and rolling of the head against that of the opponent is often punctuated by deliberate, sudden hooking motions in which the horn tips are thrust toward the opponent's head. (Photo by S. L. Woodward; used with permission.)

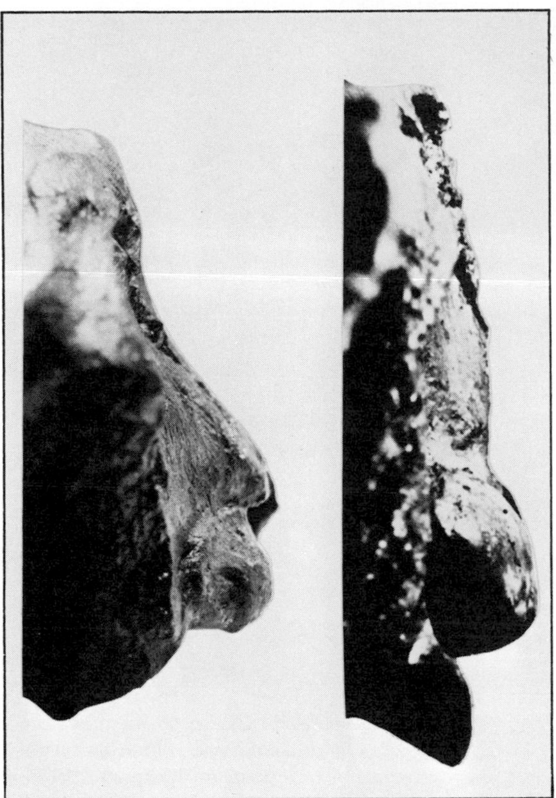

Pl. 24 Extreme differences exist in the shape of the occipital between larger and smaller bodied bison. Here, the dorso-ventral profile of the *Bison latifrons* holotype (ANSP G 12993) shows the exaggerated protrusion of the ventral occipital and the resulting proximal displacement of the occipital condyles and concavity of the occipital surface. A similar profile of a *Bison bison bison* occipital (AMNH VP F:AMSD110–2210) shows the more planar surface typical of smaller bodied bison.

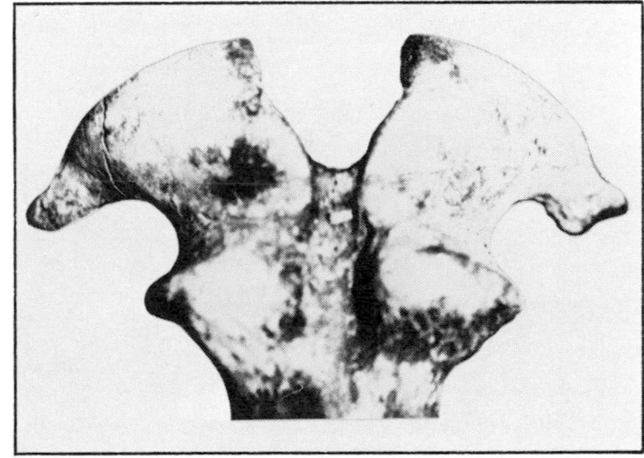

Pl. 25 The occipital condyles of larger bodied bison tend to have strongly flanged ventro-labial margins that might function to more securely fix the condyles of the large, heavy, flexible head into the articular cup of the atlas.

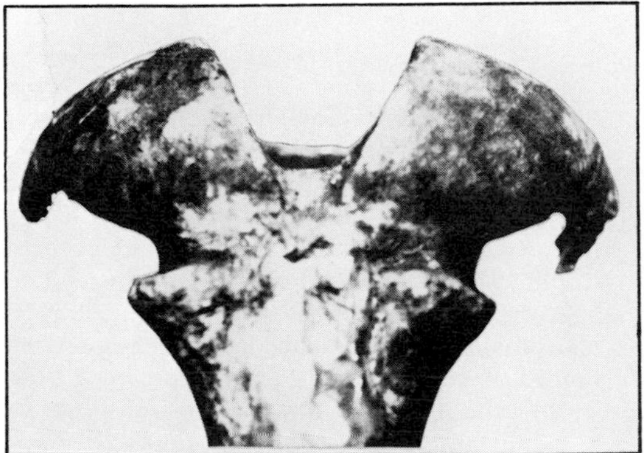

Pl. 26 The occipital condyles of smaller bodied bison are usually not flanged, although weak flanging is found in some specimens. *Bison bison* condyles show less flanging than those of any other taxon, probably because the smaller, lighter, and less flexible head precludes the necessity of such nonslippage characters.

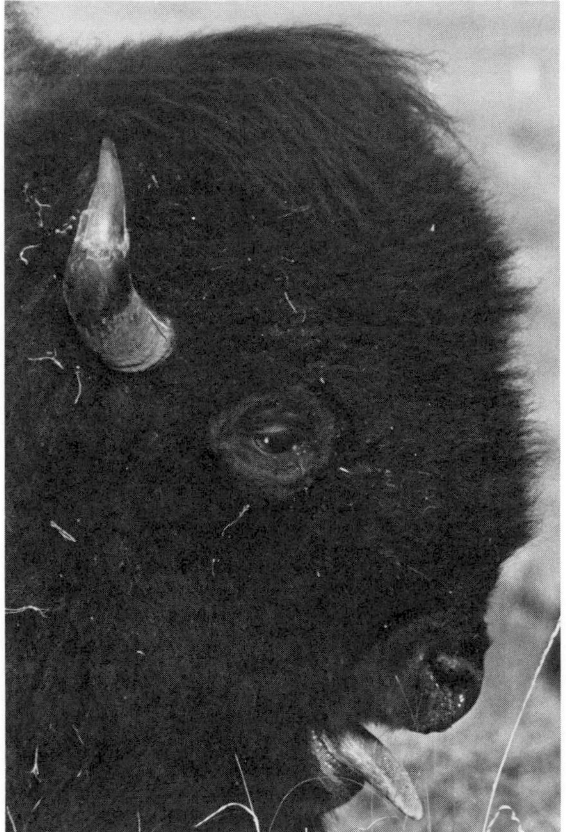

Pl. 27 The broadside display or threat posture is intended to maximize the visual impression of large body size as one bull presents the greatest silhouette area to another. Here two bulls have been threatening each other with broadside displays. The one on the left is starting to break away, whereas the one on the right maintains the threat posture.

Pl. 28 Exaggerated facial hair probably serves several adaptive functions. As a social organ it inflates apparent body size, thereby helping to establish and maintain the group dominance hierarchy. During escalated fighting, the mop of hair over the frontals cushions the impact of collisions and absorbs the impact of hooking horns (in part). Hair also shields the face during harsh winter weather, helping to conserve body heat and protect the head while animals forage in snow and ice. (Photo by H. L. Gunderson; used with permission.)

Pl. 29 Sexually mature males invest significant time and energy during the rut in "tending." A "tending bond," which may last from a few seconds to a few days, consists of a male joining and staying with a female and, to an extent, attempting to regulate the female's movement. The tending bull may or may not service the tended female. Tending lasts until the bull leaves voluntarily or is forced away by a more dominant bull. Tending probably serves to reduce tension and fighting among bulls during the rut by allowing subordinate bulls social participation in the rut short of actual insemination. The apparent head size of the bull in this photograph is exaggerated by facial hair to more than twice the width of the frontals. (Photo by H. L. Gunderson; used with permission.)

Pl. 30 The enamel border of molar fossets is normally smooth and simple (as in this M^1 and M^2, left and center), but some teeth possess reentering folds of enamel along this border (as in this M^3, right). This is an abnormal character in bison, and it might result from inbreeding or genetic drift.

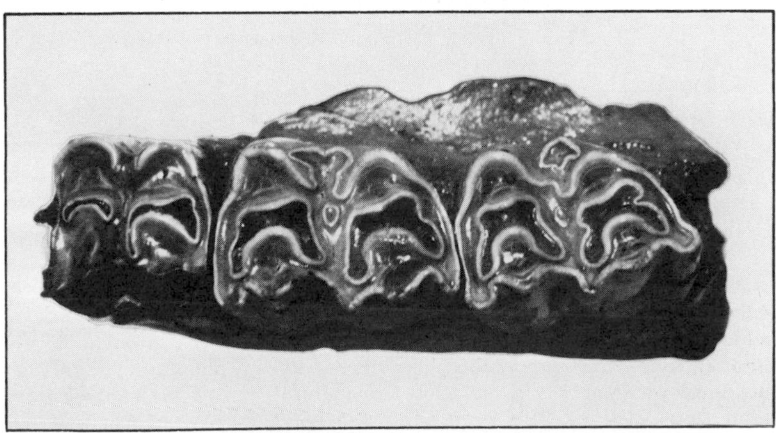

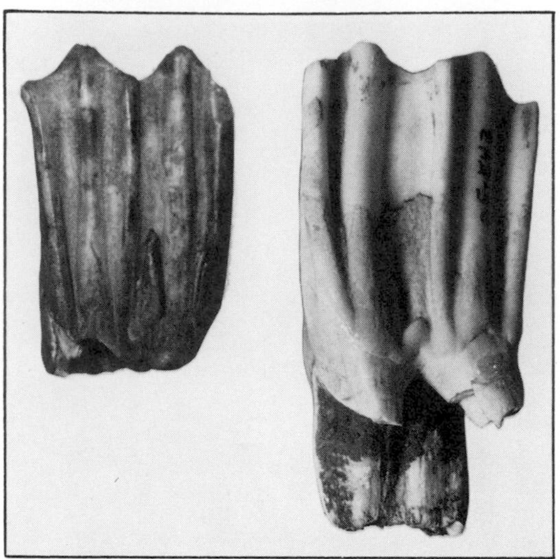

Pl. 31 The labial enamel surface of superior molars is normally characterized by three narrow and two wide alternating ridges. Occasionally, distinct auxiliary styles occur on the labial surface opposite the normal lingual style. These abnormal labial characters might also possibly identify populations experiencing inbreeding or genetic drift, especially when they occur in high frequencies or in conjunction with other abnormal characters. (Specimens are in the American Museum of Natural History and the Museum of Arid Lands Biology.)

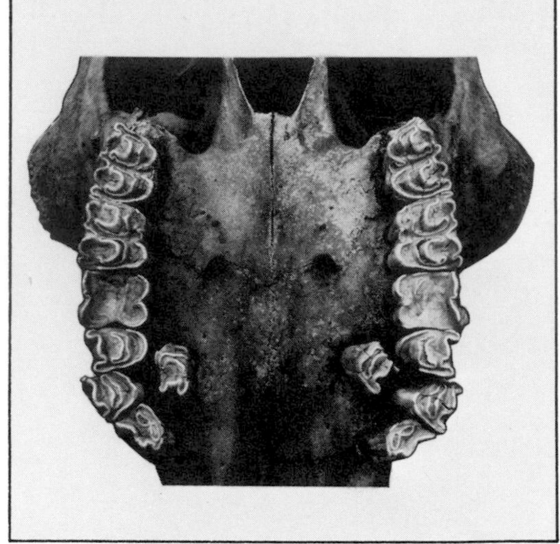

Pl. 32 Supernumerary teeth occasionally appear in bison. The holotype of *Bison bison oregonus* (USNM M 250145) contains an extra pair of premolars, and over 15% of the superior molars examined from the Malheur Lake sample had abnormal fosset borders. The Lake Malheur population possibly represents a small, inbred population from the western periphery of the *Bison bison* range.

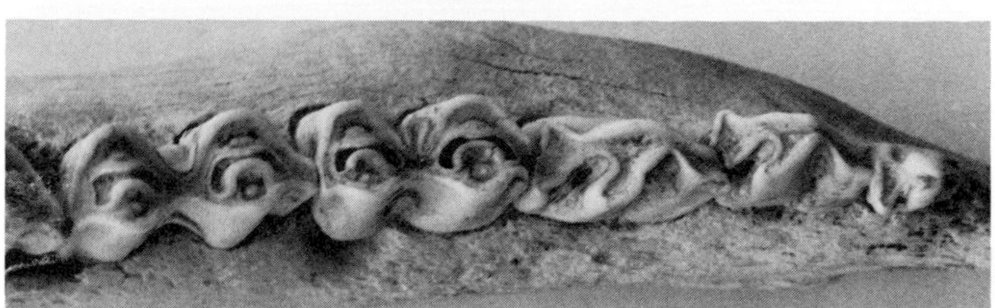

Pl. 33 The Casper archeological site, excavated near the town of Casper, Natrona County, Wyoming, yielded a bison deme possessive of numerous abnormalities—some probably genetic, others probably developmental. The incompletely formed third molar (top) pictured here is probably a genetic-based abnormality, whereas the case of "lumpy jaw" (bottom) is probably a developmental pathology, but a condition that suggests that the probably inbred population might have been abnormally susceptible to somatic traumas.

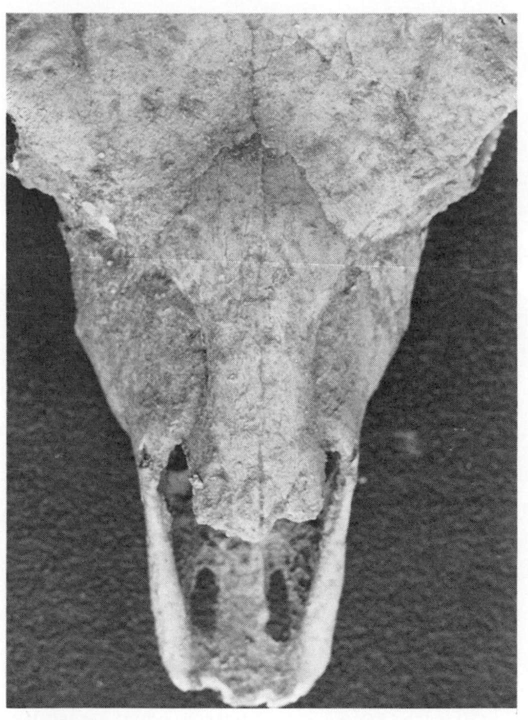

Pl. 34 Competition, or crowding during growth, between the maxillae and nasal bones frequently occurs in association with other abnormal characters. A mature female (UW NC2560) from the Casper archeological site illustrates nasal-maxillae competition. In this case the nasals have been pinched inward at the nasal-maxilla sutures, forming them into the shape of an hourglass.

Pl. 35 Radiocarbon dating has provided dependable evidencc that paleo-Indians had successfully implemented methods for killing herds of bison by at least 11,000 BP in North America. The relatively high frequency of skeletal abnormalities which are found in bison dating from the 11,000–9,000 BP period suggests that human hunting might have been sufficient to cause enough isolation of populations to initiate widespread inbreeding and genetic drift. The Hudson-Meng archeological site, in Sioux County, Nebraska, shown here during the 1975 excavation, represents the early Holocene period of presumably heavy predation on bison by paleo-Indians. For further information on this site, see Agenbroad (1978).

Pl. 36 The Vore archeological site, located on the edge of the Black Hills in northeastern Wyoming, provides a two hundred fifty year record of late prehistoric (ca. A.D. 1500–1750) bison procurement on the northwestern Great Plains. For further information on this site see Reher (1978). (Photo by C. A. Reher; used with permission.)

Pl. 37 Millions of bison were slaughtered on the Great Plains of the United States during the two decades following the Civil War, primarily for their hides. The commercial extermination of the plains bison started in the central plains, then shifted to the southern plains in the mid-1870s, and concluded in the early 1880s in the northern plains. Dodge City, Kansas, came into existence in the early 1870s as a rail station from which bison hides from Indian Territory and the Texas panhandle were shipped to eastern tanneries. This view of the Dodge City hide yard is reproduced with permission of the Kansas State Historical Society.

Pl. 38 Perhaps only a few hundred wild bison survived the slaughter of the nineteenth century, but enough were saved and protected to allow the establishment of the many private and public herds which now exist. With but few exceptions, however, modern bison exist in artifical selection regimes and are subject, in great part, to human influence upon their continued evolution. Pictured here is a portion of the herd at Fort Niobrara-Valentine National Wildlife Refuge Complex, Nebraska, where herd size, range, composition, and predation are more or less human controlled.

Relative sexual dimorphism in linear variates of the skull and limb bones follows a generally similar pattern among all taxa, yet significant differences in degrees of dimorphism are apparent for many characters. Among skull characters, the horn cores are most dimorphic, whereas the occipital-frontal-facial characters are considerably less so (fig. 61; table 46). Horn core characters collectively show the greatest range of dimorphism among the taxa and occipital characters show the least dimorphism. Among limb bones, lengths are less dimorphic than diaphysial diameters and, usually, the antero-posterior diaphysial diameters are less dimorphic than the transverse diameters (fig. 62, table 47). The magnitude of relative dimorphism in both skull and limb characters is not directly correlated with the absolute size or the absolute dimorphism of the characters.

No single taxon is unquestionably more or less sexually dimorphic in all characters than are the other taxa, yet some taxa do show tendencies toward greater or lesser relative dimorphism (figs. 61 and 62). *B. a. antiquus* is least dimorphic in more skull and limb characters than is any other taxon. *B. priscus* is most dimorphic in more skull characters than any other taxon (this observation may be biased by the small female *B. priscus* sample size). Among North American autochthons, *B. bison* is more dimorphic in more skull and limb characters than *B. latifrons* or *B. antiquus*. The great limb dimorphism observed for *B. b. athabascae* may be biased by the small sample size.

Table 46
Relative Sexual Dimorphism in Skulls

Standard measurement	*B. b. bison*	*B. a. occidentalis*	*B. a. antiquus*	*B. latifrons*	*B. priscus*	*B. alaskensis*
Spread of horn cores, tip to tip	.75	.79	.77	.69	.67	.70
Horn core length, upper curve, tip to burr	.65	.71	.73	.64	.58	.58
Straight line distance, tip to burr, dorsal horn core	.68	.74	.76	.70	.62	.63
Dorso-ventral diameter, horn core base	.63	.64	.65	.73	.64	.69
Minimum circumference, horn core base	.63	.64	.64	.66	.60	.64
Width of occipital at auditory openings	.83	.83	.86	.81	.79	.86
Width of occipital condyles	.91	.94	.93	.92	.86	.84
Depth, nuchal line to dorsal margin of foramen magnum	.86	.84	.90	.85	.88	.90
Antero-posterior diameter, horn core base	.62	.61	.62	.63	.56	.60
Least width of frontals, between horn cores and orbits	.80	.80	.84	.80	.78	.78
Greatest width of frontals at orbits	.82	.84	.85	.81	.80	.79
$M^1 - M^3$, inclusive alveolar length	.91	1.00	.91	1.02	...	...
M^3, maximum width, anterior cusp	.94	.86	.97	.93	...	...
Distance, nuchal line to tip of premaxillae	.91	.93	.92	...	...	.88
Distance, nuchal line to nasal-frontal suture	.86	.88	.87	.79	.86	.84

Note: The male mean for each character in each taxon = 1.00. Values given in the table are multiples of the male mean.

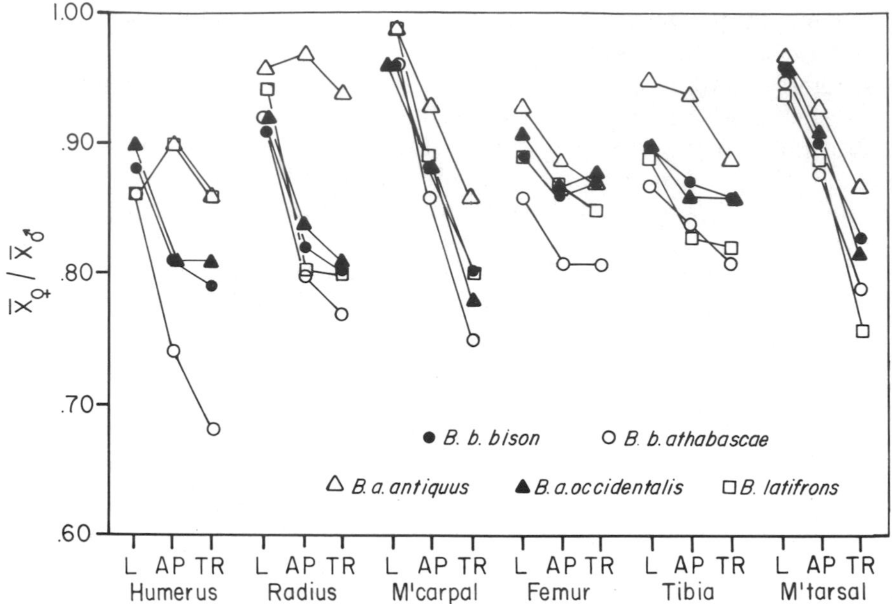

FIGURE 62 Relative sexual dimorphism among limb characters. The vertical axis expresses the female mean for each taxon as a multiple of the male mean.

Table 47
Relative Sexual Dimorphism in Limb Bones

Standard measurement	*B. b. bison*	*B. b. athabascae*	*B. a. occidentalis*	*B. a. antiquus*	*B. latifrons*
Humerus:					
Approximate rotational length of bone	.88	.86	.90	...	.86
Antero-posterior diameter of diaphysis	.81	.74	.81	.90	.90
Transverse minimum of diaphysis	.79	.68	.81	.86	.86
Radius:					
Approximate rotational length of bone	.91	.92	.92	.96	.94
Antero-posterior minimum of diaphysis	.82	.80	.84	.97	.80
Transverse minimum of diaphysis	.80	.77	.81	.94	.80
Metacarpal:					
Total length of bone	.96	.96	.96	.99	.99
Antero-posterior minimum of diaphysis	.88	.86	.88	.93	.89
Transverse minimum of diaphysis	.80	.75	.78	.86	.80
Femur:					
Approximate rotational length of bone	.89	.86	.91	.93	.89
Antero-posterior diameter of diaphysis	.86	.81	.87	.89	.87
Transverse minimum of diaphysis	.87	.81	.88	.87	.85
Tibia:					
Approximate rotational length of bone	.90	.87	.90	.95	.89
Antero-posterior minimum of diaphysis	.87	.84	.86	.94	.83
Transverse minimum of diaphysis	.86	.81	.86	.89	.82
Metatarsal:					
Total length of bone	.96	.95	.96	.97	.94
Antero-posterior minimum of diaphysis	.90	.88	.91	.93	.89
Transverse minimum of diaphysis	.83	.79	.82	.87	.76

Note: The male mean for each character in each taxon = 1.00. Values given in the table are multiples of the male mean.

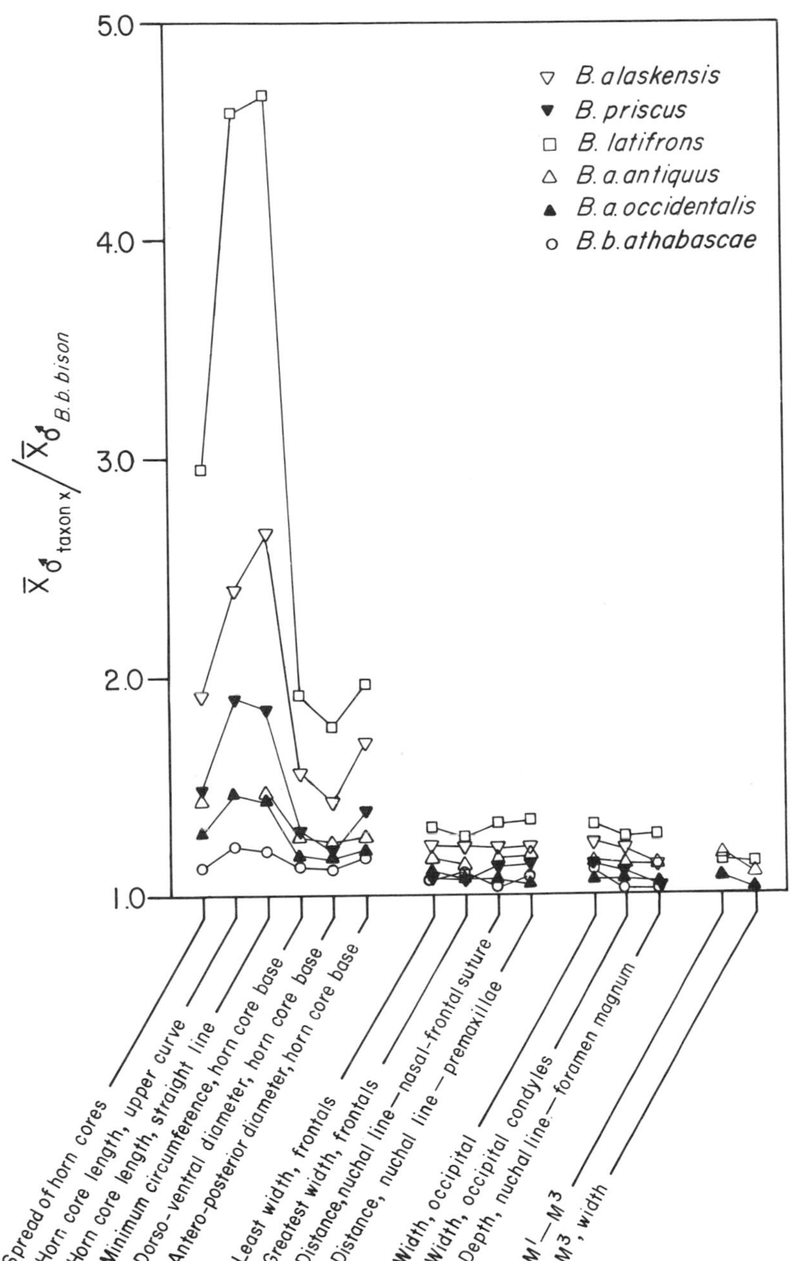

FIGURE 63 Skull allometries relative to *Bison bison bison*: males. The vertical axis expresses the male mean for each taxon as a multiple of the *Bison bison bison* male mean.

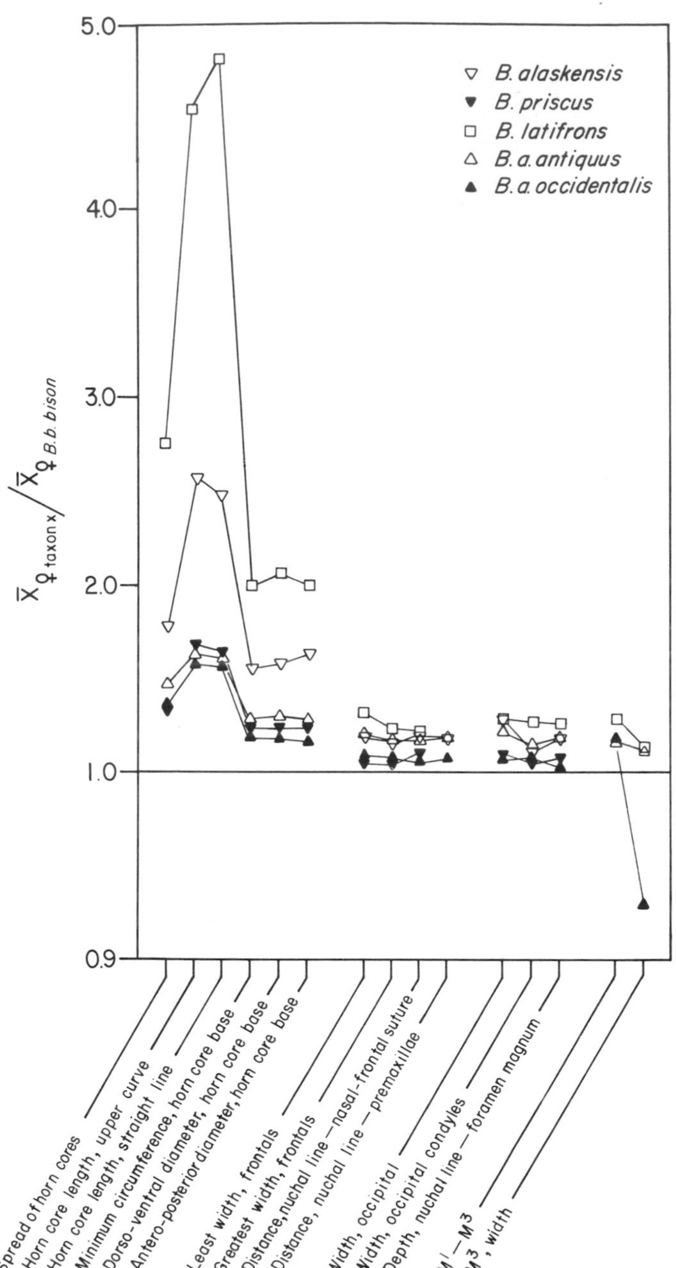

FIGURE 64 Skull allometries relative to *Bison bison bison:* females. The vertical axis expresses the female mean for each taxon as a multiple of the *Bison bison bison* female mean.

Table 48

Skull Allometries Relative to *Bison bison bison*

Standard measurement	*B. b. athabascae*	*B. a. occidentalis*	*B. a. antiquus*	*B. latifrons*	*B. priscus*	*B. alaskensis*
Males						
Spread of horn cores, tip to tip	1.13	1.29	1.44	2.96	1.47	1.91
Horn core length, upper curve, tip to burr	1.23	1.46	1.46	4.59	1.90	2.90
Straight line distance, tip to burr, dorsal horn core	1.20	1.44	1.45	4.67	1.85	2.66
Dorso-ventral diameter, horn core base	1.12	1.16	1.24	1.77	1.20	1.43
Minimum circumference, horn core base	1.13	1.18	1.27	1.92	1.29	1.56
Width of occipital at auditory openings	1.12	1.07	1.18	1.32	1.13	1.24
Width of occipital condyles	1.02	1.07	1.14	1.26	1.11	1.20
Depth, nuchal line to dorsal margin of foramen magnum	1.01	1.05	1.13	1.27	1.04	1.13
Antero-posterior diameter, horn core base	1.17	1.18	1.27	1.97	1.38	1.70
Least width of frontals, between horn cores and orbits	1.08	1.09	1.16	1.31	1.08	1.22
Greatest width of frontals at orbits	1.09	1.07	1.14	1.26	1.08	1.22
$M^1 - M^3$, inclusive alveolar length	....	1.07	1.17	1.15	....	....
M^3, maximum width, anterior cusp	....	1.03	1.08	1.14	....	....
Distance, nuchal line to tip of premaxillae	1.08	1.05	1.18	1.34	1.14	1.21
Distance, nuchal line to nasal-frontal suture	1.04	1.06	1.16	1.33	1.11	1.21
Females						
Spread of horn cores, tip to tip	....	1.36	1.48	2.75	1.33	1.79
Horn core length, upper curve, tip to burr	....	1.59	1.63	4.54	1.68	2.57
Straight line distance, tip to burr, dorsal horn core	....	1.56	1.62	4.81	1.69	2.49
Dorso-ventral diameter, horn core base	....	1.18	1.30	2.07	1.24	1.59
Minimum circumference, horn core base	....	1.18	1.29	2.00	1.23	1.56
Width of occipital at auditory openings	....	1.08	1.22	1.29	1.09	1.29
Width of occipital condyles	....	1.10	1.15	1.27	1.05	1.11
Depth, nuchal line to dorsal margin of foramen magnum	....	1.03	1.19	1.26	1.08	1.18
Antero-posterior diameter, horn core base	....	1.17	1.28	2.00	1.24	1.64
Least width of frontals, between horn cores and orbits	....	1.10	1.21	1.32	1.05	1.20
Greatest width of frontals at orbits	....	1.09	1.18	1.24	1.05	1.17
$M^1 - M^3$, inclusive alveolar length	....	1.19	1.17	1.29	....	....
M^3, maximum width, anterior cusp	....	0.93	1.11	1.13	....	....
Distance, nuchal line to tip of premaxillae	....	1.08	1.19	....	....	1.19
Distance, nuchal line to nasal-frontal suture	....	1.07	1.17	1.21	1.10	1.18

Note: The *B. b. bison* mean for each character = 1.00. Values given in the table are multiples of the *B. b. bison* mean.

ALLOMETRY

Most characters are to a greater or lesser degree allomorphic among bison taxa. The magnitude of allometry among skull characters, relative to *B. b. bison* (an arbitrarily selected norm), is illustrated in figures 63 and 64 and quantified in table 48. Horn core characters show the greatest allomorphosis among taxa, whereas occipital-frontal-facial characters show considerably less allomorphosis. Both males and females exhibit similar patterns of skull allometry.

Limb characters are less allometric than skull characters (table 49). Limb length is usually, but not always, less allomorphic than are the diaphysial diameters. There is no strong tendency for either of the diaphysial diameters to be more or less allomorphic than the other. Both sexes show similar patterns of limb allometry.

Table 49

Limb Allometries Relative to *Bison bison bison*

Standard measurement	*B. b. athabascae*	*B. a. occidentalis*	*B. a. antiquus*	*B. latifrons*
Males				
Humerus:				
Approximate rotational length of bone	1.08	1.09	....	1.23
Antero-posterior diameter of diaphysis	1.09	1.07	1.13	1.30
Transverse minimum of diaphysis	1.15	1.08	1.16	1.32
Radius:				
Approximate rotational length of bone	1.06	1.07	1.07	1.20
Antero-posterior minimum of diaphysis	1.08	1.10	1.04	1.41
Transverse minimum of diaphysis	1.10	1.08	1.01	1.40
Metacarpal:				
Total length of bone	1.10	1.06	1.07	1.21
Antero-posterior minimum of diaphysis	1.11	1.10	1.12	1.33
Transverse minimum of diaphysis	1.14	1.13	1.15	1.45
Femur:				
Approximate rotational length of bone	1.10	1.06	1.12	1.27
Antero-posterior diameter of diaphysis	1.12	1.04	1.17	1.37
Transverse minimum of diaphysis	1.09	1.04	1.24	1.33
Tibia:				
Approximate rotational length of bone	1.11	1.08	1.10	1.23
Antero-posterior minimum of diaphysis	1.09	1.12	1.22	1.43
Transverse minimum of diaphysis	1.08	1.08	1.20	1.39
Metatarsal:				
Total length of bone	1.08	1.05	1.07	1.19
Antero-posterior minimum of diaphysis	1.08	1.09	1.16	1.35
Transverse minimum of diaphysis	1.10	1.11	1.15	1.49
Females				
Humerus:				
Approximate rotational length of bone	1.06	1.12	1.19	1.21
Antero-posterior diameter of diaphysis	1.00	1.08	1.26	1.45
Transverse minimum of diaphysis	0.99	1.10	1.25	1.42
Radius:				
Approximate rotational length of bone	1.06	1.07	1.12	1.23
Antero-posterior minimum of diaphysis	1.06	1.12	1.24	1.38
Transverse minimum of diaphysis	1.06	1.09	1.18	1.40
Metacarpal:				
Total length of bone	1.10	1.06	1.10	1.25
Antero-posterior minimum of diaphysis	1.08	1.10	1.19	1.34
Transverse minimum of diaphysis	1.06	1.10	1.22	1.44
Femur:				
Approximate rotational length of bone	1.05	1.07	1.17	1.25
Antero-posterior diameter of diaphysis	1.06	1.05	1.21	1.39
Transverse minimum of diaphysis	1.01	1.04	1.23	1.29
Tibia:				
Approximate rotational length of bone	1.07	1.07	1.16	1.22
Antero-posterior minimum of diaphysis	1.05	1.10	1.32	1.37
Transverse minimum of diaphysis	1.02	1.09	1.24	1.33
Metatarsal:				
Total length of bone	1.07	1.05	1.08	1.16
Antero-posterior minimum of diaphysis	1.05	1.10	1.19	1.33
Transverse minimum of diaphysis	1.05	1.09	1.20	1.36

Note: The *B. b. bison* mean for each character = 1.00. Values given in the table are multiples of the *B. b. bison* mean.

VARIABILITY OF CHARACTERS

Pearson's coefficient of variation (V) is a measure of relative variation within samples. Higher Vs indicate greater variation among cases relative to the sample mean and imply greater variability of the respective character. Lower Vs indicate the opposite. Normally, Vs for mammalian characters are between about three and ten (Olson and Miller, 1958; Simpson, 1953; Yablokov, 1974).

The Vs for bison skull characters are greatest for horn cores and less for other characters (figs. 65 and 66; table 50). Limb Vs are higher for diaphysial diameters than for length (figs. 67 and 68; table 51). There is no correlation between Vs and absolute size of characters. Patterns of variability in skull and limb bone characters are similar for both males and females.

No single taxon shows distinctly greater or less variability than other taxa in all characters. With the exception of some horn core variates the mean coefficients of variation for all linear skull and limb variates studied fall within the general limits of variation expected in mammals (i.e., between three and ten). Coefficients of variation for horn core length characters average somewhat larger than ten (table 52). Limb character Vs range from 3.0 to 8.0 (table 53).

CHOROCLINES

Choroclines can be demonstrated among all North American bison. The *B. bison* sample represents a shorter span of geologic time than any other available sample and can, therefore, best illustrate the nature of choroclinal variation among bison. Six skull characters show different clinal gradients along the north-south axis of the primary *B. bison* range (fig. 69). Northern bison are larger than southern bison in all characters included in figure 69, but they are decidedly larger in horn core characters than in others. Also, some characters decrease in size continuously from north to south, whereas other characters increase in size from the Canadian prairies southward. Front limb bones of *B. bison* also generally decrease in length from north to south (fig. 70). Female and male limbs show parallel clinal gradients. (Female *B. bison* skull character clines are not figured because of inadequate southern plains, northern plains, and boreal samples.)

Character clines for other North American taxa are presented in figure 71. These clines reinforce the observations made from *B. bison* clines that clinal gradients vary from character to character. North-south trending clines for Pleistocene and early-to-middle Holocene bison show generally larger phenotypes to the south and smaller phenotypes to the north, whereas later Holocene *B. bison* show larger phenotypes to the north and smaller phenotypes to the south. *B. latifrons* characters (especially horn cores) are generally larger to the west and smaller to the east, whereas *B. a. antiquus* characters are larger in the east and smaller in the west. When evaluating choroclines in fossil bison, however, the fact must be kept in mind that the samples probably represent a relatively longer period of geologic time, and reflect the consequences of greater environmental fluctuations and perturbations of gene flow than does the *B. bison* sample.

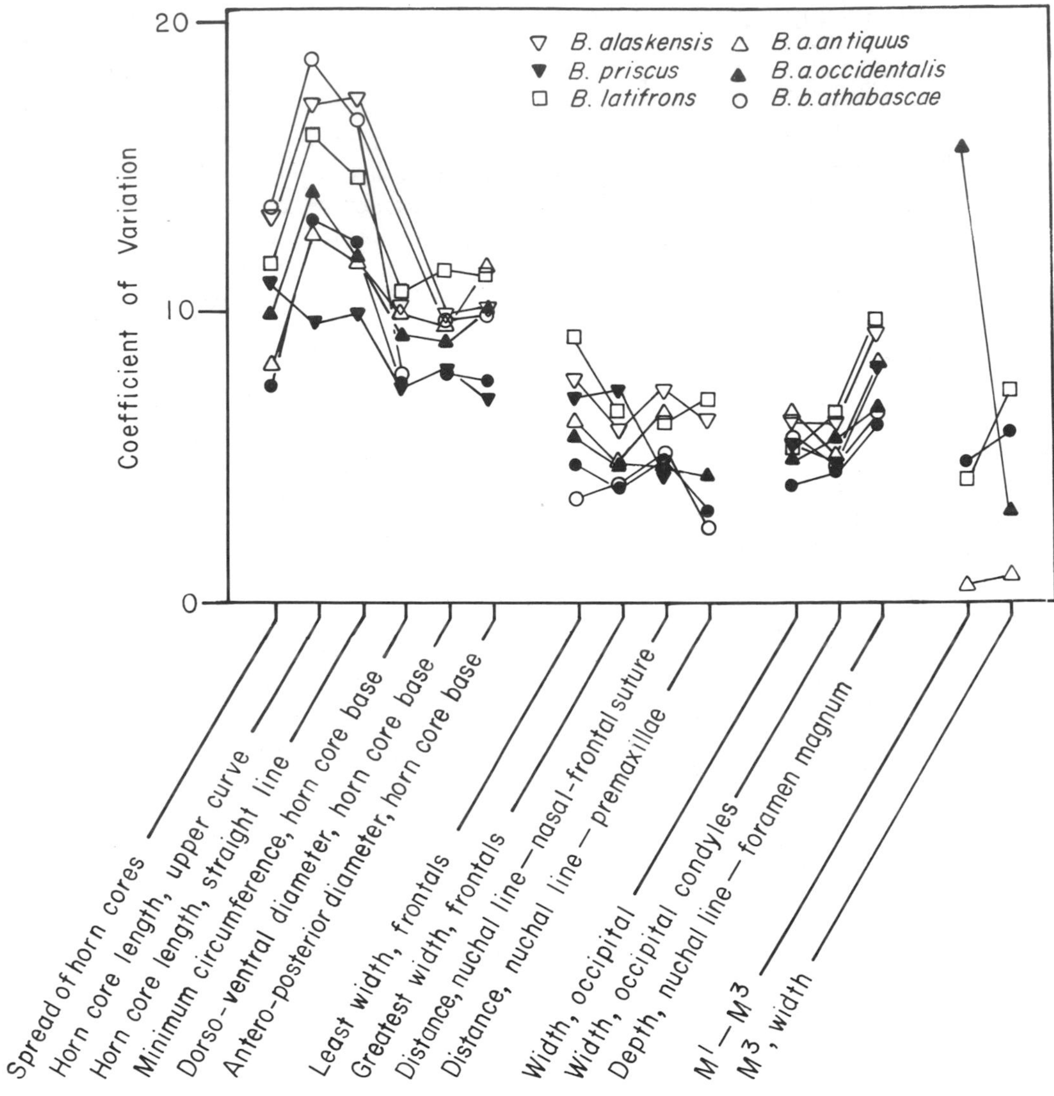

FIGURE 65 Coefficients of variation, skull characters: males.

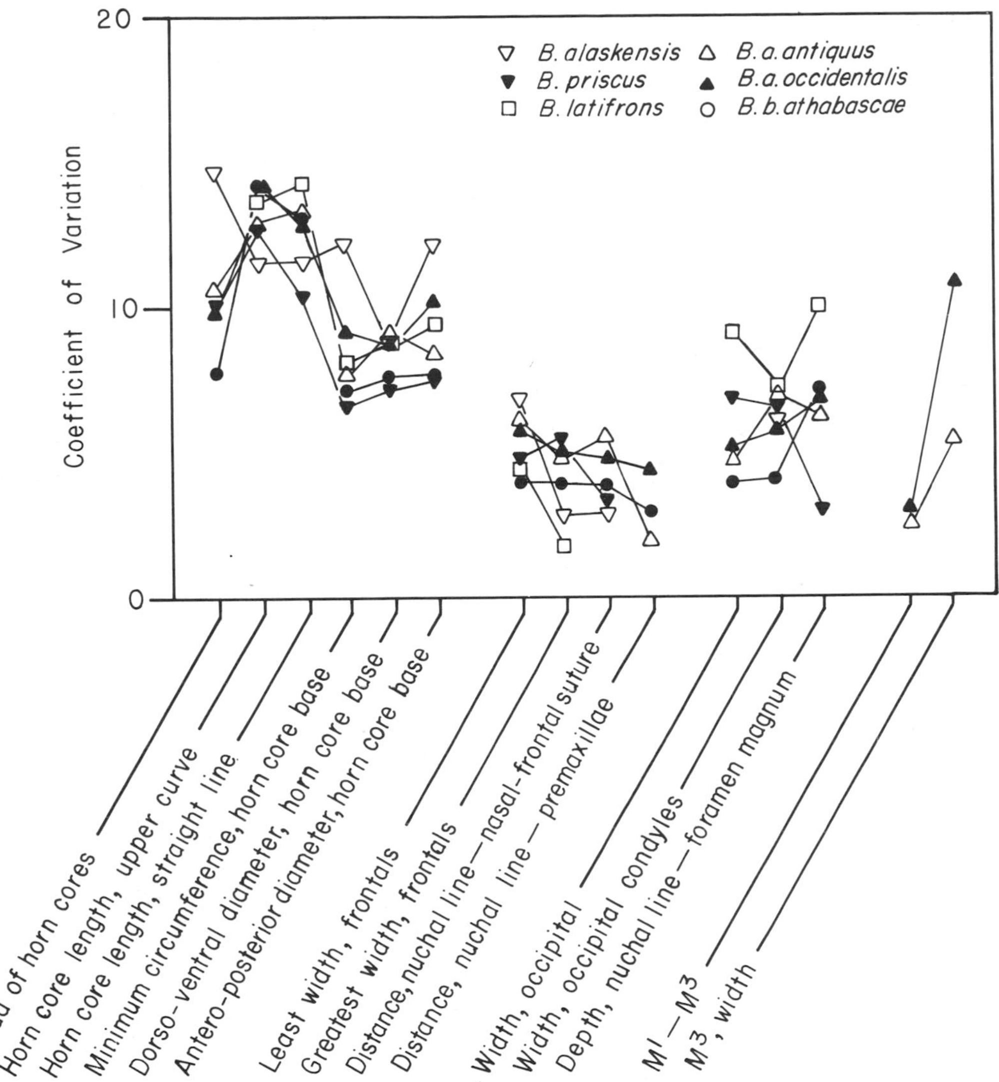

FIGURE 66 Coefficients of variation, skull characters: females.

Table 50
Coefficients of Variation: Skull Characters

Standard measurement	*B. b. bison*	*B. b. athabascae*	*B. a. occidentalis*	*B. a. antiquus*	*B. latifrons*	*B. priscus*	*B. alaskensis*
Males							
Spread of horn cores, tip to tip	7.4	13.5	9.9	8.2	11.7	11.0	13.3
Horn core length, upper curve, tip to burr	13.0	18.7	14.1	12.6	16.1	9.7	17.1
Straight line distance, tip to burr, dorsal horn core	12.4	16.6	12.8	11.7	14.6	9.9	17.4
Dorso-ventral diameter, horn core base	7.9	9.5	8.9	9.5	11.4	8.4	9.7
Minimum circumference, horn core base	7.6	7.9	9.2	10.0	10.7	6.4	10.2
Width of occipital at auditory openings	4.0	5.6	5.0	6.5	5.3	5.3	6.3
Width of occipital condyles	4.5	4.9	5.7	5.9	6.4	4.9	6.2
Depth, nuchal line to dorsal margin of foramen magnum	6.2	6.6	6.7	8.3	9.7	8.1	9.3
Antero-posterior diameter, horn core base	7.6	9.8	10.1	11.5	11.3	7.0	10.1
Least width of frontals, between horn cores and orbits	4.7	3.6	5.7	6.2	9.1	7.0	7.7
Greatest width of frontals at orbits	4.0	4.2	4.8	4.8	6.6	7.3	5.9
M^1-M^3, inclusive alveolar length	4.8	.....	.....	.....	4.0	....	....
M^3, maximum width, anterior cusp	5.9	.....	.....	.....	7.2	....	....
Distance, nuchal line to tip of premaxillae	3.2	2.6	4.4	.....	....	....	....
Distance, nuchal line to nasal-frontal suture	5.0	5.2	4.7	6.5	6.3	4.5	7.3
Angle of divergence of horn cores, forward from sagittal	6.6	7.0	7.2	6.1	7.0	5.7	7.9
Angle between foramen magnum and occipital planes	5.6	6.8	5.7	5.3	7.1	6.7	6.5
Angle between foramen magnum and basioccipital planes	4.5	5.8	4.9	6.1	6.4	5.0	6.5
Females							
Spread of horn cores, tip to tip	7.9	.....	8.4	10.6	....	10.0	....
Horn core length, upper curve, tip to burr	14.1	.....	9.6	12.8	....	12.8	11.5
Straight line distance, tip to burr, dorsal horn core	13.1	.....	77.7	13.2	....	10.4	11.5
Dorso-ventral diameter, horn core base	7.5	.....	8.0	9.1	8.6	7.1	8.6
Minimum circumference, horn core base	7.1	.....	7.0	7.7	8.1	6.6	12.1
Width of occipital at auditory openings	3.8	.....	3.7	4.7	....	....	....
Width of occipital condyles	4.0	.....	5.5	6.9	....	6.5	....
Depth, nuchal line to dorsal margin of foramen magnum	7.0	.....	6.8	6.1	....	....	....
Antero-posterior diameter, horn core base	7.6	.....	9.3	8.3	9.4	7.5	12.1
Least width of frontals, between horn cores and orbits	4.0	.....	5.2	6.0	4.3	4.7	....
Greatest width of frontals at orbits	3.3	.....	3.7	4.7	....	5.4	....
M^1-M^3, inclusive alveolar length	....	.....	3.0	2.5	....	....	....
M^3, maximum width, anterior cusp	....	.....	10.8	5.4	....	....	....
Distance, nuchal line to tip of premaxillae	2.8	.....	3.4	1.9	....	....	....
Distance, nuchal line to nasal-frontal suture	3.7	.....	6.0	5.5	....	3.4	....
Angle of divergence of horn cores, forward from sagittal	5.6	.....	5.9	6.5	....	8.8	....
Angle between foramen magnum and occipital planes	5.5	.....	5.4	3.8	....	4.0	....
Angle between foramen magnum and basioccipital planes	4.4	.....	4.9	5.8	....	5.7	....

Note: Coefficients of variation are reported only for samples in which $n \geqslant 4$. The sample size for each character is given in the skull biometric summary tables, chapter 2.

Table 51

Coefficients of Variation: Limb Characters

Standard measurement	*B. b. bison*	*B. a. occidentalis*	*B. a. antiquus*	*B. latifrons*
Males				
Humerus:				
Approximate rotational length of bone	3.4	5.9	. . .	3.7
Antero-posterior diameter of diaphysis	5.1	6.8	. . .	4.9
Transverse minimum of diaphysis	6.2	7.7	. . .	5.5
Radius:				
Approximate rotational length of bone	2.8	4.2	2.2	2.6
Antero-posterior minimum of diaphysis	5.2	5.9	2.8	5.9
Transverse minimum of diaphysis	5.2	5.6	2.3	5.7
Metacarpal:				
Total length of bone	3.4	3.9	3.0	3.3
Antero-posterior minimum of diaphysis	6.1	6.0	4.0	5.9
Transverse minimum of diaphysis	7.5	7.2	6.6	7.5
Femur:				
Approximate rotational length of bone	2.7	4.4	3.2	. . .
Antero-posterior diameter of diaphysis	5.1	6.5	7.4	5.2
Transverse minimum of diaphysis	5.1	4.7	3.6	3.0
Tibia:				
Approximate rotational length of bone	2.7	4.3	3.5	2.4
Antero-posterior minimum of diaphysis	4.2	5.8	6.9	4.5
Transverse minimum of diaphysis	4.4	5.6	5.4	5.0
Metatarsal:				
Total length of bone	3.0	3.7	3.0	2.2
Antero-posterior minimum of diaphysis	4.6	5.5	6.7	4.1
Transverse minimum of diaphysis	5.4	7.5	6.7	5.7
Females				
Humerus:				
Approximate rotational length of bone	3.0	6.6	4.0	. . .
Antero-posterior diameter of diaphysis	4.6	8.3	6.7	4.8
Transverse minimum of diaphysis	4.4	8.2	6.1	5.2
Radius:				
Approximate rotational length of bone	3.1	5.2	4.5	2.5
Antero-posterior minimum of diaphysis	6.2	7.0	10.7	2.6
Transverse minimum of diaphysis	5.8	7.2	11.2	7.6
Metacarpal:				
Total length of bone	3.4	3.5	3.1	2.5
Antero-posterior minimum of diaphysis	6.7	5.8	5.5	3.9
Transverse minimum of diaphysis	6.3	7.3	7.8	5.7
Femur:				
Approximate rotational length of bone	2.9	4.6	3.8	. . .
Antero-posterior diameter of diaphysis	6.0	8.3	4.3	1.7
Transverse minimum of diaphysis	6.0	5.3	5.7	3.9
Tibia:				
Approximate rotational length of bone	3.0	5.2	2.6	. . .
Antero-posterior minimum of diaphysis	5.8	6.4	6.1	. . .
Transverse minimum of diaphysis	5.1	6.6	4.1	. . .
Metatarsal:				
Total length of bone	3.2	3.6	3.2	. . .
Antero-posterior minimum of diaphysis	5.5	6.2	7.3	3.5
Transverse minimum of diaphysis	6.7	8.2	9.5	2.4

Note: Coefficients of variation are reported only for samples in which n ≥ 4. The sample size for each character is given in the limb biometric summary tables, chapter 2.

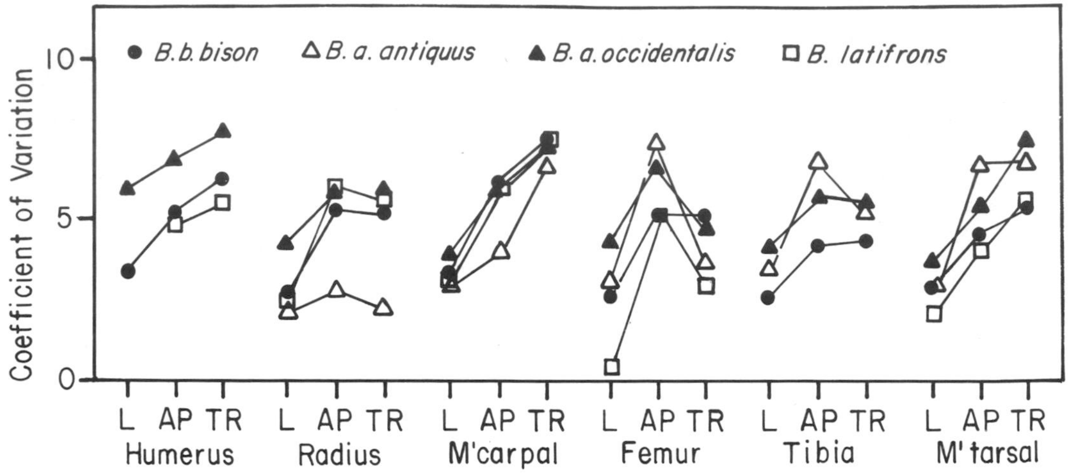

FIGURE 67 Coefficients of variation, limb characters: males.

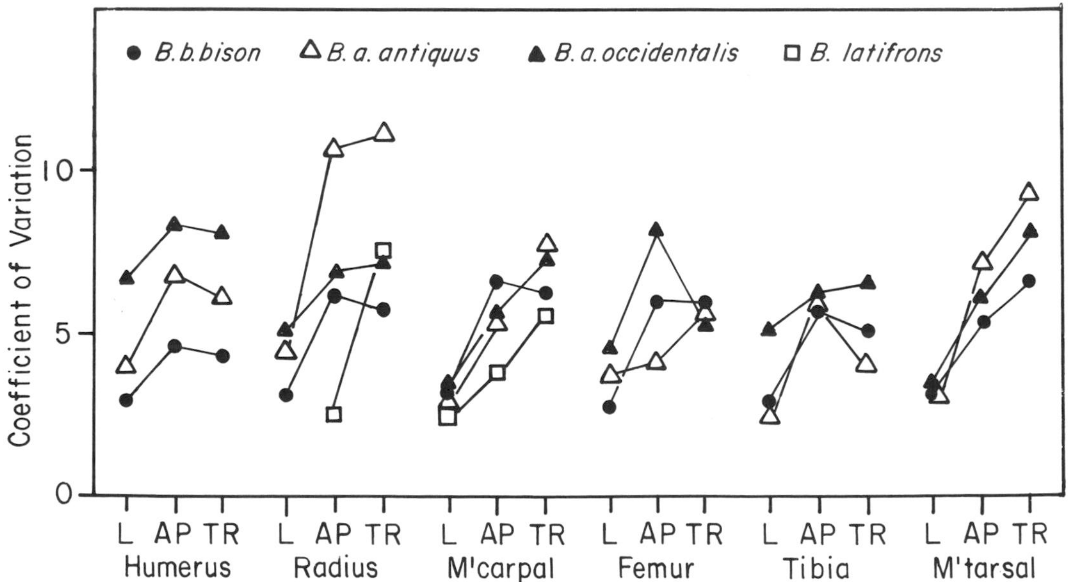

FIGURE 68 Coefficients of variation, limb characters: females.

Table 52

Mean Coefficients of Variation: Skull Characters

Standard measurement	Males	Females
Spread of horn cores, tip to tip	10.7	9.2
Horn core length, upper curve, tip to burr	14.5	12.2
Straight line distance, tip to burr, dorsal horn core	13.6	11.2
Dorso-ventral diameter, horn core base	9.3	8.2
Minimum circumference, horn core base	8.9	8.1
Width of occipital at auditory openings	5.4	4.1
Width of occipital condyles	5.5	5.7
Depth, nuchal line to dorsal margin of foramen magnum	7.8	6.6
Antero-posterior diameter, horn core base	9.6	9.0
Least width of frontals, between horn cores and orbits	6.3	4.8
Greatest width of frontals at orbits	5.4	4.4
$M^1 - M^3$, inclusive alveolar length	4.4	2.8
M^3, maximum width, anterior cusp	6.6	8.1
Distance, nuchal line to tip of premaxillae	3.4	2.7
Distance, nuchal line to nasal-frontal suture	5.6	4.7
Angle of divergence of horn cores, forward from sagittal	6.8	6.7
Angle between foramen magnum and occipital planes	6.2	4.7
Angle between foramen magnum and basioccipital planes	5.6	5.2

Note: These data have been calculated from the coefficients of variation presented in table 50.

Table 53

Mean Coefficients of Variation: Limb Characters

Standard measurement	Males	Females
Humerus:		
Approximate rotational length of bone	4.3	4.5
Antero-posterior diameter of diaphysis	5.6	6.1
Transverse minimum of diaphysis	6.5	6.0
Radius:		
Approximate rotational length of bone	3.0	3.8
Antero-posterior minimum of diaphysis	5.0	6.6
Transverse minimum of diaphysis	4.7	8.0
Metacarpal:		
Total length of bone	3.4	3.1
Antero-posterior minimum of diaphysis	5.5	5.5
Transverse minimum of diaphysis	7.2	6.8
Femur:		
Approximate rotational length of bone	3.4	3.8
Antero-posterior diameter of diaphysis	6.0	5.1
Transverse minimum of diaphysis	4.1	5.2
Tibia:		
Approximate rotational length of diaphysis	3.2	3.6
Antero-posterior minimum of diaphysis	5.4	6.1
Transverse minimum of diaphysis	5.1	5.3
Metatarsal:		
Total length of bone	3.0	3.3
Antero-posterior minimum of diaphysis	5.2	5.6
Transverse minimum of diaphysis	6.3	6.7

Note: These data have been calculated from the coefficients of variation presented in table 51.

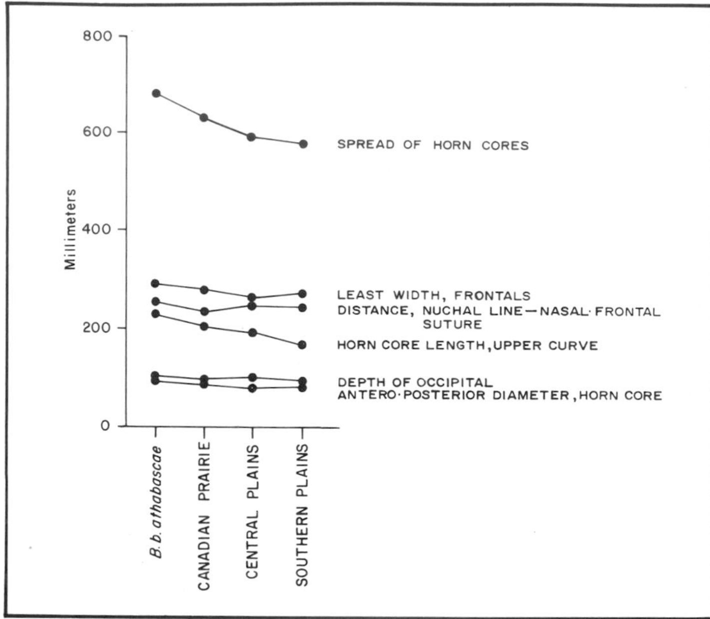

FIGURE 69 *Bison bison* choroclines: male skull characters. Six character clines document a general north–south gradient for the species, as claimed earlier by Figgins (1933). The data presented here indicate that not all characters change at the same rate, undergo the same amount of change, or continuously change in the same direction throughout the cline.

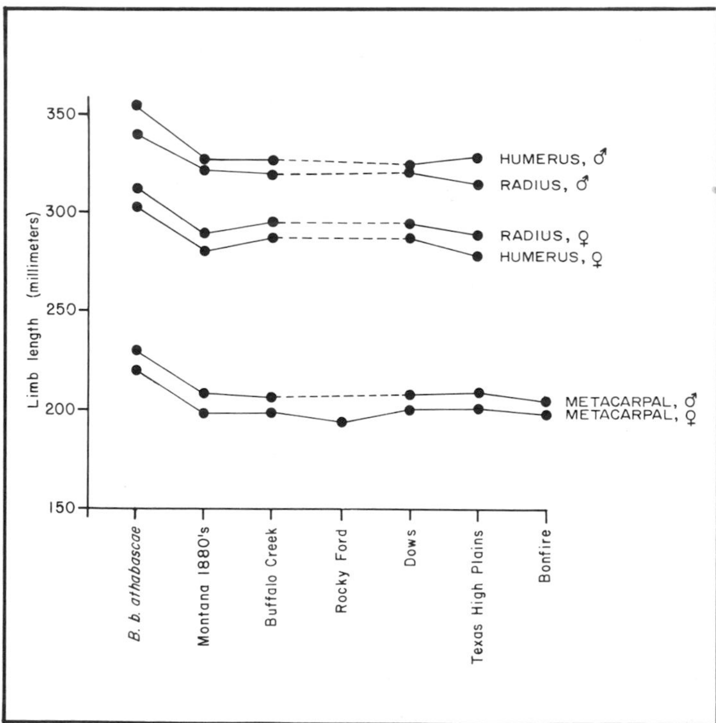

FIGURE 70 *Bison bison* choroclines: limb characters. Limb characters also decrease from north to south through the primary *Bison bison* range, but not as continuously as skull characters.

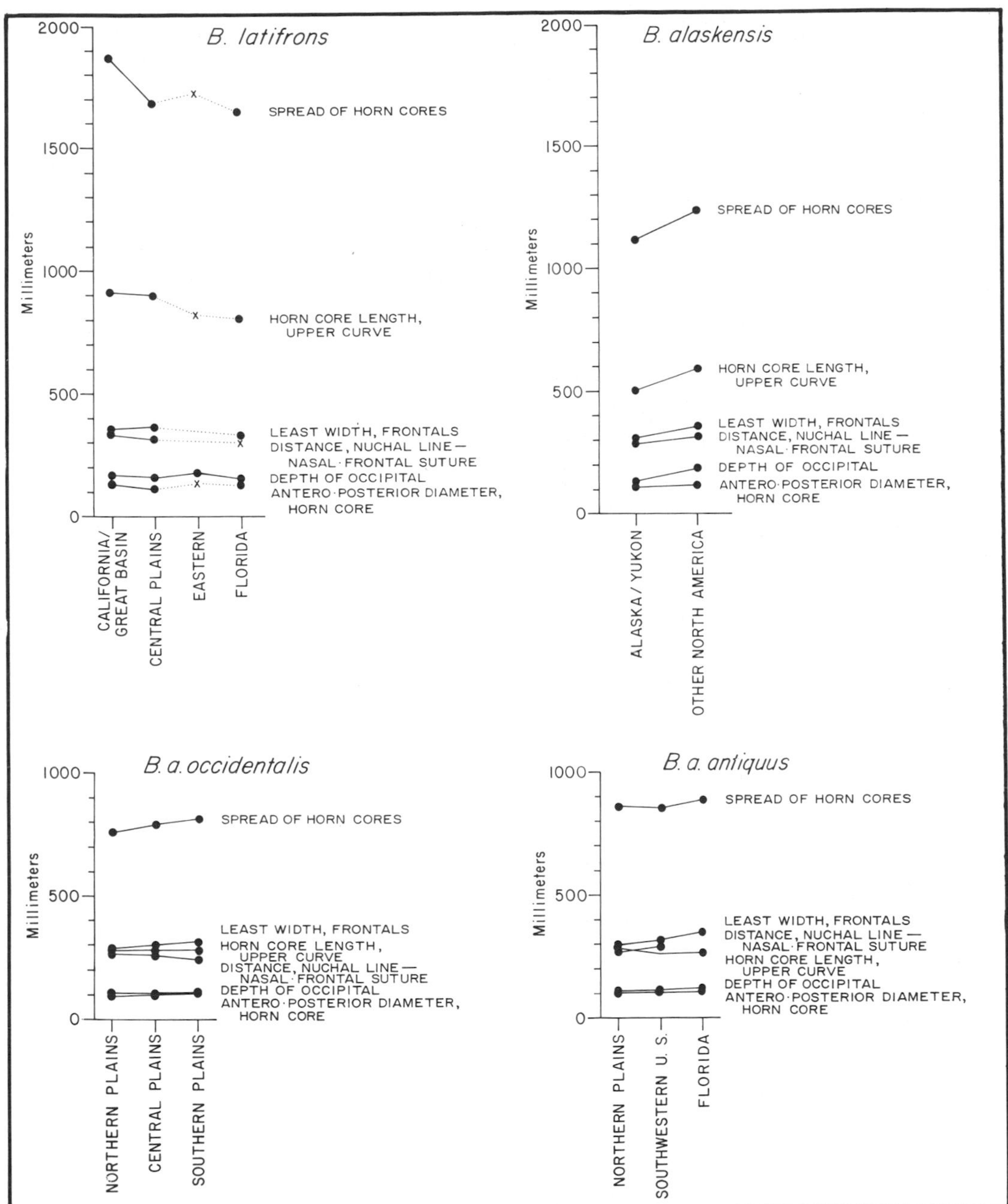

FIGURE 71 Choroclines in taxa other than *Bison bison*. These chorolines are oriented so as to represent the major axis of each taxon's range. Two axes are presented for *Bison antiquus antiquus* (southwest Florida and southwest–northern plains).

PERSPECTIVE

General aspects of variation in the bison skeleton have been documented and illustrated above. The range of variation, the distribution of cases, skewing, and the relative variability of characters are each important qualities of bison populations. These qualities are also of substantial evolutionary importance. The range over which cases within a variate sample are distributed is an unweighted measure of the phenotypic potential of a population. The pattern of case distribution is a weighted measure of relative phenotypic fitness within the population. The positive and negative tendencies (skewing) of a sample may, if the sample is truly representative of the natural population, serve as a weighted measure of (1) the taxon's adaptive potential, should a shift in selection factors occur; and (2) the direction from which an adaptive shift came, if a shift in selection factors occurred in the geologically recent past (fig. 72).

Pearson's coefficient of variation is a weighted measure of a character's plasticity. Theoretically, more variable characters will more easily respond to new forces of selection than will less variable characters. Accordingly, characters with high Vs should most readily (and extensively?) respond to differing selective forces, whereas the least variable characters should respond most slowly to differing selective forces. Less variable characters, therefore, should act as a brake on the extent of evolution (and the rate of adaptation). In newly ordered selection regimes, then, a dual phenotypic response may be envisioned where highly variable characters respond more rapidly and extensively to selection at the same time that less variable characters respond more slowly and less extensively. The intensity of new selection forces and the capacity of bison populations to respond to these forces determine whether or not adaptation will be possible and, if so, the rate at which that adaptation will take place.

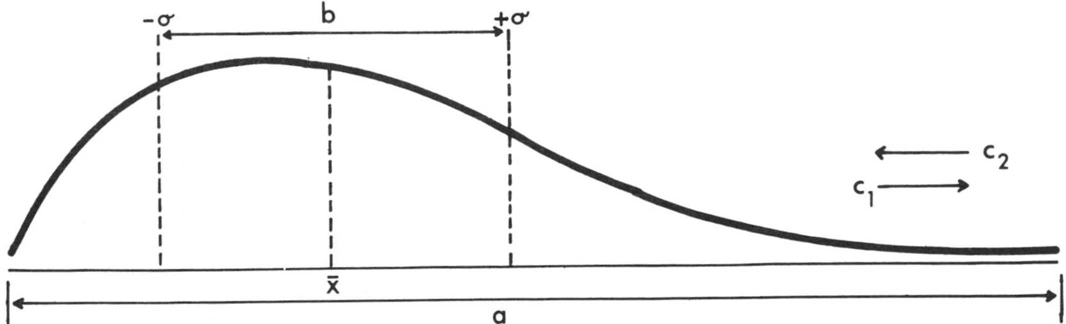

FIGURE 72 Measures of adaptive potential.

a. Range of variation: an unweighted measure of a population's potential for generating a character of such magnitude.

b. Mean and standard deviation of case distribution within the sample: a weighted measure of the population's fitness (i.e., the fitness of any given character).

c. 1. Skew indicating adaptive potential: an indicator of a population's phenotypic potential to shift toward a larger or smaller character should a shift in selective factors occur which favored a larger or smaller character.
   2. Skew indicating recent adaptive shift: an indicator of a recent phenotypic shift from a larger or smaller character to the current mean character, following upon a change in selective factors which effected the phenotypic shift.

CHAPTER 4

ADAPTIVELY DIFFERENTIATED SKELETAL CHARACTERS

The operation of different selection regimes (= environments) on populations of bison resulted, over relatively brief periods of geologic time, in the evolution of adaptively differentiated, biologically homeostatic bison species. This chapter identifies what appear to be adaptively significant morphological differences among bison taxa.

The bison skeleton, in general, changed very little during the Pleistocene and Holocene except for size. Characters that show significant shape differences among taxa are restricted to the skull and primarily consist of social organs or adjustments to feeding and visual communication in different habitat types. Accordingly, the social and ecological significance of specialized skull characters receive most attention here. Adaptation, however, comprises the complete organism, not merely isolated characters, and reflects an integrated coevolution of morphology, behavior and ecology, not a stepwise progression of one of these categories followed sequentially by others. Also, the phyletic distinctiveness of the North American and Eurasian lineages must be kept in mind; similar environments selected for parallel adaptations (= convergence) in both lineages, but exact duplication of form among vicariant taxa did not result.

HORN CORES

Living bison are dominance organized and the horns are important display paraphernalia that function in the nonfighting dominance and threat behavior of male bison. When alternate less violent forms of dominance behavior fail to establish relative ranking among competing males, actual fighting normally ensues. Fighting consists of two elements: the head-on clash, which terminates in a push, lunge, or charge directly at the opponent's head, and a hooking motion in which the horns are normally thrust upward and inward toward a vulnerable region of the opponent's body. Use of the horns in fights is normally (but not exclusively) confined to the head region where offensive and defensive maneuvering are centered. Infrequently, however, the defense maneuvering fails and injurious (sometimes fatal) wounds are delivered to the body of an opponent (Herrig and Haugen, 1969; Lott, 1972*a*, 1972*b*, 1974; McHugh, 1958). The horn is also used in living bison as a weapon of interspecific aggression and can be directed at both competitors and predators (Mahan, 1977; McHugh, 1958, 1972; Roe, 1951; pers. obscrv.). Horns may have been more important to earlier bison than to modern bison, considering the probability that body hair (which also functions as a dominance organ in living bison) may have been less prominent, that hooking may have been relatively more important than clashing in fighting behavior, and that interspecific competition may have been more intense in earlier than later bison.

Horn cores present the most obvious and extreme morphological difference among bison taxa (figs. 63 and 64). The most critical aspects of functional horn core morphology are overall horn size and tip placement. The size of horns probably determines their relative effectiveness in various types of dominance encounters. Tip placement is probably the ultimate measure of horn morphology fitness in actual fights and is also probably instrumental in determining the relative viability of fighting techniques. The length, surface area, weight (or volume), configuration, and tip placement are considered here as adaptively significant qualities of horn core morphology.

Horn core length is an important attribute of horn core size and a determinant of tip placement. The basal circumference of the horn core is strongly correlated with horn core length in both sexes, but the ratio between circumference and length is consistently greater in males than in females (fig. 73). Sexual dimorphism is also evident in the basal diameters of male and female horn cores, with males having a proportionately greater antero-posterior diameter than females (fig. 74).

Surface area and volume are important functional attributes of horns in living bison, particularly among

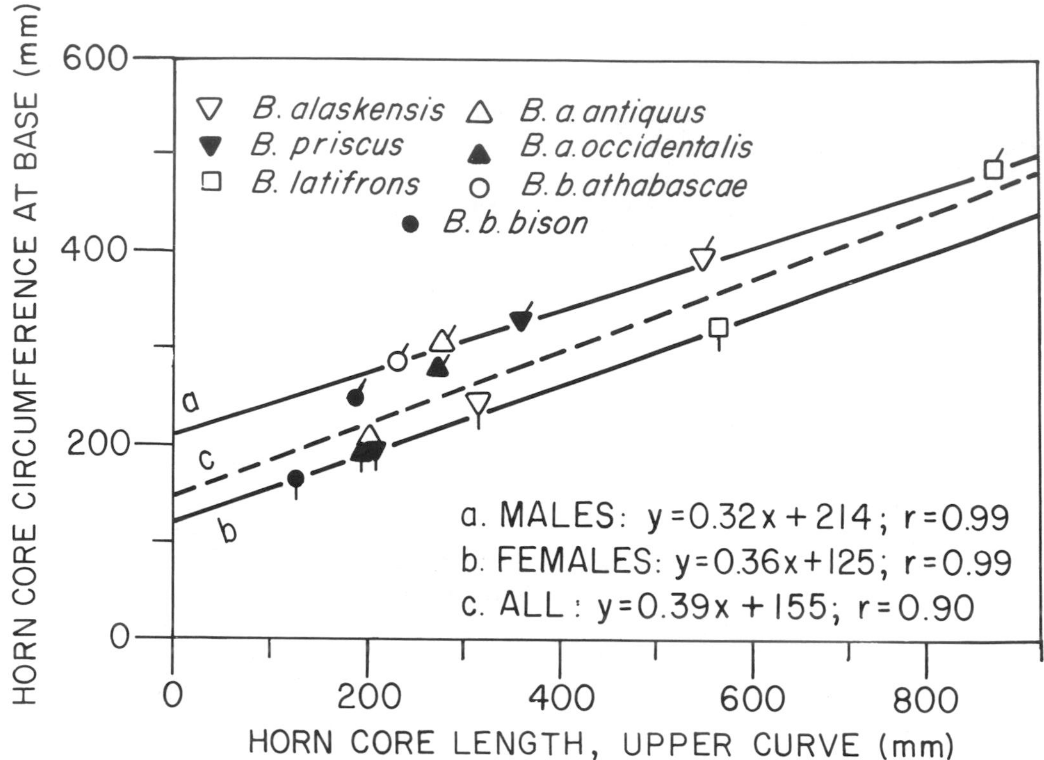

FIGURE 73 Horn core length versus basal circumference. The proportionately greater circumference: length ratio among male horn cores imparts greater surface area, volume, and weight to the male than female horn core.

males who use horns much more frequently and damagingly than females in dominance and aggressive activities. Area, particularly profile area, can conceivably function in dominance and aggression activities in two important ways. First, horn core area conveys an impression of relative size (ideally, an impression of greater than actual size), and hence relative dominance of the organism. Second, both the anterior and dorsal profile areas block, deflect, or absorb blows from an opponent's horns.

Greater horn core volume (hence weight) is an advantage in dominance and aggression activities because it presents greater surface area and can contribute to greater absolute force of delivery with a hooking blow. Larger horns, however, impose certain costs to the organism in excess of those required of bison with smaller horns. Added

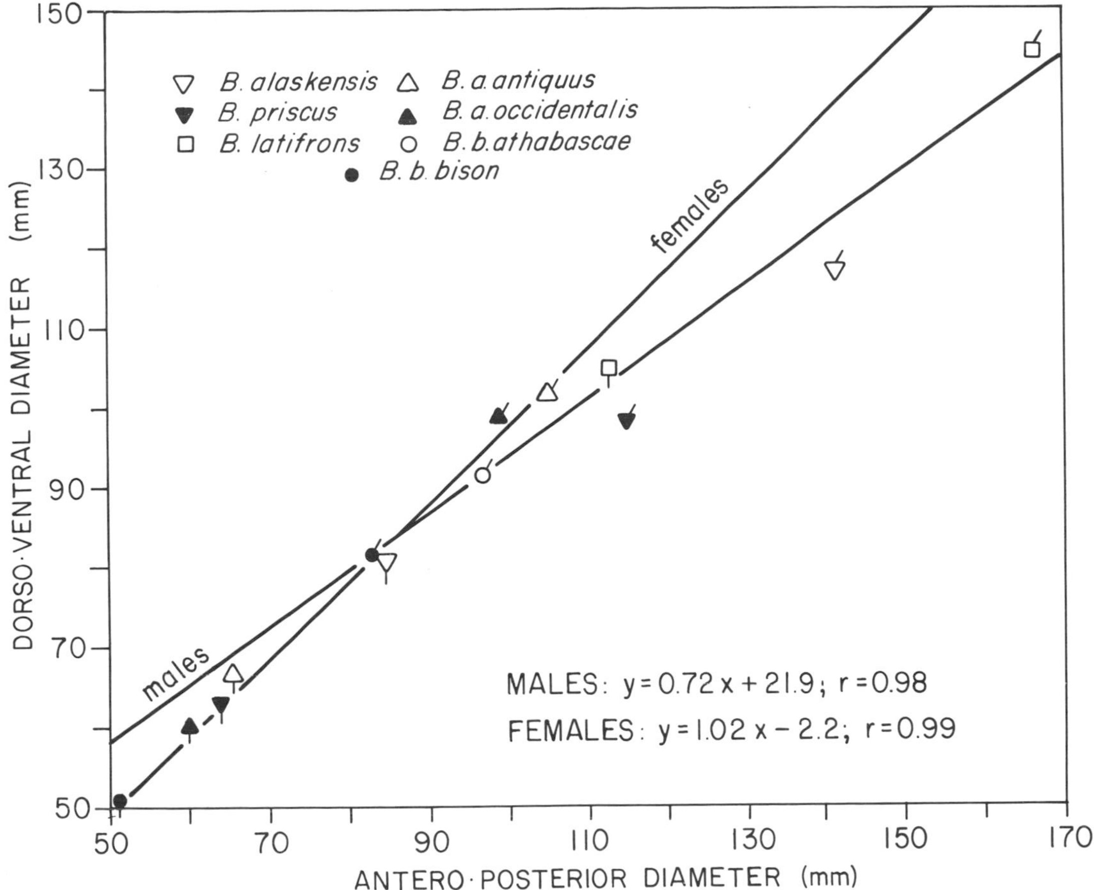

FIGURE 74 Covariation among horn core basal diameters. The proportionately greater antero-posterior diameter for males imparts relatively greater dorsal surface area to the horn.

energy demands and specialized skeletal requirements can be envisaged stemming from such routine activities as raising and lowering the head to feed, water, or check for the presence of competitors or predators; walking or running; and fighting. Smaller horns would substantially reduce the cost to the organism engaging in the activities just mentioned, but would present less surface area to intercept an opponent's offensive hooks, less weight with which to strike an opponent, and less reach with which to stab an opponent. Figure 75 illustrates the differences of scale among linear, areal, and volumetric dimensions for horn cores, and tables 54 and 55 compare the areas, volumes, and allometries of horn core characters.

Horn tips are the critical offensive attribute, so the orientation of the horns and placement of the horn

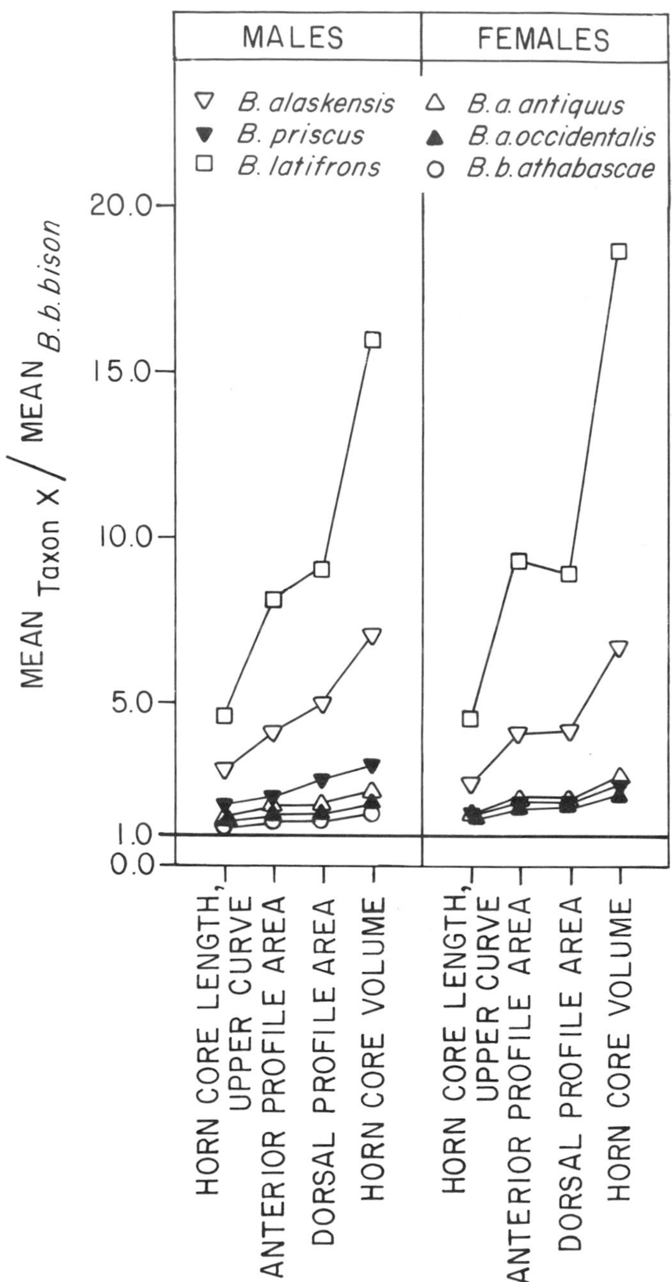

FIGURE 75 Scaled differences: horn core characters. The vertical axis expresses the mean for each taxon as a multiple of the *Bison bison bison* mean.

Table 54

Comparative Horn Core Biometrics: Areas and Volumes

Taxon	Anterior profile area (mm^2)[a]	Dorsal profile area (mm^2)[b]	Volume (mm^3)[c]
Males			
B. b. bison	7,809	7,952	341,012
B. b. athabascae	10,756	11,426	547,405
B. a. occidentalis	13,140	13,723	679,750
B. a. antiquus	14,225	14,742	786,542
B. latifrons	63,466	72,139	5,473,117
B. priscus	17,827	20,837	1,072,481
B. alaskensis	32,345	39,201	2,401,471
Females			
B. b. bison	3,177	3,195	85,668
B. b. athabascae	......	......	....
B. a. occidentalis	5,952	5,943	187,936
B. a. antiquus	6,746	6,675	232,433
B. latifrons	29,864	29,046	1,610,575
B. priscus	6,622	6,684	221,910
B. alaskensis	12,960	13,502	574,060

Note: Calculations are based on data presented in skull biometric summary tables, chapter 2. The mean value for selected standard measurements (SMs) for each taxon have been used to determine the values presented in this table.

[a]The anterior silhouette area was determined by the formula SM 3 X SM 6 X ½.

[b]The dorsal silhouette area was determined by the formula SM 3 X SM 12 X ½.

[c]The volume was determined by the formula SM 3 X SM 6 X SM 12 X 1/12.

Table 55

Allometry of Horn Core Characters Relative to *Bison bison bison*

Taxon	Horn core length, upper curve	Anterior profile area	Dorsal area	Volume
Males				
B. b. athabascae	1.23	1.37	1.44	1.61
B. a. occidentalis	1.46	1.68	1.73	1.99
B. a. antiquus	1.46	1.82	1.85	2.31
B. latifrons	4.59	8.13	9.07	16.05
B. priscus	1.90	2.28	2.62	3.14
B. alaskensis	2.90	4.14	4.93	7.04
Females				
B. b. athabascae	....	....	....	.....
B. a. occidentalis	1.59	1.87	1.86	2.19
B. a. antiquus	1.63	2.12	2.09	2.71
B. latifrons	4.54	9.40	9.09	18.80
B. priscus	1.68	2.08	2.09	2.59
B. alaskensis	2.57	4.08	4.23	6.70

Note: The *B. b. bison* mean for each character = 1.00. Values given in the table are multiples of the *B. b. bison* mean. Calculations are based on data presented in tables 48 and 54.

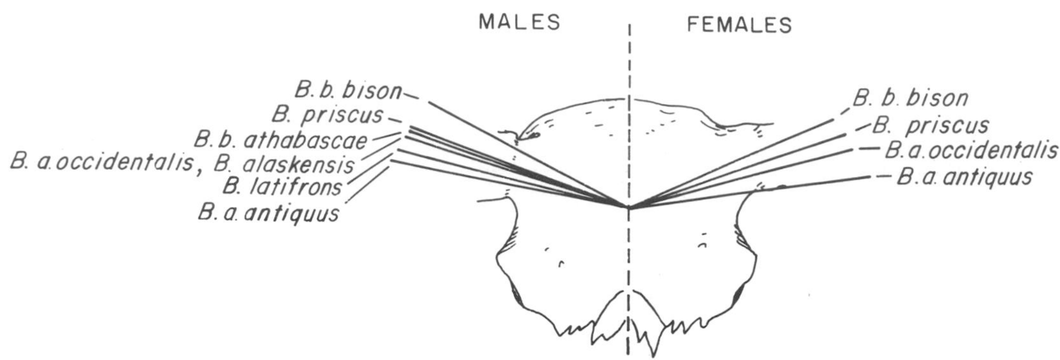

FIGURE 76 Horn core divergence forward from the sagittal plane. Longer horned taxa have greater forward divergence of horn cores than do shorter horned taxa in both the Eurasian and North American lineages.

tips is very important in determining the fitness of horn morphology and the mode of fighting behavior involving use of the horns. Three characteristics of horn tip placement are considered here, including (1) the angle at which horn cores emanate from the frontals relative to the sagittal plane; (2) the placement of the horn core tip relative to the occipital plane; and (3) the placement of the horn core tip relative to the frontal plane. The mean angle of horn core divergence forward from the sagittal for both sexes of each taxon is illustrated in figure 76, and the strong similarity of this divergence between both sexes of each taxon is illustrated in figure 77. These figures show that the angle of horn core divergence strongly affects the placement of the horn tips relative to the occipital plane, and, usually, in both the North American and Eurasian lineages the longer the horn core, the greater is the angle of divergence forward from the sagittal plane. Male *B. latifrons* and *B. a. antiquus* do not exactly fit this pattern, but both have horn cores oriented far forward, probably near the maximum functional forward orientation for horns that are deflected outward and upward.

The placement of the horn tips relative to the occipital and frontal planes is illustrated in figures 78 and 79. In the living animal the horn core would be covered with a keratin sheath, which would have a threefold effect on the placement of the horn tips relative to the horn core tips. (1) In all cases, the horn tips would be positioned rearward from the horn core tips. The extent of this rearward extension generally varies in proportion to the length of the horn core. Rearward rotation of the horn cores and posterior deflection of the horn core tips would further accentuate this rearward extension. (2) The horn tips would be positioned higher than the horn core tips. (3) The distance between the tips of the (recurved) horns would be

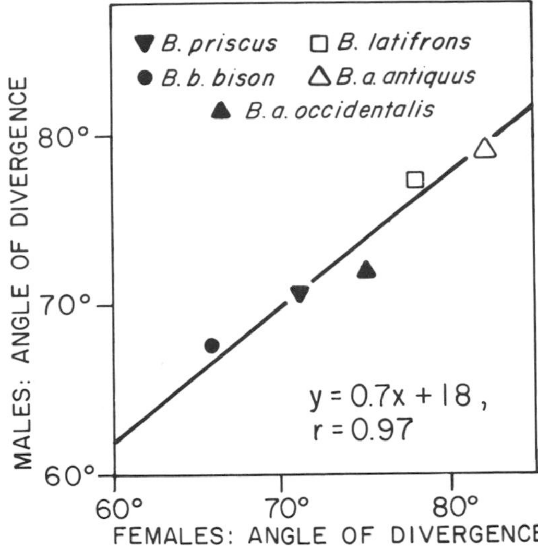

FIGURE 77 Correspondence between sexes of horn core divergence forward from sagittal. Males tend to have a slightly more posterior orientation of horn cores than do females.

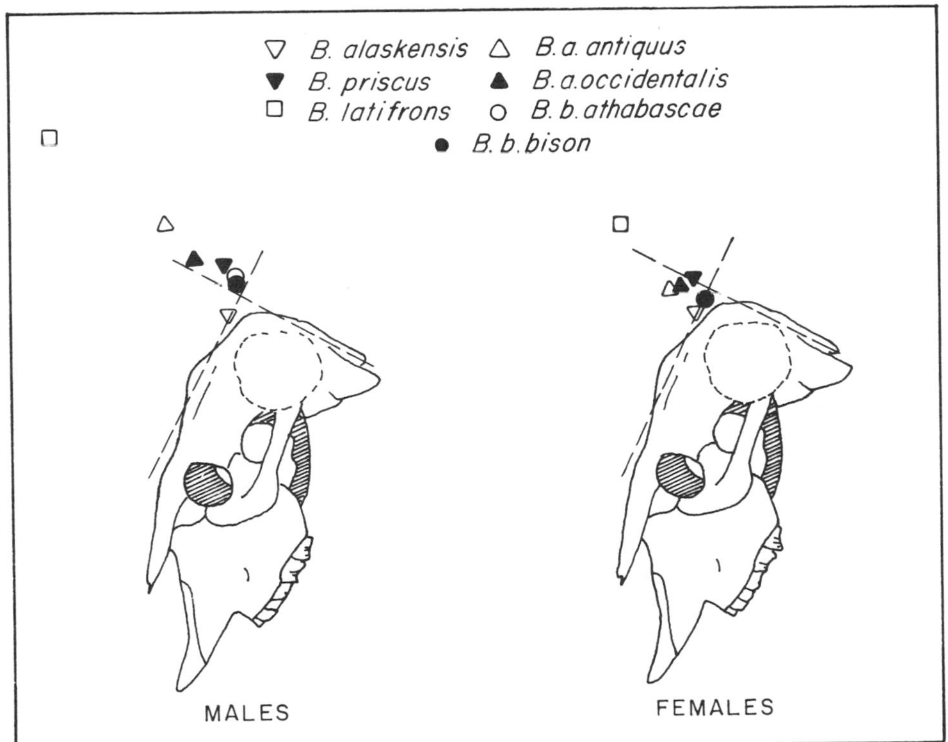

FIGURE 78 Horn core tip placement relative to the frontal plane: lateral view. Male horn core tips are placed higher and further to the rear than are female tips.

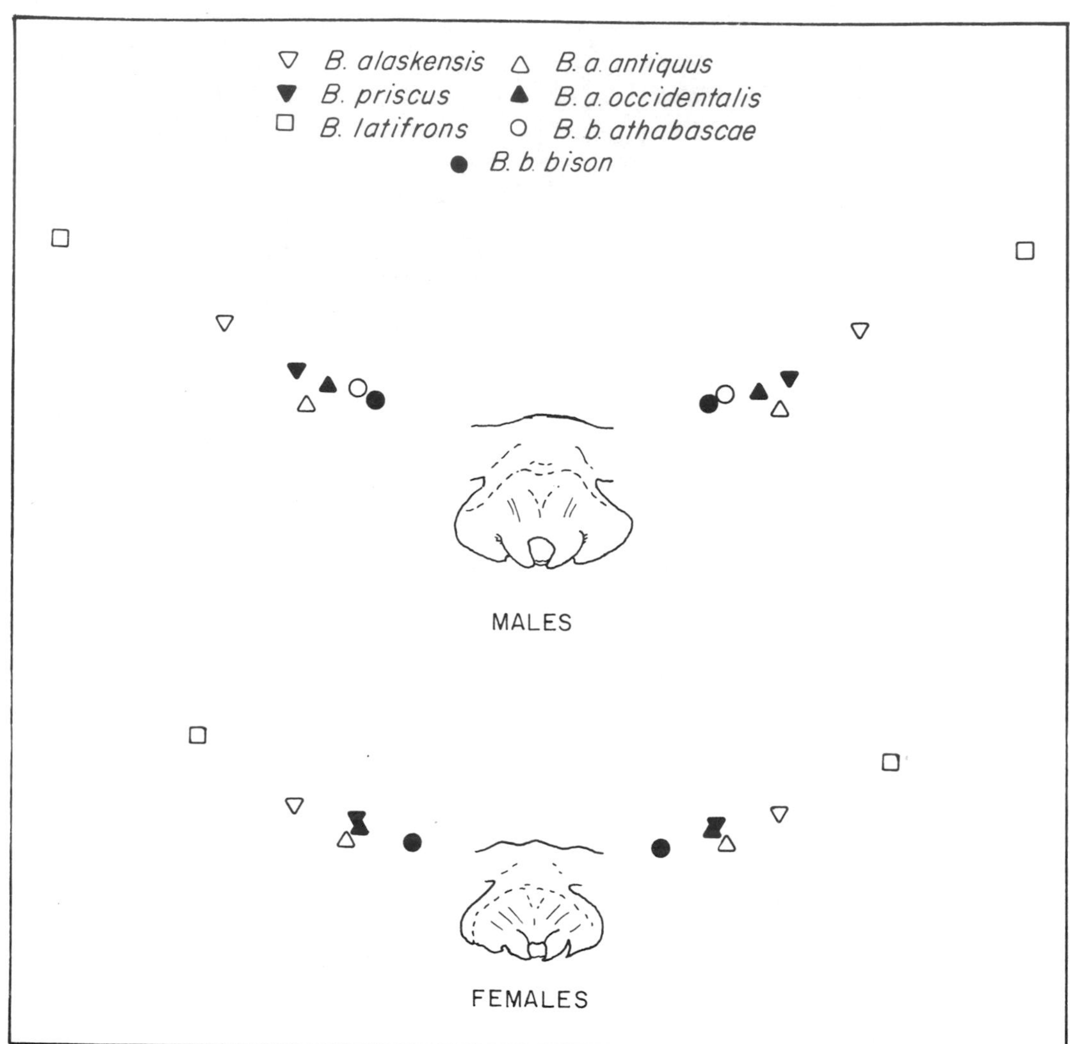

FIGURE 79 Horn core tip placement relative to the frontal plane: sagittal view. Male horn core tips are placed higher than are female tips. There is also a correlation between horn core length and height of tip placement above the frontal plane.

less than that between the tips of the horn cores. The net effect of the sheath relative to the horn cores, therefore, is to place the horn tips further above and behind the frontal-occipital region and to reduce the functional distance between the tips. The combined effects of recurved tips situated high above the frontals in larger horned bison suggests that butting was not an important aspect of dominance or fighting behavior in *B. latifrons* and *B. alaskensis*, and perhaps *B. priscus*. *B. bison*, and to a lesser extent *B. a. antiquus*, have horn tips situated lower on the skull and spaced wider relative to their height above the frontal plane, characters that are more suited to a clashing mode of fighting behavior than are the horn characters of larger bison.

In summary, the interrelated variables of length, area, volume, and tip placement are considered major characters of bison horn cores with overt adaptive significance. Larger horns are considered better adapted (1) to lower population densities where relatively few agonistic encounters (especially escalated fighting) occur; (2) to dominance and fighting behavior characterized by hooking rather than clashing; and (3) to situations of interspecific competition where size (especially where perceived size exceeds actual size) and actual aggressive ability convey net advantages in resource acquisition and defense. Smaller horns are considered better adapted (1) to higher population densities where dominance linked agonistic encounters are frequent; (2) to situations of interspecific competition where alternate ranking organs (e.g., body hair) or strategies (e.g., head-on clashes) have evolved to reduce emphasis on the ultimate and more lethal or damaging use of horns as weapons; and (3) to environments with more abundant resources and reduced absolute competition (thus less need for direct aggression between competitors).

FRONTO-FACIAL REGION

The fronto-facial region of the skull contains characters that differ sufficiently among bison taxa and which correlate with other adaptive features of the skeleton to suggest that the differences are of adaptive significance. Prominent among these characters are the nasals, the orbits, and the frontals—characters that collectively compose the dorsal surface of the bison skull. The shape and orientation of these characters bear on the phenotype's observational ability and ability to engage in and withstand clashes in fighting.

Mean size differences obviously existed among the skulls of different taxa of bison, but the linear relationships of gross fronto-facial characters remained generally constant (figs. 63 and 64, 80; table 48). These size differences appear to be more strongly correlated with size differences in the overall phenotype than to represent specific adaptive differences as such. Between sexes, the frontals of males are absolutely and proportionately wider and shorter than those of females (table 56.)

Differences in the shape of characters are the most important adaptations of the fronto-facial region. The frontals are more strongly domed transversely between the horn cores in shorter horned taxa (*B. bison* and *B.*

Table 56

Comparative Skull Biometrics: Indexes, Areas, and Volumes

Taxon	Frontal breadth: frontal length[a]	Frontal breadth: skull length[b]	Frontal area (mm^2)[c]	Dorsal skull area (mm^2)[d]	Occipital area (mm^2)[e]	Skull volume (mm^3)[f]	Molar occlusal area (mm^2)[g]	Molar occlusal area: frontal breadth[h]
Males								
B. b. bison	1.10	.51	66,609	72,560	37,814	3,371,719	2,509	9.25
B. b. athabascae	1.15	.51	75,110	84,880	42,805	4,127,837	2,558	8.72
B. a. occidentalis	1.14	.53	76,968	83,686	42,801	4,025,440	2,763	9.31
B. a. antiquus	1.11	.50	89,343	98,973	50,469	5,290,846	3,147	10.00
B. latifrons	1.09	.50	116,011	127,126	63,655	7,596,148	3,273	9.22
B. priscus	1.08	.48	80,003	89,285	44,659	4,525,520	.....	.....
B. alaskensis	1.11	.51	99,098	107,982	52,786	5,721,128	.....	.....
Females								
B. b. bison	1.02	.45	46,027	52,517	26,750	2,161,005	2,132	9.84
B. b. athabascae	....	...	......	......	......	....	.....	.....
B. a. occidentalis	1.05	.46	54,143	62,349	29,662	2,588,003	2,364	9.92
B. a. antiquus	1.06	.46	65,200	75,818	38,994	3,745,087	2,770	10.50
B. latifrons	1.12	...	73,296	......	43,750	....	3,116	10.93
B. priscus	.97	...	53,583	......	31,304	....	.....	.....
B. alaskensis	1.07	.45	65,026	74,750	40,841	3,913,901	.....	.....

Note: Calculations are based upon data presented in skull biometric summary tables, chapter 2. The mean value for selected standard measurements (SMs) for each taxon have been used to determine the values presented in this table.

[a] This index was determined by the formula SM 14÷SM ON.

[b] This index was determined by the formula SM 14÷SM OP.

[c] The frontal silhouette area was determined by the formula SM 14 X SM ON.

[d] The dorsal skull silhouette area was determined by the formula SM 14 X SM OP X ½.

[e] The occipital silhouette area was determined by the formula SM 8 X SM 10 X $\pi/2$.

[f] The skull volume was determined by the formula SM 8 X SM 10 X SM OP X $\pi/12$.

[g] The molar occlusal silhouette area was determined by the formula SM 20 X SM 20a.

[h] This index was determined by the formula (SM 20 X SM 20a) ÷ SM 14.

antiquus) than in larger horned taxa (*B. priscus, B. alaskensis* and *B. latifrons*) (fig. 81). Domed frontals are usually more strongly and completely fused along the frontal and fronto-parietal sutures than are flatter frontals, and the supraorbital nutriment foramen is usually more completely covered in domed than in flatter frontals. The pneumatic sinus protection of the brain may also be greater in skulls with domed frontals than those with flatter frontals. Relative flattening of the nasals (i.e., broadening in relation to height above the maxillae), which correlates with doming, may constitute another adaptation of the dorsal skull of shorter horned bison.

The adaptive significance of the frontal doming, nasal broadening tendency can be related to head-to-head clashes and subsequent head wrestling and hooking between fighting males (plates 22 and 23). Doming reinforces the frontal region against potentially damaging consequences of fighting in several ways. Relative to flatter frontals, domed frontals provide greater durability against the impact of clashing, more efficient dissemination of the force of impact away from the point of impact, greater cushioning of impact by allowing the skull to rotate forward a short distance after initial impact as deceleration takes place, greater protection of the brain, and greater resistance to the hooking blows of an opponent's horn tips. The reinforced dome allows an aggressor to put relatively more force behind a charge when intending to collide with an opponent head-on, and to concentrate that force into more of a point than could be possible with flat frontals. The relatively flattened nasals are able

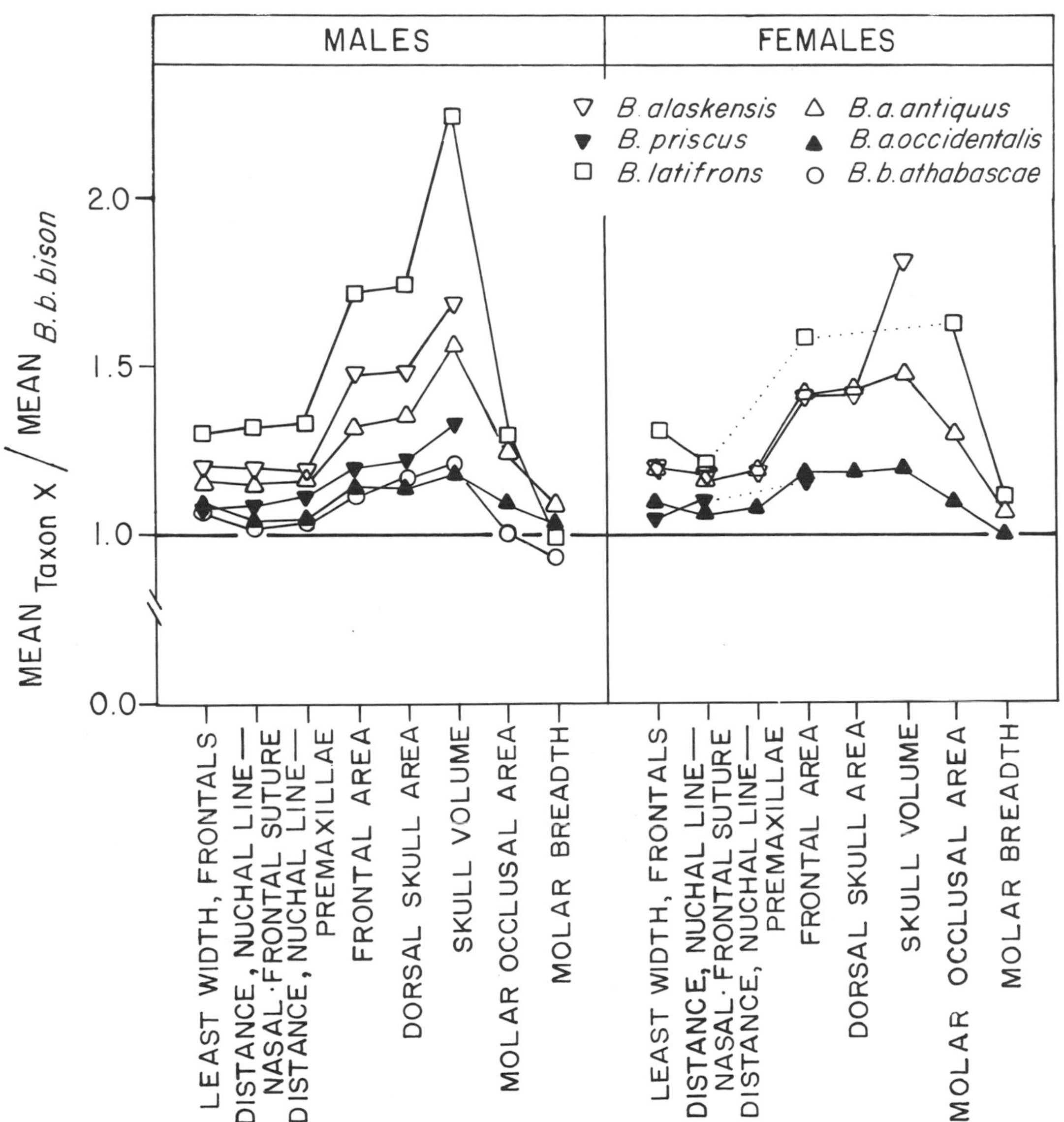

FIGURE 80 Scaled differences: fronto-facial characters. The vertical axis expresses the mean for each taxon as a multiple of the *Bison bison bison* mean.

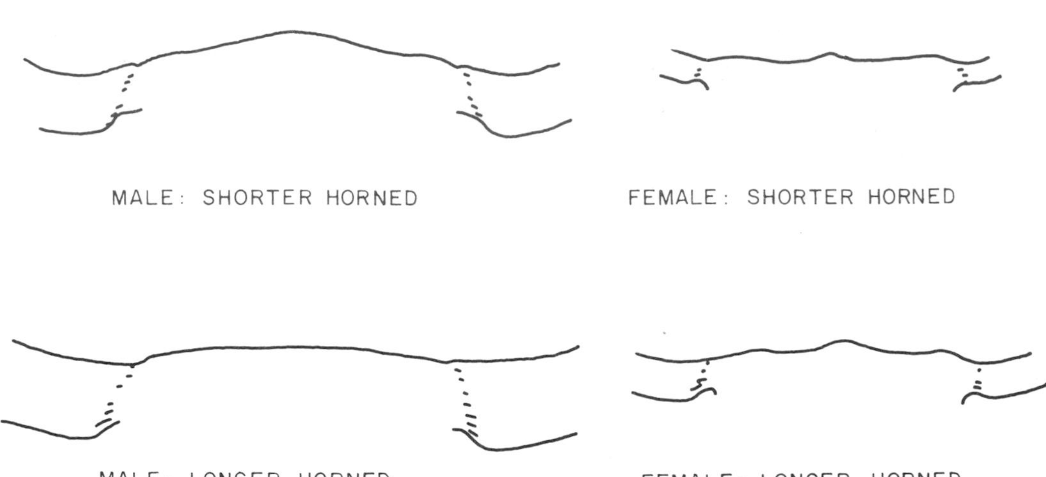

FIGURE 81 Frontal shapes: transverse profiles. The differences among frontal profiles are especially great between longer and shorter horned male bison.

to more effectively sustain the impact of collisions and deflect the hooking horns characteristic of the clashing mode of fighting behavior than are higher, narrower nasals.

Both domed frontals and lowered, broadened nasals occasionally occur in larger horned bison. Consequently, these characters have been available in all taxa and subject to selection. The extreme development of these characters in *B. bison*, and the infrequency of their occurrence in *B. latifrons* suggests that domed frontals and lowered, broadened nasals may represent adaptive differ ences that accompanied differences in fighting behavior which probably existed between longer and shorter horned bison.

Differences in orbital orientation were observed in skulls. In pre-Holocene bison and many *B. a. occidentalis* skulls the orbits tended to be placed higher (i.e., dorsally) on the cranium and to be directed more forward than is true for *B. bison*. This observation generally concurs with Flerow's (1971) observation on evolutionary change in relative orbital orientation. This observation suggests that pre-Holocene bison had relatively forward-oriented vision and vision requirements, whereas *B. bison* has more laterally oriented vision and vision requirements.

Differences in the absolute and relative protrusion of the orbits also occur among bison taxa. Absolute orbital protrusion is the maximum distance the posterior orbit margin extends laterally from the frontal region. Relative orbital protrusion expresses the narrowest distance across the frontals (standard measurement 14) as a multiple of the greatest width of the frontals at the posterior orbits (standard measurement 15). The greater the relative orbital protrusion, the less protruding are the orbits relative to the frontal width.

Mean absolute and relative orbital protrusion values for each taxon

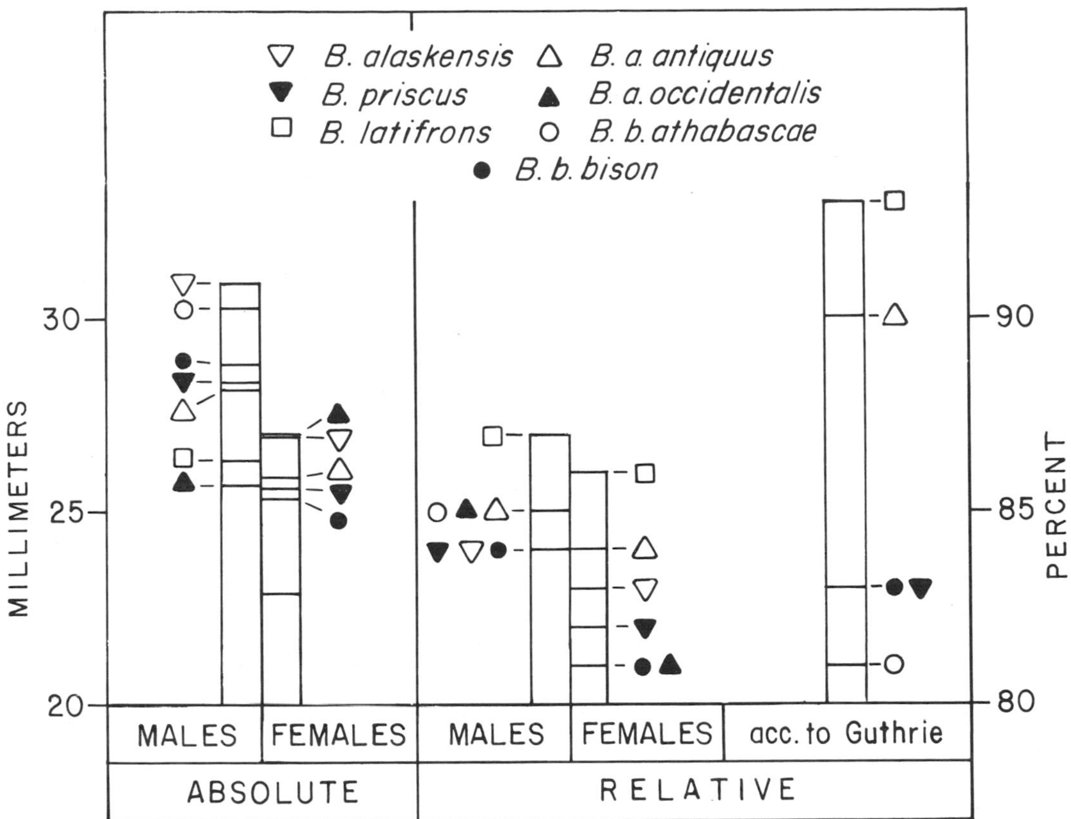

FIGURE 82 Actual and relative orbital protrusion. Actual orbital protrusion is the absolute distance the orbits extend outward from the frontals. Relative orbital protrusion, an index defined by Guthrie (1966*b*), expresses the least width of the frontals as a multiple of the greatest width of the frontals at the orbits. The greater the relative protrusion index, the less the orbit protrudes relative to the frontal width. Guthrie's original indices of relative orbital protrusion, based upon data in Skinner and Kaisen (1947), are presented for comparative purposes. Differences between Guthrie's and my indices result from differences in sample composition and bison classification.

Table 57
Absolute and Relative Orbital Protrusion

Taxon	Absolute Protrusion[a]	Relative Protrusion[b]
Males		
B. b. bison	28.8 mm	.84
B. b. athabascae	30.3 mm	.85
B. a. occidentalis	25.7 mm	.85
B. a. antiquus	28.3 mm	.85
B. latifrons	26.4 mm	.87
B. priscus	28.4 mm	.84
B. alaskensis	31.9 mm	.84
Females		
B. b. bison	25.4 mm	.81
B. b. athabascae	....	...
B. a. occidentalis	27.1 mm	.81
B. a. antiquus	25.9 mm	.84
B. latifrons	22.9 mm	.86
B. priscus	25.6 mm	.82
B. alaskensis	27.0 mm	.83

Note: Calculations are based on data presented in skull biometric summary tables, chapter 2.
[a]The absolute distance the orbit extends outward from the skull was determined by the formula (SM 15 – SM 14) X 1/2.
[b]The relative protrusion of orbits is determined by the formula SM 14 ÷ SM 15. The larger the index, the less the protrusion (see Guthrie, 1966*b*).

are listed in table 57 and illustrated in figure 82. No special adaptive meaning is apparent in the order of absolute protrusion observed among the various taxa (i.e., absolute protrusion does not correlate with mean body size, inferred behavioral clines, or inferred habitat associations). Differences in relative orbital protrusion among the taxa, however, are ordered in what appears to be an adaptively meaningful way. There is distinct ordering according to mean body size, inferred behavioral clines, and inferred habitat associations. There is no correlation between relative orbital protrusion and latitude, thus no chorocline is represented. At least four explanations can be put forth to account for differences in orbital protrusion.

1. Guthrie (1966*b*) first considered the adaptive significance of relative orbital protrusion and concluded that relative orbital protrusion could serve as an index of the length of pelage on the head, with a lower index indicating longer hair, and a higher index indicating shorter hair; that northern species had a lower index, and therefore longer facial pelage, than did southern species; and that this difference in pelage was due to climatic differences, where northern bison had evolved more pelage with which to withstand the northern climate than was required by southern bison to withstand the southern climate. Guthrie's ordering of taxa, based

on data from Skinner and Kaisen (1947), is illustrated in figure 82. *B. bison*, the obvious anomally in Guthrie's model, was considered by him to have only recently evolved from northern bison and to have not yet completely adapted to the more mild southern climate.

2. Geist (1971*a*) has suggested that there exists an inverse relationship between horn size (which is correlated with body size in *Bison*) and body hair. The magnitudes of relative orbital protrusion indices in bison correlate very well with horn and body size as predicted by Geist. Only the male *B. alaskensis* index is anomalous, equaling, but not exceeding, that of *B. priscus*. This anomally is of minor importance. Although females have proportionately lower indices than males, males have wider frontals and both sexes must accommodate a species specific (yet sexually dimorphic) amount and length of hair. In this explanation body (including facial) hair is viewed as a social organ in addition to a thermoregulatory organ, and selection for body hair is considered a social, as well as extra-social, process. Smaller bodied, shorter horned bison in which body hair is strongly selected for (perhaps as a surrogate for actual body size) would expectedly have lower relative orbital protrusion indices than larger bodied, longer horned bison in which the length of body hair, as a social organ, was probably less and may have been a consequence of stronger selection for larger body size or longer horns.
3. A third explanation for differences in relative orbital protrusion indices results from the possibility that patterns of facial hair development did not vary significantly from taxon to taxon. Assuming that the extent of orbital protrusion is causally related to the amount of facial hair present, there should be little or no meaningful difference in absolute orbital protrusion if the amount of facial hair did not vary from taxon to taxon. According to this view, larger bodied bison should have higher relative orbital protrusion indices than smaller bison, but all taxa should have about the same absolute orbital protrusion.
4. A final explanation results from shifts in the orientation of the orbits from higher, more forward directed orbits in pre-Holocene bison to lower, more laterally directed orbits in Holocene bison. The lateral shifts of the orbits, in conjunction with decreasing overall body size, could exaggerate the relative and absolute orbital protrusion in smaller bison vis-à-vis larger bison.

Any one of the foregoing explanations could conceivably account for the differences in absolute and relative orbital protrusion. All were possibly instrumental in producing the observed pattern, although the third explanation is here considered unlikely. Biometric and distribution data, however, suggest that social selection for greater or lesser body hair and morphological changes in orbital placement and orientation were the most important factors influencing orbital protrusion.

In summary, major adaptive differences in the fronto-facial region exist among bison taxa. Longer horned taxa tend to have relatively flatter and less robust frontals, higher and narrower nasals, more forward and dorsally oriented orbits, and higher relative orbital protrusion indices (i.e., less protruding orbits) than do shorter horned taxa.

MOLAR DENTITION

Molar teeth apparently differ among bison taxa only by size. The data presented here indicate that mean body size and molar occlusal area are correlated in both sexes (table 56: molar occlusal area). The ratio in males between molar occlusal area and the width of the frontals (a rough index of body size) suggests, however, that the occlusal area in larger forest/woodland taxa (*B. latifrons, B. b. athabascae*) is proportionately less than in contemporary smaller savanna/grassland taxa (*B. a. antiquus, B. b. bison*). This is not indicated for females. In all taxa the molar occlusal area:width of frontals is greater in females than males, suggesting that females have greater occlusal area per unit of body weight than males.

The differences between contemporary male forest/woodland and savanna/grassland taxa can be explained as different adaptations to the available forage. Forest/woodland bison would likely feed on less abrasive vegetation than would savanna/grassland bison, and their dentition would therefore wear less rapidly than would the dentition of the savanna/grassland bison. Less molar surface area per unit of body weight would be required in the forest/woodland bison than the savanna/grassland bison. Fossil species were probably longer lived (see chapter 5) than recent bison and therefore possessed a greater occlusal area:body weight ratio as one adaptation permitting longer functional dentition life. More data are needed to support or refute this interpretation based on admittedly limited information. The greater molar occlusal area:body size ratio in females than males can perhaps be attributed to the fact that pregnant and/or nursing females have relatively greater nutritional needs than males and simply need more food per unit of body weight to sustain these critical functions.

OCCIPITAL REGION

The occipital region consists of the posterior wall of the skull, extending from the nuchal line to the basioccipital bones. This region consists of surfaces with which the head is suspended from the neck. Interspecific differences in the contours and alignment of the occipital features appear to have adaptive significance, although some may result from direct selection for other characters. Differences in the shape of the occipital condyles and the angles formed by the occipital plane and the basioccipital axis with the foramen magnum plane are of probable adaptive significance. Intertaxa differences in the occipital area, the relative concavity or flatness of the occipital and the extent of development of the nuchal line and external occipital protuberance are probably evolutionary responses to selection for other characters, such as overall body size and head weight.

The relationship between occipital linear dimensions and area, and overall skull volume is summarized in figure 83 and tables 56, 58, and 59. Larger bodied taxa do tend to have more concave occipitals than smaller taxa, a characteristic resulting from the relatively greater development of the nuchal line/external occipital protuberance and the protrusion of this muscle/ligament attachment crescent and the occipital condyles beyond the occipital plane (plate 24). The greater development of the nuchal line and external occipital protuberance in larger bodied bison taxa indicates that muscles and ligaments normally attached along this line were more extensive than in smaller bodied

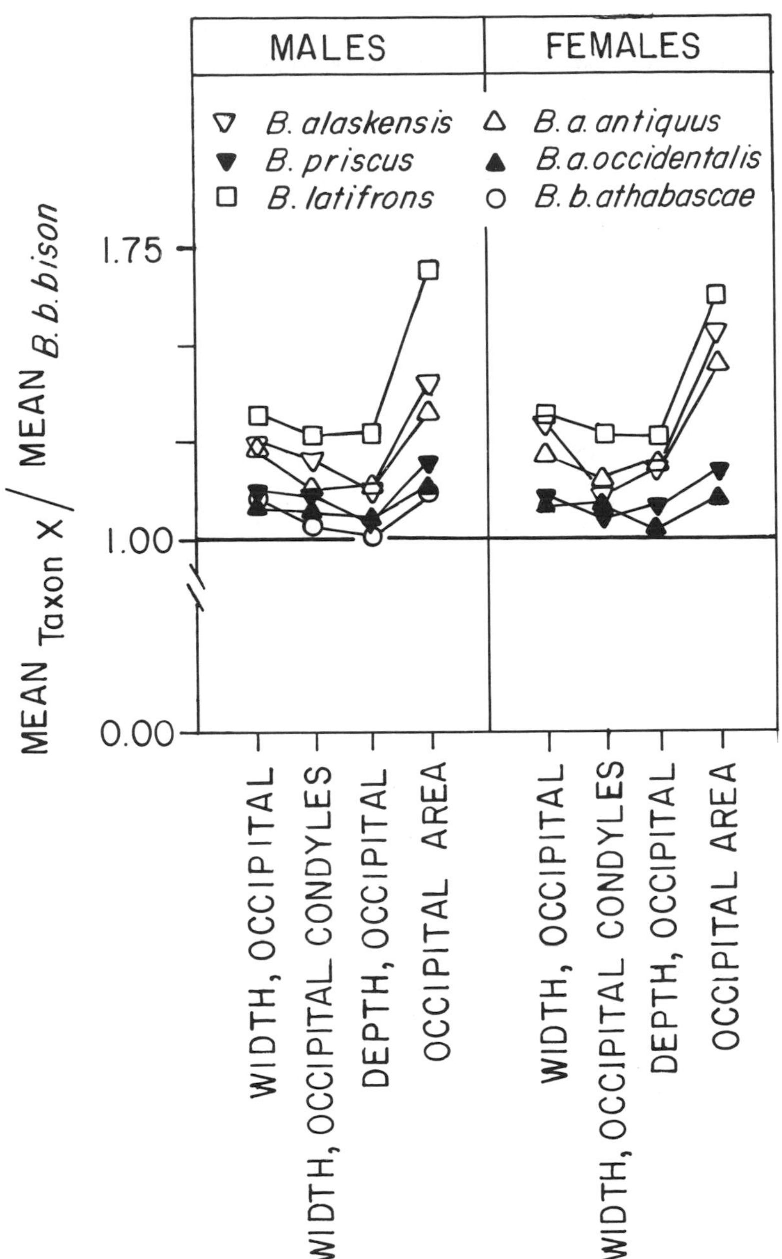

FIGURE 83 Scaled differences: occipital characters. The vertical axis expresses the mean for each taxon as a multiple of the *Bison bison bison* mean.

Table 58

Allometry of Skull Characters Relative to *Bison bison bison*: Areas and Volumes

Taxon	Molar area	Frontal area	Dorsal area	Occipital area	Skull volume
Males					
B. b. athabascae	1.02	1.13	1.17	1.13	1.22
B. a. occidentalis	1.10	1.16	1.15	1.13	1.19
B. a. antiquus	1.25	1.34	1.36	1.33	1.57
B. latifrons	1.30	1.74	1.75	1.68	2.25
B. priscus	....	1.20	1.23	1.18	1.34
B. alaskensis	....	1.49	1.49	1.40	1.70
Females					
B. b. athabascae	....	....	....	....	....
B. a. occidentalis	1.11	1.18	1.19	1.11	1.20
B. a. antiquus	1.30	1.42	1.44	1.46	1.73
B. latifrons	1.46	1.59	....	1.64	....
B. priscus	....	1.16	....	1.17	....
B. alaskensis	....	1.41	1.42	1.53	1.81

Note: The *B. b. bison* mean for each character = 1.00. Values given in the table are multiples of the *B. b. bison* mean. Calculations are based on data presented in table 56.

Table 59

Comparative Biometrics: Skulls with Horn Cores

Taxon	Overall volume (mm^3)	Horn core volume (% total volume)	Allometry relative to *B. b. bison*
Males			
B. b. bison	4,053,743	16.8	1.00
B. b. athabascae	5,222,647	21.0	1.29
B. a. occidentalis	5,384,940	25.2	1.33
B. a. antiquus	6,863,930	22.9	1.69
B. latifrons	18,542,382	59.0	4.57
B. priscus	6,670,482	32.2	1.65
B. alaskensis	10,524,070	45.6	2.60
Females			
B. b. bison	2,332,341	7.3	1.00
B. b. athabascae	....	....	....
B. a. occidentalis	2,963,875	12.7	1.27
B. a. antiquus	4,209,953	11.0	1.81
B. latifrons	....	....	....
B. priscus	....	....	....
B. alaskensis	5,062,021	22.7	2.17

Note: The overall volume is the sum of the mean skull volume and horn core volume for each taxon. The horn core volume is here expressed as a percentage of the overall volume. Allometry here expresses the overall volume as a multiple of the *B. b. bison* mean, where the *B. b. bison* mean = 1.00. Calculations are based on data presented in tables 54 and 56.

bison. This phenomenon is logically explained by the probability that more muscle and ligament mass was necessary to hold and maneuver the head. Protruding occipital condyles direct the weight of the head into more of a point at the atlas than do condyles situated more nearly flush with the occipital plane. The protruding condyles of larger bison, therefore, suggest that these species had a more maneuverable head (laterally and dorso-ventrally) than did smaller bison.

The ventro-lateral margins of occipital condyles characteristically differ in degree of development between larger and smaller bison. In larger bison, and most clearly in *B. latifrons*, the condyle margins are characteristically flanged outward, downward and backward (plate 25). In smaller bison, and most clearly in *B. bison*, the condyle margins are weakly or not at all flanged (plate 26). This flange is interpreted as an adaptation that reduced lateral slippage of the condyles in the cup of the atlas. This would be an obvious advantage to larger bodied, heavier headed bison, considering the relative weight vectors constantly imposed by the long heavy horns and the periodic torque developed when the head was moved quickly, as when startled or in fighting.

Variations in the angles that separate the foramen magnum plane from the basioccipital axis and the occipital plane among bison taxa are also of probably adaptive significance. Within both the North American and Eurasian lineages, the angle between the basioccipital axis and the foramen magnum plane ($\angle$BF) increases with increasing body size, whereas the angle between the occipital plane and the foramen magnum plane ($\angle$OF) decreases with increasing body size (fig. 84; table 60). The differing relationships among the basioccipital, foramen magnum and occipital in different taxa result in differences in the alignment of the vertebral axis (as oriented at the atlas-occipital articulation) with the skull. Smaller bison have heads rotated further downward relative to the vertebral axis than do larger bison (fig. 85). These ordered differences in alignment of the skull with the vertebral column are not easily explained by size, but they do appear to correlate with inferred differences in ecology, social organization, and dominance behavior of bison taxa. A head oriented higher relative to the vertebral column would be better suited for eye-level browsing, whereas a head oriented lower relative to the vertebral column would be better suited for grazing. The vertebral axis is more nearly perpendicular to the frontals in smaller bison than larger bison, a characteristic that would conceivably render the smaller bison better fit than larger bison for fighting with head-to-head clashes. In smaller bison with lower placed heads, the impact of collision would be more directly transferred into the postcranial skeleton and less likely to inflict injury to the neck ligaments and muscles than would be true for larger bison with higher placed heads (fig. 86).

In summary, the occipital region contains characters with probable adaptive significance which vary interspecifically, but mainly by degree. The difference in these characters can be explained in part by direct selection favoring their differential development and in part by indirect selection differentially favoring other characters with which the

Table 60
Occipital Angles

Taxon	Angle between basioccipital and foramen magnum ♂♂	Angle between basioccipital and foramen magnum ♀♀	Angle between occipital and foramen magnum ♂♂	Angle between occipital and foramen magnum ♀♀
B. b. bison	110.5°	109.7°	133.8°	131.8°
B. b. athabascae	113.8°	.....	129.4°	.....
B. a. occidentalis	113.4°	110.2°	129.6°	131.7°
B. a. antiquus	115.6°	115.0°	125.4°	123.8°
B. latifrons	127.4°	119.0°	128.5°	124.0°
B. priscus	117.2°	110.3°	133.2°	134.8°
B. alaskensis	120.5°	112.0°	129.6°	134.0°

Note: Data are from skull biometric summary tables, chapter 2.

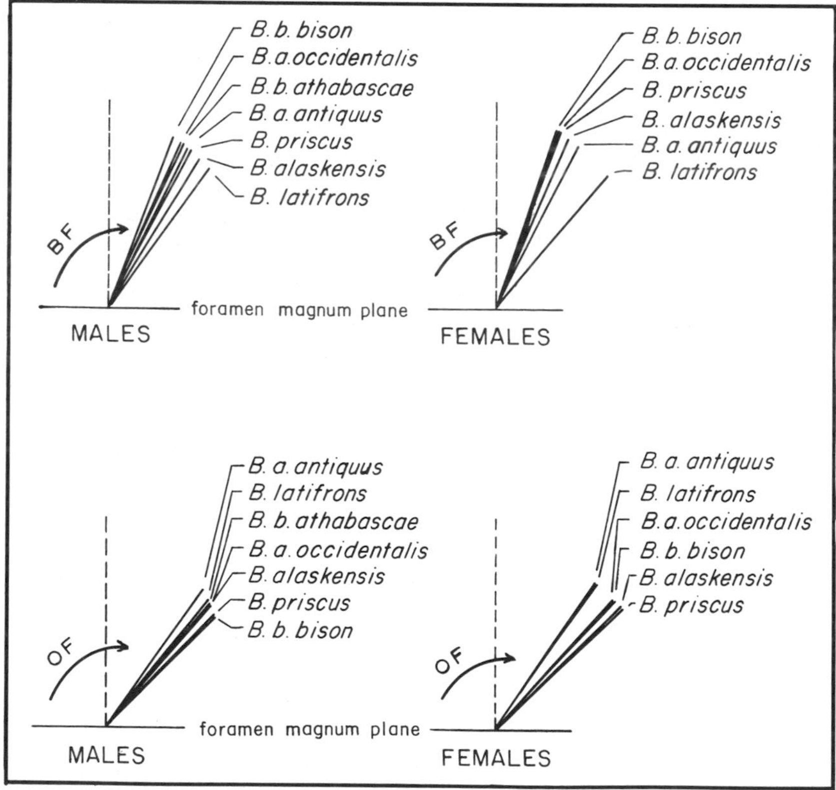

FIGURE 84 Interspecific differences in occipital angles. Top: mean angle separating the foramen magnum and basioccipital planes for each taxon. Bottom: mean angle separating the foramen magnum and occipital planes for each taxon.

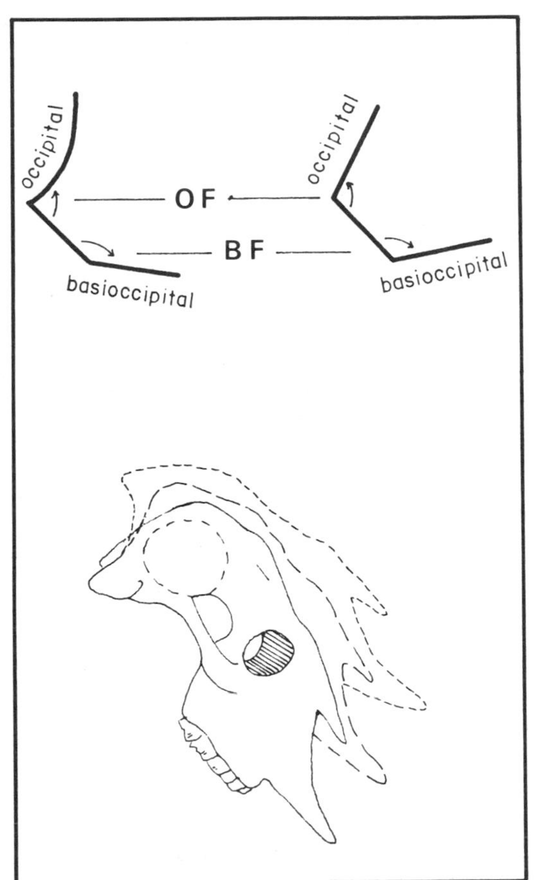

FIGURE 85 Occipital angles and placement of the skull. Top: differences in the relationship among the foramen magnum, basioccipital, and occipital planes between larger, longer horned taxa (left) and smaller, shorter horned taxa (right). Bottom: downward rotation of the head, relative to the vertebral axis, and associated changes in the shape of the occipital. The highest head position is associated with the larger, longer horned taxa while the lowest head is associated with the smaller, shorter horned taxa. The intermediate position represents the relative head placement for smaller pre-Holocene savanna/wooded steppe taxa. Notice the more concave occipital profile for larger taxa and the less concave profile for smaller taxa.

occipital characters are functionally integrated. Differences in the shape of the occipital condyles and the angular arrangement of occipital characters are probably of direct adaptive significance. Species with larger skulls have more strongly developed nuchal lines, external occipital protuberances, protruding occipital condyles, flanged occipital condylar surfaces, greater occipital areas, and heads oriented higher relative to the vertebral axis than do species with smaller skulls. These features are most strongly developed in *B. latifrons*, but they are also well developed in *B. alaskensis* and, to a lesser extent, in *B. priscus* and *B. a. antiquus*. Some of these features, such as the nuchal line and external occipital protuberance, are always obviously present and well developed, even in smaller bison, whereas other features, such as the flanged and protruding occipital condyles, are not usually present in smaller bison.

LIMBS

Limbs are the organs of mobility and elevation in ungulates. Limb length is a determinant of an organism's elevation above the ground or low growing vegetation; feeding strategy; visibility to conspecifics, competitors and

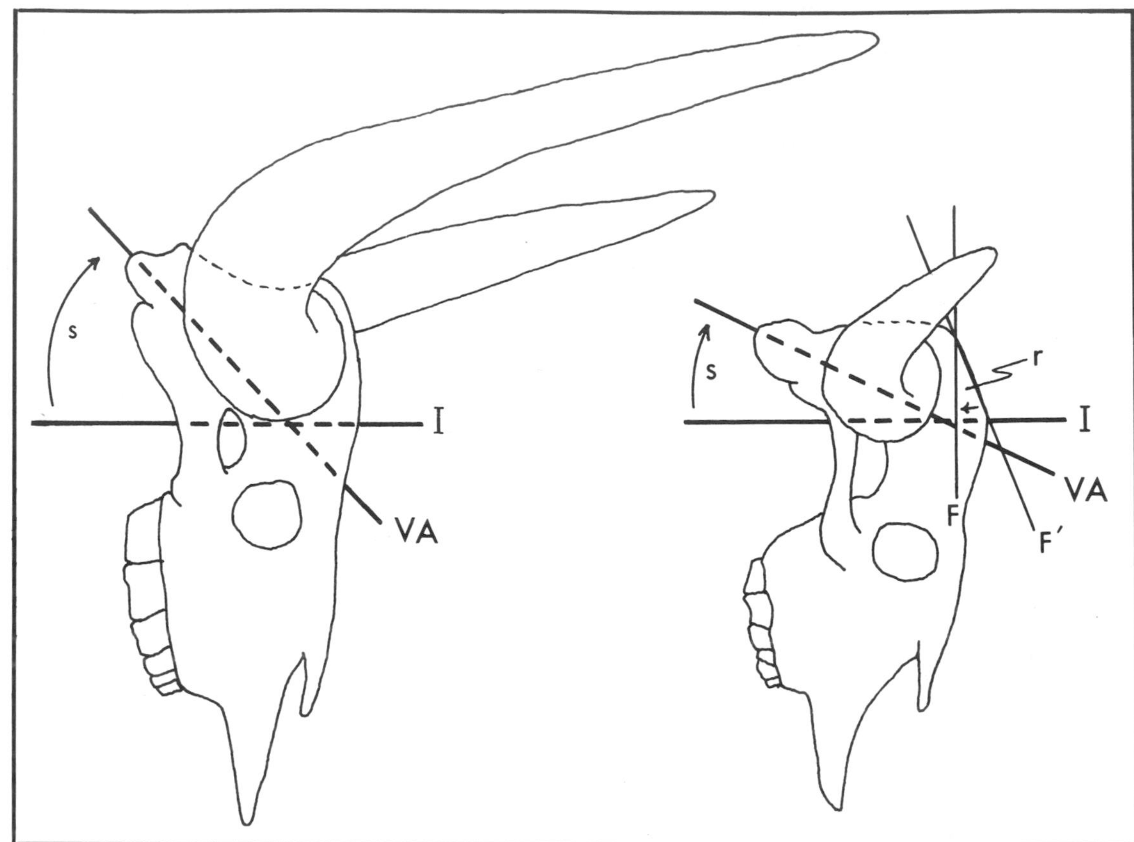

FIGURE 86 Shear and rotation angles of impact. The shear angle (∠ s) separates the vertebral axis (VA) and the impact vector (I) resulting from head-to-head collisions. The greater shear angle indicates that greater torque would result from head-to-head collisions and that the energy of impact would less efficiently be transmitted to the postcranial skeleton. A larger shear angle would increase the likelihood of neck injuries resulting from head-to-head collisions. The rotation angle (∠ r) is the maximum angle through which the skull can rotate, as deceleration occurs, while the combatants' skulls maintain frontal contact. This rotation, which would be opposed by the elastic ligaments and muscles of the neck, would absorb some of the energy of impact, thereby reducing the total energy transmitted to the skeleton. Notice the large shear angle and the absence of a rotation angle in the larger horned bison, characteristics which render the large bison less suited for head-to-head collisions than smaller bison. (Compare Schaffer [1968] and Stanley [1974] for related discussions of functional differences in skull structure among closely related taxa.)

predators; and mobility, including agility, acceleration and stride length. Limb massiveness is a gross index of the relative weights of organisms.

Among autochthonous North American bison, longer limbed taxa generally have more massive limbs than shorter limbed taxa (tables 49 and 61). *B. latifrons* has the longest limbs and *B. b. bison* has the shortest limbs among

Table 61

Mean Limb Lengths, North American Autochthons

Taxon	Humerus	Radius	Metacarpal	Total front limb	Femur	Tibia	Metatarsal	Total rear limb	Rear limb total: front limb total	Allometry relative to *B. b. bison* Front limb	Rear limb
Males											
B. b. bison	325.9 mm 38.2%	320.5 mm 37.6%	206.3 mm 24.2%	852.7 mm	380.7 mm 37.8%	371.9 mm 36.9%	255.2 mm 25.3%	1007.8 mm	1.18	1.00	1.00
B. b. athabascae	354.5 mm 38.4%	341.3 mm 36.9%	228.7 mm 24.7%	924.5 mm	421.3 mm 37.9%	414.3 mm 37.3%	275.7 mm 24.8%	1111.3 mm	1.20	1.08	1.10
B. a. occidentalis	356.9 mm 38.8%	343.8 mm 37.3%	219.8 mm 23.9%	920.5 mm	403.7 mm 37.5%	403.5 mm 37.4%	270.3 mm 25.1%	1077.5 mm	1.17	1.08	1.07
B. a. antiquus	(355.0 mm) (38.2%)	343.0 mm 37.0%	222.0 mm 23.9%	(927.0 mm)	429.0 mm 38.5%	412.0 mm 36.9%	273.9 mm 24.6%	1114.9 mm	(1.20)	1.09	1.11
B. latifrons	401.7 mm 38.9%	383.2 mm 37.1%	248.6 mm 24.0%	1033.5 mm	481.7 mm 38.7%	458.1 mm 36.8%	303.6 mm 24.4%	1243.4 mm	1.20	1.21	1.23
Females											
B. b. bison	286.3 mm 36.8%	293.0 mm 37.7%	198.9 mm 25.5%	778.2 mm	342.2 mm 37.1%	335.9 mm 36.4%	245.2 mm 26.5%	923.3 mm	1.19	1.00	1.00
B. b. athabascae	304.0 mm 36.3%	313.0 mm 37.4%	220.0 mm 26.3%	837.0 mm	362.0 mm 36.7%	361.0 mm 36.6%	264.0 mm 26.7%	987.0 mm	1.18	1.06	1.07
B. a. occidentalis	321.7 mm 38.0%	315.2 mm 37.1%	211.2 mm 24.9%	848.1 mm	367.6 mm 37.1%	362.2 mm 36.6%	259.9 mm 26.3%	989.7 mm	1.17	1.09	1.07
B. a. antiquus	343.5 mm 38.5%	329.0 mm 36.8%	219.5 mm 24.6%	892.0 mm	400.7 mm 37.9%	390.2 mm 36.9%	266.0 mm 25.2%	1056.9 mm	1.18	1.15	1.14
B. latifrons	345.0 mm 36.2%	359.2 mm 37.7%	248.1 mm 26.1%	952.3 mm	428.7 mm 38.2%	409.5 mm 36.4%	285.3 mm 25.4%	1123.5 mm	1.18	1.22	1.22

Note: Calculations are based on data presented in limb biometric summary tables, chapter 2. Values in parentheses are estimates. Percentage figures beneath mean lengths for each element indicate the percentage of the three-bone total represented by that element.

North American autochthons. The front limb of male *B. latifrons* averages 21 percent longer than the front limb in *B. b. bison*; the hind limb averages 23 percent longer in male *B. latifrons* than in male *B. b. bison* (table 61). Diaphysial diameters of the *B. latifrons* limbs are proportionately much larger than limb lengths vis-à-vis these same characters in *B. b. bison*. Diaphysial diameters range from 18 percent to 43 percent, and average about 30 percent, greater in male *B. latifrons* than *B. b. bison*. Male *B. a. antiquus* have proportionately shorter and less massive limbs than do *B. latifrons*, averaging only about 9 percent longer and 15 percent wider in the diaphyses than *B. b. bison*. Male *B. a. occidentalis* are smaller yet, averaging about 7 percent longer and 9 percent wider in the diaphyses than *B. b. bison*. Male *B. b. athabascae* have larger limbs than *B. a. occidentalis*, with limbs averaging about 9 percent longer and 10 percent wider in the diaphyses than *B. b. bison*. The hind limbs of *B. b. athabascae* are proportionately longer than the front limbs vis-à-bis *B. b. bison* (8.4 percent longer front limbs, 10.2 percent longer hind limbs). The pattern of limb length and diaphysial diameter allometry observed among males is generally the same for females. *B. b. athabascae* female limbs, the only exception to this pattern, average about the same length as female *B. a. occidentalis* limbs but are considerably more slender (4.1 percent in *B. b. athabascae* vs. 8.8 percent in *B. a. occidentalis* of the *B. b. bison* diaphysial diameter means).

The information available for limb dimensions of North American autochthons indicates that *B. latifrons* was a tall and heavy bison not built for sudden movements or prolonged running. The long and massive legs suggest that *B. latifrons* was probably primarily an eyelevel browser, well adapted to feed on shrubs and smaller trees. *B. a. antiquus* was about midway between *B. b. bison* and *B. latifrons* in limb size. The relatively shorter and less massive legs suggest that *B. a. antiquus* was more adapted to feeding on lower growing vegetation (herbs and shrubs), running and agile movements. *B. b. bison*, with the shortest and most slender limbs, appears best adapted of all North American bison to grazing, agile movements, rapid acceleration, and prolonged running. The relative differences of limb length and massiveness between *B. a. occidentalis* and *B. b. athabascae* are probably consequences of their phyletic ancestry and evolutionary directions. *B. a. occidentalis* evolved from the longer and more massive legged *B. a. antiquus* to the shorter and less massive legged *B. b. bison*, whereas *B. b. athabascae* probably evolved to a larger size from *B. b. bison* or, alternatively, has probably been influenced by gene flow from the smaller *B. b. bison*.

Limb bones of Eurasian autochthons were not available for study.

BODY SIZE

Overall body size was not determined for any bison taxon, but implications from those portions of the skeleton studied clearly indicate that differences of overall body size existed among North American bison. Ranking the seven recognized taxa according to inferred body size, using skull length as a guide, *B. latifrons* was largest followed by *B. alaskensis, B. priscus, B. a. antiquus, B. a. occidentalis, B. b. athabascae,* and *B. b. bison*. Difference in body size is of adaptive significance itself, but it also correlates with other adaptive differences, especially differences in longevity, maturation rates, and reproductive potential (Western, 1979).

Generally predictable differences exist between larger bodied and smaller bodied mammals, especially within a closely related and behaviorally similar taxonomic group. Larger-bodied taxa generally live longer, mature more slowly, have fewer offspring per unit of time, have greater absolute energy requirements, consume absolutely more forage (if equalized for nutrition and energy), and have larger space requirements per individual than do smaller bodied taxa (Crandall, 1964; Estes, 1974; Gould, 1977; Hediger, 1964; Morris, 1965; Western, 1979; E. Wilson, 1975). Large body size can be considered adaptive when prolonged sexual maturation, low biotic potential, and available resources combine successfully to maintain population viability in any given selection regime. Larger body size can convey a competitive advantage in the acquisition of food, water, and space; dominance; and defense; and can be considered adaptive so long as the advantages of large body size exceed the disadvantages. Smaller body size can be considered adaptive if more rapid sexual maturation, greater biotic potential, and greater specialization in resource use combine to successfully maintain the species population.

The adaptive significance of differences in maturation rates is closely tied to the nature of selection regimes operating upon populations. The extremes in maturation rates vis-à-vis their adaptiveness to environments are commonly termed r-selected and K-selected (E. Wilson, 1975). Generally speaking, K-selected maturation rates characterize populations that inhabit relatively stable environments and have relatively slow maturation rates and lower biotic potential. These characteristics can result in larger bodied, longer lived individuals with lower overall numbers and lower population densities maintained at or near carrying capacity. R-selected maturation rates characterize populations that inhabit relatively unstable environments and have relatively fast maturation rates and higher biotic potential. These characteristics result in smaller bodied, shorter lived individuals with higher population numbers and densities maintained well below carrying capacity. The significances of r-selected adaptations are that rapidly maturing individuals can potentially contribute to the reproductive effort sooner than more slowly maturing individuals, rapidly expanding populations gain access to available resources before more slowly expanding populations, and the more fecund population can better absorb and rebound from periodic decimation by environmental catastrophies, including predation. Both K- and r-selected maturation rates are, of course, extremes along a continuum and represent only one level of the overall adaptation hierarchy.

CHAPTER 5

MODELS OF BISON ADAPTATION TO DIFFERENT ENVIRONMENTS

Certain characters of the skull, limb bones, and overall body size of presumed adaptive significance were described in the preceding chapter. These adaptations will now be integrated as taxon-level adaptive complexes with the environments to which they were presumably adapted (tables 62 and 63). The temporal and spatial distribution of North American autochthons implies that three general habitat types were occupied during the late Quaternary: forests and woodlands, savannas and wooded steppes, and open grasslands. Limited populations of each species no doubt occupied more than one habitat type (as is true for *B. bison*), and these limited populations probably differed from the main species populations morphologically, behaviorally, and ecologically according to the potentials or limitations of the specific environments they occupied. The models presented here, however, are intended only to relate the selective environments and adaptive responses of typical species populations. The function of many characters and behavioral and ecologic patterns of living bison have been determined by direct observation (e.g., Borowski, Krasinski, and Milkowski, 1967; Fuller, 1960; Gunderson, 1975; Krasinski, 1967; Lott, 1974; Mahan, 1978; McHugh, 1958; Shackleton, 1968; Shult, 1972; pers. observ.), whereas others have been inferred either from analogs in other ungulates (e.g., Bell, 1971; Dalimier, 1955; Dubost, 1979; Eisenberg and McKay, 1974; Estes, 1974; Geist, 1971*a*, 1971*b*, 1978; Gould, 1974; Halder, 1976; Hirth, 1977; Schaller, 1977; Schloeth, 1958, 1961; Sinclair, 1977; Tulloch, 1978; Walther, 1974; Western, 1979; Wilson and Hirst, 1977; Woodward, 1976, 1979), or by deduction. General evolutionary and sociobiological theory must also be considered in modeling adaptive complexes of North American bison taxa.

Table 62

Adaptive Differences Among North American Bison Species

Character	*B. bison*	*B. antiquus*	*B. latifrons*	*B. priscus*	*B. alaskensis*
Horn cores:					
Length	shortest	shorter	longest	shorter	longer
Volume/weight	least	less	greatest	less	great
Divergence	rearward	forward	forward	intermediate	intermediate
Tip placement	low, back	low, forward	high intermediate	intermediate, back	intermediate, back
Fronto-facial area:					
Domed frontals	most	intermediate	least	intermediate	intermediate
Orbit orientation	lower, laterally	intermediate	higher, forward	higher, forward	higher, forward
Relative orbital protrusion[a]	low	intermediate	high	intermediate	low
Dentition:[b]					
Molar occlusal area	smallest	large	largest	....	....
Area:body size	intermediate	high	low	....	....
Occipital condyles:					
Protrusion	least	low	greatest	intermediate	great
Flanging	weakest	weak	strongest	intermediate	strong
Skull:					
Volume	least	intermediate	greatest	intermediate	great
Downward rotation	greatest	intermediate	least	intermediate	intermediate
Limb length:[c]	shortest	intermediate	longest	(intermediate)	(long)
Body size:	smallest	intermediate	largest	intermediate	large

Note: Information presented in this table has been extracted from chapters 3 and 4.

[a]The lower the relative orbital protrusion index value, the greater the actual protrusion of the orbit outward from the skull (see Guthrie, 1966*b*).

[b]Teeth for the Eurasian autochthons were not studied.

[c]Limbs representing these taxa have not been positively identified from North American localities. The characteristics in parentheses are assumed.

Those facets of adaptation to be considered include intraspecific and interspecific competition, intraspecific dominance strategies and activities, defense, social or grazing tendencies, and individual and group mobility. All are interdependent to some degree. The objective is to generate logical models of possible adaptation to the environments that North American bison inhabited during the late Quaternary. This effort unifies specific adaptations, organisms, populations, and their environments. These models are intentionally generalized; the available data on bison distribution and dispersal history do not permit more detailed resolution at this time.

FOREST/WOODLAND MODEL

Much of Eurasia, Beringia, and North America were covered with forests or woodlands (table 3) during the Pleistocene (chapter 1). Although the distribution and density of forests and woodlands fluctuated during this period (figs. 2–5), the habitats nonetheless remained and were available for occupation by terrestrial herbivores. Forests and woodlands were much more persistent in North America than Eurasia during the Quaternary. The extent to which these habitats could be occupied by terrestrial herbivores depended on the amount and quality of usable forage available. Forest openings and edges and woodlands, therefore, could potentially have supported a greater terrestrial megafaunal herbivore biomass than could closed forests because more utilizable biomass is nearer the ground in open and early seral than in late seral or climax forests. Parklands, meadows, and other herb-rich seral phases in the early to intermediate stages of succession would be habitats especially suited to large herbivores. Such seral stages would, however, tend to be patchy in

Table 63

Habitat and Adaptive Characteristics of Large Herbivores in Selected North American Environments

Parameter	Pre-Holocene forest openings/woodlands	Pre-Holocene savannas/ wooded steppes	Holocene midlatitude grasslands
Habitat:			
Stability	least stable	relatively stable	most stable
Forage diversity	intermediate	most diverse	least diverse
Megafaunal carrying capacity	lowest	high	high
Megaherbivores in general:			
Species diversity	lower	higher	low
Carrying capacity (for each adapted species)	low	intermediate	high
Feeding strategy	browsing/grazing	grazing/browsing	grazing
Interspecific competition	highest	high	low
Intraspecific competition	low	intermediate	high
Predator pressure	lower, simple	higher, complex	higher, simple
Defense by size/strength	more reliance	more reliance	less reliance
Defense by evasion/flight	less reliance	more reliance	more reliance
Social cohesiveness	weak	intermediate	high
Dispersal tendency	high – both sexes	moderate – males low – females	moderate – males low – females
Group size	smallest	intermediate	largest
Maturation/development	slower	intermediate	more rapid

Note: Both Pre-Holocene habitats are considered to represent faunally diverse, relatively K-type selection regimes. The Holocene grassland is considered to represent a faunally simple, relatively r-type selection regime. Only those megaherbivores that followed a generalist adaptive strategy are considered here; those following a specialist adaptive strategy are excluded.

distribution and/or prone to succession to woody vegetation. Frequent destruction or impaired development of dominant woody vegetation by such disruptive processes as fire, snowfall, windfall, rockfall, flooding, sedimentation, edaphic factors, grazing and browsing, and so on, and arrested succession following disruption or impairment would tend to enhance the suitability of forests and woodlands for use by large herbivores. Alternatively, infrequent disruption and rapid seral succession would tend to reduce the suitability of forests and woodlands to large herbivores. Resources accessible to large herbivores would be relatively scarce in closed forests and, as a result, competition would potentially be intense and immediate—less so as more openings were created or their succession delayed, more so as fewer openings were created or their succession accelerated.

Forest/woodland seral habitats would be most efficiently utilized by large terrestrial herbivores characterized by the following traits. (1) Feeding strategy could include browsing and/or grazing. A grazing tendency would permit use of the usually more nutritious herbaceous vegetation, but total dependence on herbs would normally result in the individual's resource base being eliminated as succession proceeded (and, simultaneously, competition for herbs from other grazers would intensify). A browsing tendency would permit use of the usually more enduring woody vegetation (shrubs and small trees) which would normally be available soon after disruption of the climax vegetation occurred and, as the shrubs and trees migrated into the sere, would remain available as the herb phase of succession passed. Both grazing and browsing tendencies (i.e., a generalist feeding strategy) would permit greater flexibility in utilizing either herbs or woody vegetation. (2) Competition for limited resources would be actually or potentially intense and immediate. Successful competitors would have to obtain ample resources when and where they were available. With resources limited in quantity, alternative means of obtaining sufficient resources would be few. Intraspecific partitioning of habitat would be advantageous. The competitive advantage among animals with large body size (generalist feeders) would lie with assertive, belligerent, large individuals capable of backing up threats by direct aggression if less physical methods failed. Male-like characters in females would make females more competitive with conspecifics and interspecifics. (3) Dispersal tendencies, the inclination and ability to search out new and dispersed sources of food, water and space when necessary would be favored over philopatry (adherence to a home range even as resources declined) or social cohesiveness (groups would more quickly exhaust the limited resources). This dispersal tendency would reduce intraspecific competition, would facilitate the location of new or more ample resources as older resources became depleted, and would probably reflect greater cursoriality than was characteristic of taxa adapted to more open habitats. (4) Individuality or selfishness would convey greater fitness than would social cohesiveness and tolerance in the resource limited, intensively competitive forest opening/woodland environment. (5) Defense from predators would be mainly by factors of size (intimidation, strength, and counterattacks), rather than flight.

FIGURE 87 The inverse relationship between horn size and body hair in male bison. Top: *Bison latifrons*, as inferred from skeletal morphology and the assumed inverse relationship between horn size and body hair. Center left: *Bison priscus*, as reconstructed from Paleolithic cave drawings (after Geist, 1971*a*). Center right: *Bison antiquus antiquus*, as inferred from skeletal morphology and the assumed inverse relationship between horn size and body hair. Bottom: *Bison bison bison*, from photographs of living bison. Notice the differences in orientation of the head relative to the vertebral axis.

B. latifrons, the largest species of bison known, was probably adapted to forest openings and woodlands. Based on the inverse relationship between body (display) hair and horns which occurs in some ungulates, including bison (Geist, 1971*a*, 1971*b*; Geist and Karsten, 1977), *B. latifrons* is assumed to have had less exaggerated pelage on the forebody and head than any other species of bison (figs. 87 and 88). *B. latifrons* was probably the most slowly maturing and longest lived bison, and possibly had the lowest biotic potential of all bison. The large overall body size, hypermorphic horns, and other specialized characters of the skull are attributable to a relatively K-selected maturation rate. Female *B. latifrons* were relatively male-like in many characters (figs. 61 and 62; table 46), a quality that probably improved their competitiveness in the forest opening or woodland environment.

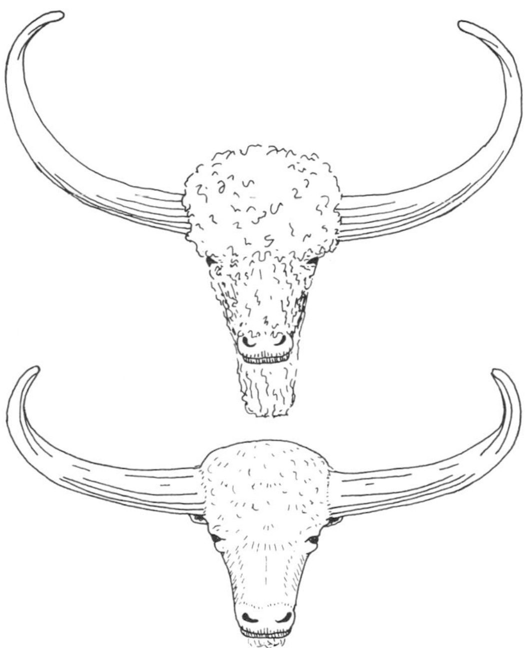

FIGURE 88 Two renditions of the *Bison latifrons* head. Top: Based on the restoration in Vanderhoof (1942), this head was modeled after *Bison bison bison* and as a result probably has too much facial hair and the horns are probably rotated too far to the rear. Bottom: *Bison latifrons* with horns rotated further forward and with most display hair removed, a rendition which probably more nearly approximates the actual appearance of *Bison latifrons* males.

Overall large size was probably the most generally important differentiating adaptation of *B. latifrons*. This characteristic conveyed the competitive advantage necessary for *B. latifrons* to coexist successfully with other large Pleistocene herbivores (*Camelops, Cervalces, Equus, Mammut, Mammuthus, Symbos* et al.) in resource-limited forest opening and woodland seral habitats of midlatitude North America. Larger horns were probably selected for, among other possible alternate morphologies, because they both conveyed or exaggerated the impression of large body size and they were functional tools with which to back up threats assertively if necessary. Longer and heavier limbs contributed to and supported the large body size, but they also elevated the body, thereby making the individual more visible amidst the low shrubs, tree seedlings, and broken background of the forest opening/woodland environment.

B. latifrons was probably characterized by a browsing and grazing feeding strategy, with greater emphasis on browsing than was characteristic of savanna or grassland bison. Generalized feeding would have permitted greater utilization of the variety of available energy and nutrition sources and greater tolerance of the vegetation changes which accompanied forest/woodland seral succession. The shape of the *B. latifrons* occipital suggests that the head in this species was held higher and oriented more forward than the head of any other species of bison (figs. 85 and 87). The great size of the head suggests that eye-level browsing would simply have been more mechanically efficient and less costly than would grazing.

Individual *B. latifrons* were probably relatively unsociable. Seral habitats simply could not support large populations of any large herbivore for long, nor could the forest openings or woodlands as a whole continuously support large numbers of large herbivores. Energy limitations, therefore, constituted an important external restriction on the size and density of populations. Selection favored individualistic phenotypes, those which were capable of surviving alone, or, perhaps, with relatively few other individuals in a simple, clearly structured dominance hierarchy. Male *B. latifrons* may have been more territorial than savanna or grassland bison, actively holding or attempting to hold openings in the forest to the exclusion of others as long as usable resources were available. (Territoriality and dominance are but extremes of a continuum; a slight shift toward the territorial extreme is suggested here. This does not conflict with the dispersal tendency attributed to individuals, an adaptation that would facilitate the search for new resources. Territoriality would permit the retention of resources once located.) Asociality in both sexes permitted effective dispersal of individuals or small groups, a characteristic that permitted a high level of gene flow throughout the dispersed population.

Rutting was probably a summer activity, but it may have differed from the rut of *B. bison* in several ways. The role of vocalizations was probably more important to *B. latifrons* as a way of announcing the presence, availability, and assertion of a male willing to participate in the reproductive effort. Female vocalization also may have been important in reproductive behavior. The relatively closed habitat would have reduced (but not eliminated) the visual role in locating a potential mate. Male dominance encounters were probably relatively infrequent, especially those in which escalated fighting took place. This situation existed simply because the population was dispersed and the probability of two males of approximately the same dominance capabilities meeting was relatively low. Relative dominance was probably determined by initial impressions of size. Displays probably included the basic bovine broadside exhibition (plate 27) with a limited assortment of other postures and threat gestures, such as the presentation of horns, nodding, and hooking inanimate objects—all of which are common in living bison (Lott, 1974; McHugh, 1958) and other living Bovini (Sinclair, 1977). These activities were, however, probably conducted in more of a face-to-face position than the side-to-side position common in living bison, in large part because of the probable lack of display hair (which strongly exaggerates the lateral profile of the body); the forward-oriented orbits; and the orientation and development of the horns, which makes their full size more apparent from the front than the side (cf. figs. 87 and 88). Competing males could have turned their heads sideways to face the opponent in order to present both broadside and full-head views simultaneously, a dominance threat posture common in ungulates (Walther, 1974). The forward presentation of horns and nose-up threat known from living Bovini (Sinclair, 1977) may also have been important aspects of *B. latifrons* dominance behavior.

If visual assessment of each other failed to rank opponents, combat probably ensued. Escalated fighting probably entailed hooking both head-to-head and at an opponent's side. The long, heavy, recurved horns, with tips extending high above and slightly behind the frontals, were probably swung upward and inward (offensively) or downward (defensively) at an opponent from close range, normally while facing each other. The length and weight of the horns and the maneuverable head probably allowed the generation of considerable force with each hooking stroke. Circling of the hind quarters about a central pivot may have accompanied the horn swinging as both a defensive and aggressive strategy. This circling would have consisted of limited movement to gain position without the necessity of excessive, costly body movement.

Charges with head-on collisions and head wrestling were probably not parts of the *B. latifrons* fighting strategy. The skull and horns of *B. latifrons* were not well structured to execute or withstand the impact of head-on collisions (fig. 86). The horns were high and placed relatively forward, presenting a narrow passage between their tips which an opponent would have to penetrate with both his head and horns in order to execute a clash (fig. 88). Such a feat would be nearly impossible, considering the coordinated weaving motion that would have been necessary for opponents to maneuver their horns around the opponent's horns to effect a clash.

Frontal shape and structure and skull-vertebral axis alignment suggest that the impact of head-to-head clashes between *B. latifrons* bulls would have been more potentially damaging and less effectively absorbed and disseminated than in the *B. bison* skull. The fact that head-on clashing is an important aspect of fighting behavior in living *B. bison* does not require that all antecedent bison behaved similarly. *Homoioceras*, a late Pleistocene African Bovini that may be considered analogous to *B. latifrons* insofar as general horn core morphology and head positioning are concerned, apparently fought by hooking. Figure 89 clearly shows the narrowness of the corridor between the horn tips through which an opponent's head would have had to pass to execute a clash.

Fights in *B. latifrons* were probably relatively rare (i.e., relative to *B. bison*, for example), brief, and decisive. Fighting with large, heavy blows would be tiring and dangerous, so greater efficiency and decisive conclusions in dominance threats, perhaps accomplished by clear presentation of fighting weapons and greater assertiveness, would be favored. Once relative dominance had been established, it would likely be relatively lasting, especially insofar as any one season's reproductive effort was concerned. Also, dominant males would have had relatively few females to tend and service. They would, in effect, devote more rutting energy to the tending and servicing

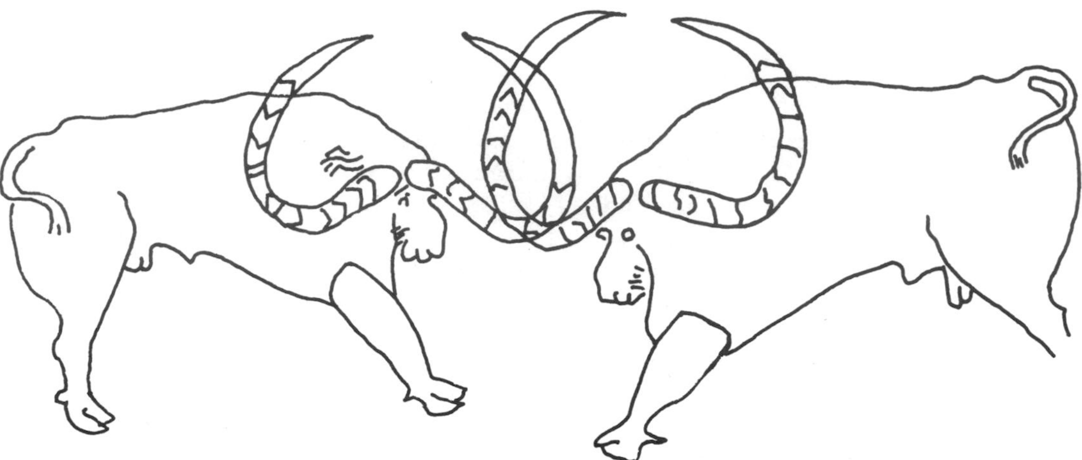

FIGURE 89 Male *Homoioceras*, assumed to be fighting. The facts that both animals are male, that both are leaning forward braced on their forequarters, that the head of the individual on the right is turned further to the left than is the head of the individual on the left, and that the horns intersect suggest that these bulls are fighting by hooking actions. The narrow distance between the tips of the horns would have precluded head-to-head clashes. The fighting behavior illustrated here probably characterized *Bison latifrons* (after a rock engraving located at Enfouss, Atlas Mountains, Algeria, illustrated in Frobenius and Obermaier, 1965).

of one or a few females, thus assuring their contribution to the gene pool continuum by fertilizing the available females when they reached estrous, rather than investing extra energy searching for a larger number of dispersed females. By investing more rutting energy in tending activity than promiscuity, fewer rutting fights would have occurred and proportionately more males would have participated in the annual reproductive effort, thereby maintaining relatively high genetic variation within the species.

B. alaskensis is considered a Eurasian vicar of *B. latifrons*. *B. alaskensis* probably adapted to the forest opening/woodland seral stages and dense savanna habitats of Eurasia by evolving many morphological, physiological, and behavioral characters similar to, but less exaggerated than, those evolved by *B. latifrons*. The presence of domed and thickened frontals and greater relative orbital protrusion are not unexpected, considering the probability that *B. alaskensis* evolved from *B. priscus* (in which these characters occur) and Eurasian forest/woodland habitats were less enduring than similar habitats in North America (see chapter 1), which precluded the same degree of adaptive differentiation and isolation from the savanna congener in *B. alaskensis* (less differentiated, less isolated) than in *B. latifrons* (more differentiated, more isolated).

The nearest modern analog to the *B. latifrons* adaptation model is probably the moose-forest opening system. Among living species, moose (*Alces alces*) most nearly parallels *B. latifrons* in having enlarged antlers, in being the largest extant cervid, and in its forest/woodland association. Moose are dependent on transient forest openings to maintain relatively high population levels. As forest openings succeed to advanced seral stages or reach climax, the moose population is either reduced or displaced. Feeding is by both browsing and grazing. Moose are also individualistic, dispersal-oriented animals. Males are not strongly polygamous, if polygamous at all. Vocalizations are important elements of the courtship effort. (For details of moose behavior and ecology, see Geist [1963, 1971*b*], Houston [1974], Knight [1968], Markgren [1974], and Peterson [1955].)

When comparing moose behavior and ecology with *B. latifrons* adaptation, however, certain important differentiating facts must be kept in mind. Perhaps most important is the more simplified megafauna of the moose system relative to that which characterized the *B. latifrons* system. Direct competition from herbivores is much reduced for moose, which allows a relatively greater moose population than would otherwise be possible. Human disruption of the boreal and alpine forests has also created more and probably larger openings than would have been created by natural processes, a condition that increases the potential carrying capacity for moose. Humans also hunt moose; there is no evidence that man hunted *B. latifrons*. This hunting pressure has worked in various ways to stimulate dispersals, reduce populations, and possibly to select for accelerated maturation rates. Morphological, maturation, and longevity consequences of a shift from relative K- toward relative r-selection regimes could expectedly parallel those postulated above for North American bison. Living moose are smaller than pre-Holocene moose; possibly they mature more quickly and have a higher biotic potential and population levels as well.

In summary, pre-Holocene forest opening/woodland bison were large bodied, long legged, and large horned, all factors that conveyed greater competitiveness to individuals in a resource limited environment. *B. latifrons* is considered a good example of a relatively K-selected species; it probably occurred in relatively small numbers, was dispersal-oriented, individualistic, and morphologically specialized, characters reflecting the intense selection in such environments.

SAVANNA/WOODED STEPPE MODEL

Savanna and steppe habitats (table 3) were common throughout Eurasia, Beringia, and North America during the Pleistocene (figs. 2–5). Like forests and woodlands, the distribution of savannas also fluctuated during this period as they waxed and waned with both large-scale global climatic oscillations and more local oscillations of climate and other regulatory or disruptive factors. Steppes are generally more stable than savannas. The pattern of woody and herbaceous vegetation resulted in a habitat of relatively great structural diversity. Ample resources would normally be available to large herbivores. Competition for these resources would be potentially high, but not immediate—that is, spatial displacement by competitors could be tolerated as long as sufficient resources could be acquired elsewhere in the same generally continuous habitat. Pre-Holocene savanna and wooded steppe habitats with their diverse megafaunal associates (*Bison, Camelops, Equus, Mammuthus*, et al.) would have constituted relatively

K-selection regimes in which larger and more cohesive numbers of conspecifics (relative to a forest/woodland selection regime) existed in a near carrying-capacity situation. Thus, larger bodied, slower maturing, longer lived individuals with greater social tolerance (relative to forest/woodland adapted populations) would be selected for.

Savanna and wooded steppe habitats could most effectively be occupied and utilized by herbivores characterized by the following traits. (1) Feeding strategy could include both grazing and browsing, with advantages accruing to a generalist or grazing strategy rather than a browsing strategy. Resources would be sufficiently abundant and vegetation diversity sufficiently stable that both feeding strategies would be successful, yet generalists or grazers would in most cases have access to more nutrients and energy than would browsers. (2) Interspecific and intraspecific competition would be sufficiently high that selection would favor large individuals, but not necessarily as assertive or belligerent as the forest opening or woodland situations favored. Immediate competitive success in resource control would not be as critical in the savanna or wooded steppe habitat as in the forest opening/woodland habitats. Displacement of an individual or group from resources in a savanna or wooded steppe habitat would simply require a change of location, not a denial of access to limited resources. (3) Dispersal tendencies are not as adaptive to savanna/wooded steppe environments as to transient seral openings in forests or woodlands. Resources are continuous and abundant in savanna/wooded steppe environments and dispersal to seek new resources would not be necessary. (4) Sociality is adaptive in that some aggregation of conspecifics permits greater interspecific security and competitiveness for all individuals within the group—especially females and juveniles—than would be permitted by individuals living alone, and improves fitness by maximizing male dominance and permitting only more dominant males to fertilize females. Sociality of small, discrete groups requires a more complex social organization and dominance hierarchy than would be required in the more individualistic social system of forest/woodland species. Intraspecific competition is greater than in individualistic populations, so tolerance for conspecifics must be developed. Dispersal, especially of males, would increase gene flow among demes and would relax psychological stresses derived from high population levels. (5) Defense and evasion would be by either flight or factors of individual or group size (intimidation, strength, or counterattack), with flight being more important than in forest/woodland species.

B. a. antiquus was the North American pre-Holocene savanna bison. The overall large body size is typical of relatively K-selected maturation rates. Facial and forebody hair was probably more exaggerated in *B. a. antiquus* than in *B. latifrons*, although this is not certain (fig. 87). The relatively greater doming of the frontals and wider, lower placements of the horns suggests that *B. a. antiquus* may have behaved more like *B. bison* than *B. latifrons*. *B. a. antiquus* females are more male-like than the females of any other taxon (fig. 61; table 46).

Large body size was advantageous to *B. a. antiquus* in that it conveyed greater competitiveness and defensive fitness than would be true for smaller body size. Relatively open habitat and a group formation with consequent reduction in interspecific competition mollified the intensity of selection for larger body size and, perhaps, longer maturation rates, which probably produced the size extremes in forest/woodland bison. Savanna and wooded steppe bison, in effect, could afford to be more flexible than forest/woodland bison in acquiring resources. There was more latitude for bluffing in interspecific competition without the critical need to back it up with absolute force if necessary. Extra body hair could function to inflate apparent size beyond the individual's actual size by accentuating certain body zones or points. Unlike large horns in forest/woodland bison, however, body hair itself could do no physical damage in escalated fighting with an unyielding opponent. If the inflated image of size failed to dislodge a competitor the unsuccessful bluffer could simply move elsewhere to feed.

B. a. antiquus was a generalized feeder, feeding by both grazing and browsing. The occipital region of the skull suggests that the head was oriented higher than in grassland bison, but not as high as in forest/woodland bison (fig. 85; table 60).

The smaller body size of *B. a. antiquus* (relative to *B. latifrons*) probably conveyed less individual competitiveness, but the size difference was probably compensated for by the establishment of small, socially cohesive groups. The formation of small groups would have necessitated relatively greater tolerance of conspecifics by members of the group and the formation of a more complex social structure, with cohesiveness, leadership, and a clearly established dominance hierarchy to instill stability. Females and younger males probably formed and retained small kin groups throughout the year, with females—as relatively permanent members of the comparatively stable groups—assuming leadership roles. Mature males apparently retained dispersal tendencies (expressed by voluntary segregation from the larger female-juvenile male social groups for most of the year), which allowed them to retain somewhat more belligerent personalities than females or younger males. This belligerence was a general social disadvantage, but a definite dominance advantage. Selection, nonetheless, probably favored reduced intraspecific belligerence and aggressive tendencies, especially less ultimately damaging dominance activities on the part of the males than was characteristic of forest bison.

Rutting was probably a summer activity. The visual role in rutting was probably more important with savanna/wooded steppe than forest/woodland bison, and the audial role in rutting behavior relatively less significant. Dominance-linked encounters between males were probably more frequent than in forest/woodland bison, simply because the savanna/wooded steppe bison population was larger and more dense and the chances of contesting males coming together was greater. Still, the population density of savanna and wooded steppe bison was probably much lower than that of late Holocene grassland bison, so dominance encounters were probably less frequent and selection for ranking organs was probably less intense in savanna than in grassland bison. Dominance could be established and maintained in a small group in which all members actually or potentially contending for dominance could be mutually recognized and their relative ranking remembered.

Dominance among male *B. a. antiquus* was probably established in typical Bovini fashion with body displays or, ultimately, combat. The absence of large horns probably resulted in more emphasis being placed on body hair as a ranking organ. Escalated fighting probably included head-to-head clashes, head wrestling, and hooking. Lowering of the head, movement of the frontals into a more perpendicular alignment with the vertebral column, moderate doming and greater fusion of the frontal sutures, and smaller horns were morphological adjustments that made clashing and head wrestling first of all possible, and thereafter safer and more efficient while simultaneously reducing the likelihood of skull or neck injuries as a result of these activities.

According to this savanna/wooded steppe model, relatively fewer males would have contributed to the reproductive effort than was true for forest/woodland bison. The social cohesiveness of savanna/wooded steppe bison resulted in several females being together as a unit. The small size of the group was reasonable assurance that estrous for all females would probably be spaced so that perhaps as few as one or two dominant male(s) could inseminate all the females of the group. Small group size also assured that relative dominance was established and remembered by all to whom it was of importance, and, thus, the dominant male(s) could concentrate more on tending and servicing females than on maintaining dominance over the other mature males which had joined the female-juvenile herd for the rut.

B. priscus inhabited savanna, wooded and open steppe, and tundra environments, but probably retained relatively K-selected characteristics (populations were probably near the carrying capacity most of the time, except when populations dispersed into newly accessible habitats), which functionally and morphologically resembled those of *B. a. antiquus*. The open steppe and tundra, at least until the late Wisconsin, were environments in which selection was similar to that of the savanna. The only major adjustments to the savanna/wooded steppe adaptation model required by a pre-Holocene open steppe/tundra model would be shifts toward total grazing as a feeding strategy and social tendencies to accommodate somewhat larger groups of conspecifics. Interspecific competition would remain high, carrying capacity would probably be approximated, and a relatively K-type selection regime would probably exist.

The nearest modern analog to the *B. a. antiquus* adaptation model proposed here is probably the African buffalo-savanna system. The African buffalo (*Syncerus caffer*) resembles *B. a. antiquus*, as conceived herein, in being, in part of its range, a savanna adapted bovine associated with a diverse megafauna. *Syncerus* forms herds that vary from large (ca. 1,000 individuals) to small (one or a few individuals), with herd size generally determined by habitat and abundance of resources (herds are large in grassy savannas, small in closed savannas, smallest in forests). Herding results in relatively less predation and thus greater survivorship than is experienced by isolated individuals or very small groups. Population density appears to fluctuate with herd size and available resources, being highest in large herds during periods when abundant resources are available. Adult males established a strong dominance hierarchy, with relatively few males participating in the reproductive effort. (Dalimier [1955], Kruuk [1972], Schaller [1972], and especially Sinclair [1977] have presented details on the behavior and ecology of *Syncerus caffer*).

As was the case in the *B. latifrons*-moose comparison, certain differences between late Pleistocene *B. a. antiquus* and late Holocene *S. caffer* and their environments must be kept in mind when comparing the two. Differences between Holocene tropical and subtropical environments of *S. caffer* and pre-Holocene midlatitude and high latitude environments of *B. a. antiquus* and *B. priscus* are significant insofar as annual patterns of energy production and biological reproduction are concerned. Human agency is another important difference. Human burning and hunting have long been practiced in southern and eastern Africa. These activities have probably resulted in greater disruptions of the woody vegetation (thereby increasing the region's carrying capacity by keeping it in subclimax) and reduction of the larger and presumably longer maturing, longer lived taxa. This has resulted in large numbers of *Syncerus* adapted to primitive hunting pressures, which is a parallel with *B. bison* but not with *B. antiquus*. Also, modern *Syncerus* exhibits a direct relationship between body size and openness of habitat; large bodied *S. c. caffer* are typically found in savanna habitat whereas smaller bodied *S. c. brachyceros* and *S. c. nanus* are found in forested regions (Sinclair, 1977). This pattern, at first glance, conflicts with the bison adaptation models presented here, but I believe that it represents, instead, one of two possible general adaptive strategies to forest/woodland environments. The generalist strategy results in larger bodied individuals with more elaborate ornamentation, such as *B. latifrons*, and the specialist strategy results in smaller bodied individuals with less elaborate ornamentation (Estes, 1974), such as *S. c. nanus*. Human hunting, probably more so than any other single factor, has nullified the generalist strategy of adaptation to forests and woodlands.

In summary, pre-Holocene savanna and wooded steppe bison (*B. a. antiquus* and *B. priscus*) were smaller bodied, shorter horned, relatively more hairy individuals showing stronger herding tendencies and greater social tolerances than contemporary forest/woodland bison. Savanna/wooded steppe bison were examples of relatively K-selected populations adapted to a species rich, resource abundant environment.

GRASSLAND ADAPTATION MODEL

North American midlatitude grasslands were apparently much less extensive before than during the Holocene, whereas Arctic steppe and tundra seems to have been widespread before the Holocene, especially during glacial periods (chapter 1; figs. 2–5). Open grasslands of regional extent are generally more stable over longer periods of time than either forest/woodland seral phases or savannas. These grasslands, normally the result of both climatic factors and fire, are structurally simple and offer an abundance of space and continuously distributed herbaceous vegetation readily accessible to large herbivores. The pre-Holocene Arctic steppe and tundra fauna was diverse (*Mammuthus, Equus, Bison, Coelodonta* et al.), and interspecific competition was probably high. As a result of late Wisconsin-early Holocene megafaunal extinctions (Martin, 1967), however, considerably fewer species were present in the Holocene North American and Eurasian grasslands, and consequently interspecific competition in these habitats was low.

Grassland habitats could most effectively be occupied and utilized by large bodied herbivores characterized by the following traits. (1) Grazing would be the only feeding strategy capable of supporting a viable population of large herbivores. (2) Intraspecific competition would be more important than interspecific competition, assuming that fewer species would adapt to the structurally simple grassland environment than to the structurally more complex savanna environment. As a result, conspecifics would constitute the majority of direct competitors. (3) High social cohesiveness and tolerance with a relatively complex social organization would be desirable. (4) Defense and evasion would be by factors of size or flight, with greater emphasis on flight than was characteristic of the forest/woodland or savanna/wooded steppe adaptation models.

B. b. bison is probably the only North American bison to have adapted to pure grasslands. Exaggerated body hair is probably more developed and more socially important in *B. bison* than in any other species of North American bison (fig. 87). The relatively small body size and locally high population densities of North American grassland bison are characteristics expected of relatively r-selected species, species with relatively rapid maturation rates (hence relatively small size), and relatively high biotic potential (hence potentially high population densities). *B. bison* evolved under a more r-selection regime (relative to the forest/woodland and savanna/wooded steppe regimes) characterized by heavy predation, abundant resources, and low interspecific competition. Continued periodic disruption of the *B. bison* population by human hunting has prevented the species population from coming as near the habitat's carrying capacity as was realized by pre-Holocene bison and other megafauna.

The grassland bison probably are more socially cohesive and have a more complex social organization than any other North American bison taxon. Prior to the midnineteenth century, large herds of *B. bison* probably contained hundreds—or sometimes thousands—of individuals. Smaller, more persistent groups of females and younger males, however, probably constituted the basic social unit of *B. bison*. These groups may have been organized hierarchically with female leadership, at least within some or all age and sex groups of the basic herd. These were possibly kin groups, although continuous gene flow among such groups is anticipated as a result of male dispersal. Mature males, retaining primitive dispersal tendencies, separated from the main herd and remained apart for much of the year as individuals or in small bull groups, usually returning to the larger group only for the summer rut, if at all. Tolerance of conspecifics and the amelioration of physically damaging dominance behavior has been more strongly selected for in grassland bison than in any other bison taxon. Females are relatively passive, whereas males have developed complex rituals by which relative equality is assessed. Threats and bluffs can be made, received and reacted to in ways that normally preclude actual physical combat (Fuller, 1960; Lott, 1974; McHugh, 1958). The social disadvantage of male belligerence in nonrutting periods is mollified by male dispersal between ruts.

Grassland bison are essentially grazing animals, although they will browse if shrubs and trees are available. Morphological features of the skull which are adaptive to grazing in an open habitat include the downward rotation of the head relative to the vertebral column (fig. 85; table 60); the lateral placement of the orbits, which facilitate maintaining herd contact and watching for predators or other environmental information; and shorter limbs (table 61).

Rutting is a summer activity. The auditory element is very much present and functional, but its relative significance in the rutting behavior of grassland bison is probably less than it was to earlier bison. Visual stimuli during the rut are probably more complex and meaningful in grassland bison than in savanna/wooded steppe bison and certainly more than in the forest/woodland bison. The exaggerated development of forebody and head pelage with strategically placed salients at the beard, forelimb, penis, and tail tip presents an inflated image of the actual size of the individual. The fundamental bovine broadside display in which much of the body hair is visible is most successful in exaggerating the size of the individual. This again reveals selection for apparent size, but in the form of hair—a less lethal organ than horns. The greater development of body hair in *B. bison* than in its Eurasian vicar, *B. bonasus*, may be attributed to two circumstances. The larger population levels of *B. bison* resulted in a larger number of phenotypes (= potentially more morphological variety) from which selection could choose and exaggerate more fit characters and more intense selection that resulted from the larger *B. bison* population densities. *B. bonasus* is best known from the forests of eastern Europe, but it occurred as far east as Siberia (based on specimens in the U.S. National Museum). The available record on *B. bonasus*, therefore, must be viewed as a record of a relict population surviving in what was probably a marginal portion of the species' former range—which was largely grassland habitat.

Fighting in grassland bison consists of head-to-head collisions, head wrestling, and hooking. Strongly domed and strongly fused frontals, lowering and broadening of the nasals, more perpendicular alignment of the frontals with the vertebral column and lowering of the head are morphological adaptations that increase the effectiveness of and ability to withstand frequent fights (figs. 81, 85, and 86). Externally, the exaggerated mop of hair covering the frontals functions as a cushion to reduce the force of a collision before the impact reaches the skeleton (plate 28). The horns have been reduced in length and the tips are located further to the rear than in forest/woodland or savanna/wooded steppe bison (figs. 78 and 87).

Dominance linked activities are probably more frequent in grassland bison than they were in earlier bison. This is in part a result of the relatively prolonged rut during which several females may come into estrous at about the same time, and in part a result of relatively large numbers of males contending for the privilege of servicing those females. Male-dominance hierarchies are probably not as easily, as clearly, or as permanently established in larger groups as in smaller groups. Consequently, dominant males in grassland bison spend more rutting energy in maintaining dominance than in tending and servicing, relative to savanna/wooded steppe or forest/woodland bison. This results in proportionately more males contributing to the annual reproductive effort, a process that maintains somewhat greater variability than would be the case in smaller groups by diversifying the overall genetic contributions of males to the calf crop. In relatively r-selected populations, however, as Gould (1977) points out, quantity, not quality, is generally the product.

In summary, Holocene grassland bison (*B. b. bison*) are small bodied, short legged, short horned, herding individuals with exaggerated body hair and with a cranium specialized for grazing, maintaining herd contact in an open environment, and fighting in such a way as to minimize actual damage to individuals of the group while frequently reordering the dominance hierarchy. The small body size is attributable to accelerated sexual maturation, the failure of the species population to approximate the carrying capacity of its environment, and the absence of selective factors requiring larger body size. *B. b. athabascae*, which has been subject to most of the same Holocene selective forces as has *B. b. bison*, differs from the grassland bison mainly as a result of the somewhat more intense forest opening selection regime favoring larger body size and an inverse horn-to-body hair ratio. Geist and Karsten (1977) have compared and differentiated the size and external appearance of the two *B. bison* subspecies, and attributed at least some of these differences to the selective effect of different environments.

SUMMARY

The foregoing models have presented hypothetical views on the possible nature of bison adaptation to pre-Holocene forest/woodland and savanna/wooded steppe habitats and Holocene grassland habitats, as developed around inferences made from the identification of morphological characteristics of the various taxa occupying these habitats (table 63).

Bison adaptations to three habitats, as hypothesized above, consist of a series of adaptive complexes, each characterized by relatively differentiated emphasis on shared characters. Adaptations to the forest/woodland habitat, at one extreme, is considered to emphasize individualistic or selfish tendencies in the context of scarce resources, enhanced by absolute size and strength that are attained by a generally prolonged period of somatic development and sexual maturation. Adaptation to the grassland habitat, at the other extreme, is considered to emphasize social cohesiveness amidst abundant resources and low interspecific competition, enhanced by small size, more complex social organization, and greater tolerance of others. Adaptation to savanna and wooded steppe habitats was characterized by intermediacy in the development and function of adaptive characters, relative to those of the forest/woodland and grassland habitats. The point is stressed that the preceding models are simply logical interpretations of bison adaptations to different environments.

CHAPTER *6*

UNUSUAL MORPHOLOGIES: POSSIBLE EVIDENCE OF HYBRIDIZATION AND GENETIC DRIFT

Many species of large mammals are composed of numerous demes distributed throughout the species' range. Some demes exist on the periphery of the species' range, and individuals from these peripheral demes are likely to disperse into new habitats or regions when dispersal is permitted by environmental changes or instigated by population pressures. Such dispersal has several potential consequences besides range expansion, including hybridization with individuals of other species or isolation from the species' gene pool and the subsequent development of unbalanced polymorphism as a result of genetic drift and inbreeding. Isolation of demes within an established range can also occur if surrounding demes are destroyed or if barriers to interdemic gene flow develop.

Both hybridization and genetic drift are known or inferred to exist among natural populations (e.g., R. Doughty, written comm., August 12, 1979; Heptner, Nasimovitsch, and Bannikov, 1961; Lande, 1976, 1979; McDonnell, Gartside, and Littlejohn, 1978; Palmieri, 1968; Patton, 1973; Schaller, 1977; Schueler and Rising, 1976; Valdez, Nadler, and Bunch, 1978; E. Wilson, 1975; Wint, 1960; Woodruff, 1973; pers. observ.). These processes have received little attention insofar as their roles in accounting for variation in bison morphology are concerned, but hybridization between *B. b. bison* and *B. b. athabascae* has been assumed for bison in Wood Buffalo National Park since the large-scale introduction of *B. b. bison* into the park after 1924, and Wilson (1969, 1974*b*) has invoked both hybridization and genetic drift to explain the high variability and morphological abnormalities that characterize late Wisconsin-early Holocene bison from the western Great Plains. Both processes are important considerations in assessing morphological variation in bison; they simply and logically account for some of the most enduring problems in the study of bison taxonomy and morphological variation, including intermediate phenotypes, abnormal specimens, and the origin of *B. a. occidentalis*.

HYBRIDS AND HYBRIDIZATION

I consider horn cores possessing characters diagnostic of both North American and Eurasian autochthons to be hybrid specimens. Hybrids of only North American (e.g., *B. latifrons* X *B. a. antiquus*) or only Eurasian (e.g., *B. alaskensis* X *B. priscus*) autochthons were not recognized among fossil specimens although such hybrid specimens undoubtedly exist. Adult *B. b. athabascae* collected after 1928 are considered to represent *B. b. athabascae* X *B. b. bison* hybrids. Here I will present information on the nature of variation characterizing what are considered hybrid populations, document the distribution of hybrid specimens, and designate hybrid zones which appear to have existed in the past.

B. b. athabascae X B. b. bison

B. b. athabascae had been nearly exterminated when it was protected by law in 1891. Wood Buffalo reserve was created in 1893 as a refuge for the remaining population. By the early 1920s the *B. b. athabascae* population numbered about 1500 animals. From 1925 to 1928, however, 6,673 *B. b. bison* were released in the Wood Buffalo National Park (Banfield and Novakowski, 1960). The introduced bison freely interbred with the numerically fewer *B. b. athabascae* and, as a result, few true wood bison now remain. Most bison in Wood Buffalo National Park are assumed to be *athabascae* X *bison* hybrids, although a pure population survived in one remote portion of the park (Banfield and Novakowski, 1960). Available biometric data support the contention that widespread hybridization occurred. Bison from Wood Buffalo National Park collected after 1928 are phenotypically intermediate between *B. b. bison* from south of the boreal forest and pre-1929 adult *B. b. athabascae* (tables 64–68).

Table 64

Summary of *Bison bison athabascae* (Post-1928 Adults) Skull Biometrics

Standard measurement	n	Range	Mean ± SE	σ	V
Males					
Spread of horn cores, tip to tip	8	555 – 778 mm	636.1 ± 24.0 mm	67.9 mm	10.6
Horn core length, upper curve, tip to burr	8	183 – 249 mm	219.3 ± 7.0 mm	20.0 mm	9.1
Straight line distance, tip to burr, dorsal horn core	8	167 – 230 mm	195.4 ± 6.2 mm	17.6 mm	9.0
Dorso-ventral diameter, horn core base	8	81 – 109 mm	91.5 ± 3.3 mm	9.5 mm	10.4
Minimum circumference, horn core base	8	233 – 336 mm	280.3 ± 10.2 mm	29.0 mm	10.3
Width of occipital at auditory openings	11	243 – 276 mm	266.2 ± 2.8 mm	9.4 mm	3.5
Width of occipital condyles	10	117 – 142 mm	131.1 ± 2.0 mm	6.5 mm	5.0
Depth, nuchal line to dorsal margin of foramen magnum	10	96 – 114 mm	106.2 ± 1.6 mm	5.0 mm	4.8
Antero-posterior diameter, horn core base	8	77 – 112 mm	91.5 ± 3.6 mm	10.2 mm	11.1
Least width of frontals, between horn cores and orbits	11	264 – 303 mm	284.7 ± 4.0 mm	13.4 mm	4.7
Greatest width of frontals at orbits	11	311 – 365 mm	345.4 ± 5.9 mm	19.6 mm	5.6
M^1–M^3, inclusive alveolar length	0	....	....	....	....
M^3, maximum width, anterior cusp	0	....	....	....	....
Distance, nuchal line to tip of premaxillae	11	512 – 602 mm	555.6 ± 8.6 mm	28.6 mm	5.1
Distance, nuchal line to nasal-frontal suture	11	233 – 268 mm	251.8 ± 3.5 mm	11.9 mm	4.7
Angle of divergence of horn cores, forward from sagittal	7	64° – 69°	66.8° ± 0.9°	2.4°	3.6
Angle between foramen magnum and occipital planes	10	115° – 136°	125.4° ± 2.3°	7.3°	5.8
Angle between foramen magnum and basioccipital planes	10	109° – 127°	117.8° ± 1.8°	5.8°	4.9
Females					
Spread of horn cores, tip to tip	3	429 – 457 mm	440.0 ± 8.6 mm	14.9 mm	3.3
Horn core length, upper curve, tip to burr	3	131 – 177 mm	156.0 ± 13.4 mm	23.2 mm	14.9
Straight line distance, tip to burr, dorsal horn core	3	127 – 144 mm	137.6 ± 5.3 mm	9.2 mm	6.7
Dorso-ventral diameter, horn core base	3	53 – 56 mm	54.3 ± 0.8 mm	1.5 mm	2.8
Minimum circumference, horn core base	3	164 – 180 mm	171.3 ± 4.6 mm	8.0 mm	4.7
Width of occipital at auditory openings	3	205 – 220 mm	213.7 ± 4.4 mm	7.7 mm	3.6
Width of occipital condyles	3	109 – 123 mm	115.3 ± 4.0 mm	7.0 mm	6.1
Depth, nuchal line to dorsal margin of foramen magnum	3	93 – 100 mm	95.3 ± 2.3 mm	4.0 mm	4.2
Antero-posterior diameter, horn core base	3	53 – 59 mm	56.0 ± 1.7 mm	3.0 mm	5.3
Least width of frontals, between horn cores and orbits	3	227 – 232 mm	228.6 ± 1.6 mm	2.8 mm	1.2
Greatest width of frontals at orbits	3	275 – 280 mm	277.0 ± 1.5 mm	2.6 mm	0.9
M^1–M^3, inclusive alveolar length	0	....	....	....	....
M^3, maximum width, anterior cusp	0	....	....	....	....
Distance, nuchal line to tip of premaxillae	3	495 – 514 mm	501.3 ± 6.3 mm	10.9 mm	2.1
Distance, nuchal line to nasal-frontal suture	3	210 – 221 mm	214.0 ± 3.5 mm	6.0 mm	2.8
Angle of divergence of horn cores, forward from sagittal	3	61° – 70°	66.3° ± 2.7°	4.7°	7.1
Angle between foramen magnum and occipital planes	3	124° – 131°	128.0° ± 2.0°	3.6°	2.8
Angle between foramen magnum and basioccipital planes	3	109° – 113°	111.3° ± 1.2°	2.0°	1.8

Note: The above sample was collected after 1928 and is assumed to represent *B. b. athabascae* X *B. b. bison* hybrids.

Table 65

Skull Allometries in *Bison bison athabascae* (Post-1928 Adults)

Standard measurement	Relative to *B. b. bison* ♂♂	Relative to *B. b. bison* ♀♀	Relative to *B. b. athabascae* ♂♂	Relative to *B. b. athabascae* ♀♀
Spread of horn cores, tip to tip	1.05	0.98	0.93	....
Horn core length, upper curve, tip to burr	1.15	1.26	0.93	....
Straight line distance, tip to burr, dorsal horn core	1.13	1.18	0.94	....
Dorso-ventral diameter, horn core base	1.12	1.06	1.00	....
Minimum circumference, horn core base	1.10	1.05	0.97	....
Width of occipital at auditory openings	1.09	1.06	0.97	....
Width of occipital condyles	1.04	1.00	1.00	....
Depth, nuchal line to dorsal margin of foramen magnum	1.08	1.13	1.08	....
Antero-posterior diameter, horn core base	1.10	1.09	0.94	....
Least width of frontals, between horn cores and orbits	1.05	1.05	0.97	....
Greatest width of frontals at orbits	1.06	1.04	0.97	....
M^1 – M^3, inclusive alveolar length	1.00	1.15	0.99	....
M^3, maximum width, anterior cusp	0.98	1.00	0.98	....
Distance, nuchal line to tip of premaxillae	1.04	1.03	0.97	....
Distance, nuchal line to nasal-frontal suture	1.02	1.01	0.98	....

Note: Calculations are based on data presented in tables 29, 34, and 64. The *B. b. athabascae* (post-1928 adults) mean for each character = 1.00. Values reported in this table are multiples of the *B. b. athabascae* (post-1928 adults) mean for each character. Here, post-1928 adult *B. b. athabascae* are synonymous with *B. b. athabascae* X *B. b. bison* hybrids.

Table 66

Summary of *Bison bison athabascae* (Post-1928 Adults) Limb Biometrics

Standard measurement	n	Range	Mean ± SE	σ	V
Males					
Humerus:					
Approximate rotational length of bone	4	324 – 357 mm	346.0 ± 7.5 mm	15.0 mm	4.3
Antero-posterior diameter of diaphysis	4	60 – 65 mm	63.3 ± 1.1 mm	2.3 mm	3.7
Transverse minimum of diaphysis	4	48 – 55 mm	52.3 ± 1.5 mm	3.1 mm	5.9
Radius:					
Approximate rotational length of bone	4	318 – 348 mm	336.3 ± 7.1 mm	14.2 mm	4.2
Antero-posterior minimum of diaphysis	4	32 – 36 mm	34.0 ± 0.9 mm	1.8 mm	5.3
Transverse minimum of diaphysis	4	53 – 59 mm	57.5 ± 1.5 mm	3.0 mm	5.2
Metacarpal:					
Total length of bone	5	205 – 230 mm	220.0 ± 4.1 mm	9.3 mm	4.2
Antero-posterior minimum of diaphysis	5	27 – 31 mm	29.2 ± 0.6 mm	1.4 mm	5.0
Transverse minimum of diaphysis	5	44 – 52 mm	48.6 ± 1.2 mm	2.9 mm	6.1
Femur:					
Approximate rotational length of bone	4	372 – 433 mm	409.3 ± 13.0 mm	26.1 mm	6.3
Antero-posterior diameter of diaphysis	4	47 – 55 mm	51.0 ± 1.8 mm	3.6 mm	7.1
Transverse minimum of diaphysis	4	48 – 51 mm	49.5 ± 0.6 mm	1.2 mm	2.6
Tibia:					
Approximate rotational length of bone	4	362 – 411 mm	389.8 ± 10.4 mm	20.9 mm	5.3
Antero-posterior minimum of diaphysis	4	36 – 38 mm	37.0 ± 0.4 mm	0.8 mm	2.2
Transverse minimum of diaphysis	4	48 – 53 mm	51.0 ± 1.0 mm	2.1 mm	4.2
Metatarsal:					
Total length of bone	4	256 – 279 mm	268.0 ± 5.4 mm	10.8 mm	4.0
Antero-posterior minimum of diaphysis	5	30 – 35 mm	32.6 ± 0.8 mm	1.9 mm	5.9
Transverse minimum of diaphysis	5	34 – 39 mm	37.6 ± 0.8 mm	2.0 mm	5.5
Females					
Humerus:					
Approximate rotational length of bone	1	....	303.0 mm	....	...
Antero-posterior diameter of diaphysis	1	....	48.0 mm	....	...
Transverse minimum of diaphysis	1	....	41.0 mm	....	...
Radius:					
Approximate rotational length of bone	1	....	310.0 mm	....	...
Antero-posterior minimum of diaphysis	1	....	26.0 mm	....	...
Transverse minimum of diaphysis	1	....	45.0 mm	....	...
Metacarpal:					
Total length of bone	2	210 – 218 mm	214.0 ± 3.9 mm	5.6 mm	2.6
Antero-posterior minimum of diaphysis	2	24 – 25 mm	24.5 ± 0.4 mm	0.7 mm	2.8
Transverse minimum of diaphysis	2	37 – 40 mm	38.5 ± 1.4 mm	2.1 mm	5.5
Femur:					
Approximate rotational length of bone	1	....	360.0 mm	....	...
Antero-posterior diameter of diaphysis	1	....	43.0 mm	....	...
Transverse minimum of diaphysis	1	....	41.0 mm	....	...
Tibia:					
Approximate rotational length of bone	1	....	348.0 mm	....	...
Antero-posterior minimum of diaphysis	1	....	30.0 mm	....	...
Transverse minimum of diaphysis	1	....	42.0 mm	....	...
Metatarsal:					
Total length of bone	2	256 – 260 mm	258.0 ± 1.9 mm	2.8 mm	7.1
Antero-posterior minimum of diaphysis	2	....	28.0 mm	....	...
Transverse minimum of diaphysis	2	29 – 33 mm	31.0 ± 1.9 mm	2.8 mm	9.1

Note: The above sample was collected after 1928 and is assumed to represent *B. b. athabascae* X *B. b. bison* hybrids.

Table 67

Limb Bone Allometries in *Bison bison athabascae* (Post-1928 Adults) Relative to True *Bison bison bison* and *Bison bison athabascae*

Standard measurement	Relative to *B. b. bison* ♂♂	Relative to *B. b. bison* ♀♀	Relative to *B. b. athabascae* ♂♂	Relative to *B. b. athabascae* ♀♀
Humerus:				
Approximate rotational length of bone	1.06	1.06	.98	.99
Antero-posterior diameter of diaphysis	1.03	.96	.93	.96
Transverse minimum of diaphysis	1.03	1.02	.89	1.02
Radius:				
Approximate rotational length of bone	1.05	1.06	.98	.99
Antero-posterior minimum of diaphysis	1.06	.98	.97	.92
Transverse minimum of diaphysis	1.07	1.04	.96	.97
Metacarpal:				
Total length of bone	1.07	1.08	.96	.97
Antero-posterior minimum of diaphysis	1.07	1.02	.96	.94
Transverse minimum of diaphysis	1.07	1.05	.93	.98
Femur:				
Approximate rotational length of bone	1.07	1.05	.97	.99
Antero-posterior diameter of diaphysis	1.06	1.04	.94	.98
Transverse minimum of diaphysis	1.05	1.01	.96	1.00
Tibia:				
Approximate rotational length of bone	1.05	1.04	.97	.96
Antero-posterior minimum of diaphysis	1.06	.99	.97	.94
Transverse minimum of diaphysis	1.04	1.00	.95	.98
Metatarsal:				
Total length of bone	1.05	1.05	.97	.98
Antero-posterior minimum of diaphysis	1.08	1.02	.98	.97
Transverse minimum of diaphysis	1.07	1.06	.96	1.00

Note: Calculations are based on data presented in tables 30, 35, and 66. The ***B. b. athabascae*** (post-1928 adults) mean for each character = 1.00. Values reported in the table are multiples of the ***B. b. athabascae*** (post-1928 adults) mean for each character. ***B. b. athabascae*** (post-1928 adults) are assumed to be ***B. b. athabascae*** × *B. b. bison* hybrids.

Table 68

Bison bison athabascae (Post-1928 Adults): Referred Specimens

Institution and identification number		Sex	Provenience
AMNH	M 86950	♂	Wood Buffalo National Park (1935)
AMNH	M 98228	♂	10 miles from Slave River, Alberta (1935)
AMNH	M 98229	♂	10 miles from Slave River, Alberta (1935)
AMNH	M 98953	♂	Salt Lick, Five Acre Wallow, Wood Buffalo National Park (1936)
AMNH	M 98954	♀	Salt Plain, 7 mile camp, Alberta (1934)
AMNH	M 98957	♀	Salt Plain, 7 mile camp, Alberta (1934)
AMNH	M 130171	♂	Salt River, Wood Buffalo National Park (1936)
NMC	M 11436	♂	Cabin 7, Murdock Creek, Wood Buffalo National Park
NMC	M 24026	♂	Nyarling River, Frying Pan Lake, Wood Buffalo National Park
NMC	M 39876	♀	Needle Lake Area, Wood Buffalo National Park (1968)
PMA	Z 68.31.1	♂	Wood Buffalo National Park (1930s)
PMA	Z 73.82.1	♂	Elk Island National Park, Alberta (1973)
PMA	Z 74.86.1	♂	Alberta Game Farm, Alberta (1974)
USNM	M 263390	♂	Fort Fitzgerald, Alberta (1936-7)

Note: These specimens were collected after 1928; known dates of collection are given in parentheses.

Male *athabascae* X *bison* hybrids average about 5 percent larger than *B. b. bison* in frontal and occipital characters, but about 11 percent larger than *B. b. bison* in horn core characters (table 65). *Athabascae* X *bison* hybrids average about 1.5 percent smaller than pre-1929 adult *B. b. athabascae* in frontal and occipital characters, but almost 5 percent smaller than true *B. b. athabascae* in horn core characters (table 65). The limbs of crosses are about 6 percent larger than *B. b. bison* and about 4 percent smaller than true *B. b. athabascae* (tables 66 and 67). Female crosses show similar size differences. *Athabascae* X *bison* crosses are, therefore, phenotypically intermediate between *B. b. bison* and true *B. b. athabascae*, but they are nearer *B. b. athabascae* than *B. b. bison* in size. Male horn cores are proportionately larger than other skeletal characters in the crosses, which suggests the nature of selection in the parkland environment. The numerical superiority of *B. b. bison* to *B. b. athabascae* immediately following the releases of plains bison into Wood Buffalo National Park during the late 1920s could have resulted in the complete swamping of wood bison by plains bison. The fact that *athabascae* X *bison* crosses are nearer true *B. b. athabascae* than *B. b. bison* in size suggests that selection has been relatively stronger than simple genotypic frequency (in late 1920s) in determining the resultant phenotypic characters of hybrids. The forest/woodland adaptation model presented in chapter 5 emphasized the intensity of selection characteristic of these habitat types and the adaptive value of both larger body size and exaggerated dominance organs. Larger horns and larger body size would predictably convey greater competitive and dominance fitness than would smaller characters. Larger phenotypes, with emphasis on larger horns, would, therefore, be selectively favored by the relatively wooded environment of Wood Buffalo National Park.

In contrast, size increase, especially of horn cores, has been attributed to an abundance of nutritious forage (Geist, 1971*a*, oral comm., September 2, 1978; Guthrie, 1966*a*, 1970, 1978). If environmental benevolence alone were the determinant of organic and character size, however, the *athabascae* X *bison* hybrids should be as large as true *B. b. athabascae*; they are not.

Bison "crassicornis"

The term *B. crassicornis* was introduced by Richardson (1852–1854) as the name for a new species of heavy horned bison from Alaska. Skinner and Kaisen (1947) recognized *B. crassicornis* as a species in their revision, and the term is still used in publications, although normally as a subspecies or synonym of *B. priscus*. Herein, *B. "crassicornis"* is used in reference to all hybrids of North American X Eurasian autochthons from the Beringian region regardless of the species involved (which in most cases can only be guessed anyway). Specimens identified as *B. "crassicornis"* constitute a majority of the skulls studied from Beringia. The fact that many specimens from Beringia show diagnostic characters of both North American and Eurasian autochthons, that these specimens occur in the most likely zone of contact between dispersing populations from both continents (fig. 90), and that dispersals of taxa into Beringia from both continents have been confirmed by nonhybrid specimens collected from Beringia, lead to the conclusion that the specimens that possess shared characters can reasonably be considered hybrids.

B. "crassicornis" averages smaller than other pre-Holocene North American bison in most skull characters, but the minimum observed size for most characters in *B. "crassicornis"* is usually very near or greater than the minima observed for the same characters in other pre-Holocene North American bison. Horn core characters in *B. "crassicornis,"* however, tend to be either larger than, or intermediate to, those of *B. priscus* or *B. a. antiquus*, the two pre-Holocene taxa most nearly resembling *B. "crassicornis"* and the two taxa that, probably, most frequently interbred to give rise to these hybrids. Limb evidence from Old Crow River and Lost Chicken Creek suggests that *B. "crassicornis"* was about the same size as *B. a. antiquus* (figs. 45–56) or perhaps a little smaller, considering that the smaller limb elements from the Fairbanks and Dawson areas probably represent *B. "crassicornis"* as well as *B. priscus* (and, perhaps, *B. a. antiquus* and *B. bison* as well). The distribution of cases within character samples is normal or near normal for both sexes and the ranges of variation for samples are about what is expected (fig. 57; table 69). Sexual dimorphism is somewhat greater than is observed in other samples (tables 46 and 70), and the coefficients of variation tend to be lower than those observed in other samples (tables 50 and 69).

The fact that *B. "crassicornis"* horn cores are larger than, or intermediate to, those of *B. a. antiquus* or *B. priscus*, and that other skull characters average smaller than those of *B. a. antiquus* or *B. priscus*, presents a problem that can be resolved to a degree by consulting the analyses of variation among bison (chapter 3) and the various adaptation models developed earlier (chapter 5). Mean coefficients of variation for bison skull and limb characters (tables 52 and 53) show that horn cores are more variable than other skeletal characters. The differences between r- and K-selection regimes were noted in chapter 5. Beringia was probably characterized by a relatively r-type selection regime for large herbivores during the transition to glacial or stadial periods. This occurred because available habitat and carrying capacity increased as ocean levels dropped and interglacial/interstadial forests and woodlands withdrew to more suitable environments or disintegrated. At other times the subaerial Beringian environment was probably characterized by a K-type selection regime. This is because once the habitat made available by the onset of glacial/stadial conditions became completely occupied by the megaherbivores, populations would tend to approximate the region's carrying capacity.

The small mean size of *B. "crassicornis"* is explained by the spread of *B. a. antiquus* and *B. priscus* populations into Beringia during the onset of glacial/stadial conditions. Dispersal through this relatively r-type selection regime permitted greater representation of small phenotypes in the population as the formerly dominant K-type selection regimes, prevailing on both sides of Beringia, were relaxed. Smaller phenotypes, then, would characterize the r-type, or

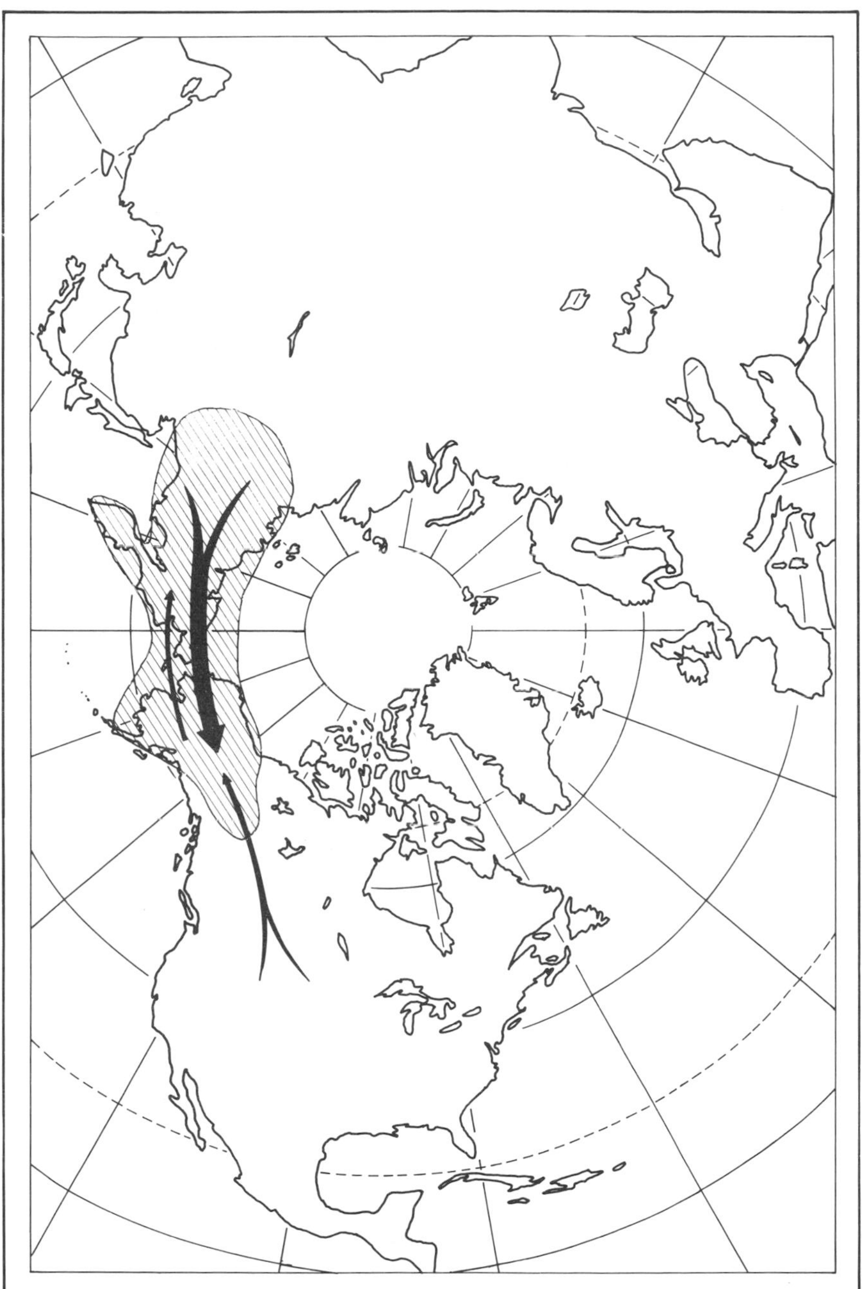

FIGURE 90 The Beringian hybrid zone. The heavy arrow represents the strong dispersal of Eurasian bison into glacial/stadial eastern Beringia. The lighter arrows represent the moderate dispersal of North American bison into (a) interglacial/interstadial Alaska and (b) glacial/stadial Siberia.

Table 69

Summary of *"Bison crassicornis"* Skull Biometrics

Standard measurement	n	Range	Mean ± SE	σ	V
Males					
Spread of horn cores, tip to tip	46	635 – 1192 mm	928.8 ± 17.4 mm	118.4 mm	12.7
Horn core length, upper curve, tip to burr	63	292 – 512 mm	390.8 ± 5.9 mm	47.0 mm	12.0
Straight line distance, tip to burr, dorsal horn core	63	252 – 455 mm	350.7 ± 5.1 mm	40.8 mm	11.6
Dorso-ventral diameter, horn core base	73	90 – 123 mm	102.1 ± 0.9 mm	7.8 mm	7.6
Minimum circumference, horn core base	73	297 – 375 mm	325.2 ± 2.3 mm	20.4 mm	6.2
Width of occipital at auditory openings	50	252 – 292 mm	271.6 ± 1.3 mm	9.6 mm	3.5
Width of occipital condyles	55	126 – 151 mm	139.8 ± 0.8 mm	6.1 mm	4.4
Depth, nuchal line to dorsal margin of foramen magnum	57	86 – 115 mm	101.1 ± 0.8 mm	6.2 mm	6.1
Antero-posterior diameter, horn core base	73	96 – 135 mm	110.1 ± 0.9 mm	8.4 mm	7.6
Least width of frontals, between horn cores and orbits	60	251 – 311 mm	286.3 ± 1.8 mm	14.0 mm	4.9
Greatest width of frontals at orbits	46	320 – 397 mm	346.0 ± 2.1 mm	14.6 mm	4.2
M^1–M^3, inclusive alveolar length	0	....	....	....	....
M^3, maximum width, anterior cusp	0	....	....	....	....
Distance, nuchal line to tip of premaxillae	8	582 – 629 mm	600.1 ± 5.6 mm	15.9 mm	2.6
Distance, nuchal line to nasal-frontal suture	15	248 – 313 mm	274.8 ± 4.0 mm	15.8 mm	5.7
Angle of divergence of horn cores, forward from sagittal	20	64° – 81°	73.1° ± 0.9°	4.1°	5.6
Angle between foramen magnum and occipital planes	59	115° – 150°	133.2° ± 0.9°	7.4°	5.5
Angle between foramen magnum and basioccipital planes	59	108° – 130°	116.2° ± 0.7°	5.4°	4.7
Females					
Spread of horn cores, tip to tip	14	517 – 667 mm	591.6 ± 10.6 mm	397.7 mm	6.7
Horn core length, upper curve, tip to burr	13	171 – 275 mm	217.9 ± 7.9 mm	28.8 mm	13.2
Straight line distance, tip to burr, dorsal horn core	13	165 – 251 mm	208.5 ± 6.7 mm	24.4 mm	11.7
Dorso-ventral diameter, horn core base	19	57 – 72 mm	63.0 ± 0.8 mm	3.5 mm	5.6
Minimum circumference, horn core base	18	183 – 223 mm	198.9 ± 2.4 mm	10.5 mm	5.2
Width of occipital at auditory openings	3	218 – 225 mm	220.7 ± 2.1 mm	3.7 mm	1.7
Width of occipital condyles	8	120 – 134 mm	125.5 ± 1.6 mm	4.8 mm	3.8
Depth, nuchal line to dorsal margin of foramen magnum	4	86 – 91 mm	87.7 ± 1.9 mm	3.9 mm	4.5
Antero-posterior diameter, horn core base	19	57 – 69 mm	62.6 ± 0.7 mm	3.4 mm	5.4
Least width of frontals, between horn cores and orbits	19	202 – 242 mm	222.4 ± 2.1 mm	9.4 mm	4.2
Greatest width of frontals at orbits	14	254 – 308 mm	277.9 ± 3.5 mm	13.7 mm	4.9
M^1–M^3, inclusive alveolar length	0	....	....	....	....
M^3, maximum width, anterior cusp	0	....	....	....	....
Distance, nuchal line to tip of premaxillae	1	....	561.0 mm	....	....
Distance, nuchal line to nasal-frontal suture	5	218 – 240 mm	232.0 ± 3.8 mm	8.6 mm	3.7
Angle of divergence of horn cores, forward from sagittal	19	61° – 77°	68.8° ± 1.0°	4.4°	6.3
Angle between foramen magnum and occipital planes	6	123° – 132°	127.0° ± 1.3°	3.3°	2.6
Angle between foramen magnum and basioccipital planes	7	107° – 116°	111.6° ± 1.3°	3.6°	3.3

Table 70

Absolute and Relative Sexual Dimorphism in *"Bison crassicornis"* Skulls

Standard measurement	Absolute (mm)	Relative[a]
Spread of horn cores, tip to tip	337	.63
Horn core length, upper curve, tip to burr	173	.55
Straight line distance, tip to burr, dorsal horn core	142	.59
Dorso-ventral diameter, horn core base	39	.61
Minimum circumference, horn core base	126	.61
Width of occipital at auditory openings	51	.81
Width of occipital condyles	14	.89
Depth, nuchal line to dorsal margin of foramen magnum	13	.86
Antero-posterior diameter, horn core base	47	.56
Least width of frontals, between horn cores and orbits	64	.77
Greatest width of frontals at orbits	68	.80
M^1 – M^3, inclusive alveolar length	...	...
M^3, maximum width, anterior cusp	...	...
Distance, nuchal line to tip of premaxillae	39	.93
Distance, nuchal line to nasal-frontal suture	43	.84

[a]The male mean for each character = 1.00. Values given are multiples of the male mean.

Note: Calculations are based on data presented in table 69.

colonization, phase of the megaherbivore occupation of glacial/stadial Beringia. (See Geist [1971*b*, 1974], for a sharply contrasting point of view.) The meeting of eastward dispersing Eurasian autochthons and westward dispersing North American autochthons (fig. 90) would result in hybridization between the two.

Once the available habitat was completely occupied and a more K-selection regime had developed, selection for large size would again occur. Presumably, the most variable characters would be the first to respond to selection, and the least variable characters would be the last to respond to selection. Thus, according to this reasoning, K-selection would result in larger horns appearing in phenotypes before other, less variable characters showed size increases. Also, with the passing of time, the K-selected *B. priscus* population would probably expand into Beringia and swamp the hybrid populations.

An alternate point of view, and the traditional one, maintains that what is here considered *B. "crassicornis"* was ancestral to Holocene North American bison. According to this point of view Beringian bison became smaller near the end of the Wisconsin and some populations of these dwarfing bison dispersed southward into midlatitude North America and there gave rise to *B. occidentalis* (= *B. a. occidentalis* of this classification) and *B. bison*. Evidence presented in chapters 3 and 5, however, suggests that horn cores should perhaps be the first and most obvious of characters to change, not the last. A further difficulty is that the *B. "crassicornis"* sample (table 71) probably represents a longer period of time than just the late Wisconsin, and it probably overlaps with *B. priscus*, *B. a. antiquus*, and *B. alaskensis* samples as well. *B. "crassicornis"* is better considered a hybrid than a discrete regional taxon because it is not morphologically distinct from *B. priscus* and *B. a. antiquus*, but it is instead generally intermediate, and the environment of late Quaternary Beringia was too dynamic, habitats were too varied over several glacial and interglacial oscillations, and other species of bison were too frequently present to convincingly argue that *B. "crassicornis"* was specifically distinct and homeostatic. Beringia was probably much more important as a megafaunal crossroad and mixing tank than a source area for new taxa.

Hybrid Zones

Eurasian autochthons occur widely over Beringia and midlatitude North America (figs. 26 and 27). So, too, do specimens that possess diagnostic characters of both North American and Eurasian autochthons, and, although these are relatively rare south of Beringia, they serve to identify zones where hybridization seems to have taken place between North American and Eurasian autochthons.

The most intense zone of hybridization was Beringia, which alternated between being an easterly extension of the Palearctic during glacial/stadial periods and a Nearctic outlier during interglacial/interstadial periods. Over half of the skulls studied from Alaska and northwest Canada shared North American and Eurasian characters. This zone probably formed as eastern Siberian bison dispersed eastward into Beringia during the onset of

Table 71
"Bison crassicornis": Referred Specimens

Institution and identification number	Sex	Provenience
AMNH VP F:AM13721	♂	Fox Gulch, Yukon Territory
AMNH VP F:AMA173-8103	♀	Ester Creek, Alaska
AMNH VP F:AMA256-7389	♀	Goldstream at Fox, Alaska
AMNH VP F:AMA284-6594	♀	Goldstream at Fox, Alaska
AMNH VP F:AMA284-6595	♀	Goldstream at Fox, Alaska
AMNH VP F:AMA284-6600	♀	Goldstream at Fox, Alaska
AMNH VP F:AMA437-1348	♂	Fairbanks area, Alaska
AMNH VP F:AMA449-1041	♀	Fairbanks Creek, Alaska
AMNH VP F:AMA502-1013	♀	Fairbanks Creek, Alaska
AMNH VP F:AMA579-2034	♀	Fairbanks Creek, Alaska
AMNH VP F:AMA579-2035	♀	Fairbanks Creek, Alaska
AMNH VP F:AMA580-2037	♀	Fairbanks Creek, Alaska
AMNH VP F:AMA580-2040	♀	Goldstream at Fox, Alaska
AMNH VP F:AMA605-1003	♂	Fairbanks Creek, Alaska
AMNH VP F:AMA605-1022	♀	Fairbanks Creek, Alaska
AMNH VP F:AMA605-3002	♂	Dome Creek, Alaska
AMNH VP F:AMA606-1082	♂	Fairbanks Creek, Alaska
AMNH VP F:AMA606-1103	♀	Fairbanks Creek, Alaska
AMNH VP F:AMA683-3244	♂	Dome Creek, Alaska
AMNH VP F:AMA683-3251	♂	Dome Creek, Alaska
AMNH VP F:AMA683-3256	♂	Dome Creek, Alaska
AMNH VP F:AMA683-8090	♂	Chatam Creek, Alaska
AMNH VP F:AMA684-3253	♀	Dome Creek, Alaska
AMNH VP F:AMA685-3262	♀	Dome Creek, Alaska
AMNH VP F:AMA685-3265	♂	Dome Creek, Alaska
AMNH VP F:AMA698-2026	♂	Fairbanks Creek, Alaska
AMNH VP F:AMA698-2031	♀	Fairbanks Creek, Alaska
AMNH VP F:AMA862-	♀	Cleary Creek, Alaska
AMNH VP F:AMA863-	♀	Cleary Creek, Alaska
AMNH VP F:AM30512	♂	Cleary Creek, Alaska
AMNH VP F:AM30517	♂	Fairbanks area, Alaska
AMNH VP F:AM30523	♂	Goldstream, Alaska
AMNH VP F:AM30524	♂	Fairbanks area, Alaska
AMNH VP F:AM30526	♂	Fairbanks area, Alaska
AMNH VP F:AM30529	♂	Fairbanks area, Alaska
AMNH VP F:AM30530	♂	Cleary Creek, Alaska
AMNH VP F:AM30533	♂	Fairbanks area, Alaska
AMNH VP F:AM30541	♂	Fairbanks area, Alaska
AMNH VP F:AM30551	♂	Lillian Creek, Alaska
AMNH VP F:AM30558	♂	17 miles north of Fairbanks, Alaska
AMNH VP F:AM30567	♂	Cleary Creek, Alaska
AMNH VP F:AM30571	♂	Cleary Creek, Alaska
AMNH VP F:AM30582	♂	Cleary Creek, Alaska
AMNH VP F:AM30583	♂	Cleary Creek, Alaska
AMNH VP F:AM30588	♂	Goldstream, Alaska
AMNH VP F:AM30595	♂	Cleary Creek, Alaska
AMNH VP F:AM30601	♂	Goldstream, Alaska
AMNH VP F:AM30617	♂	Cleary Creek, Alaska
AMNH VP F:AM30618	♂	Cleary Creek, Alaska
AMNH VP F:AM30622	♂	Gilmore Creek, Alaska
AMNH VP F:AM30629	♂	Fairbanks area, Alaska

Table 71

"Bison crassicornis": Referred Specimens (Continued)

Institution and identification number			Sex	Provenience
AMNH	VP	F:AM30631	♂	Goldstream, Alaska
AMNH	VP	F:AM30644	♂	Goldstream, Alaska
AMNH	VP	F:AM30648	♂	Upper Cleary Creek, Alaska
AMNH	VP	F:AM30651	♂	Cleary Creek, Alaska
AMNH	VP	F:AM30653	♂	Goldstream at Fox, Alaska
AMNH	VP	F:AM32761	♂	Alder Creek, Seward Peninsula, Alaska
AMNH	VP	F:AM46884	♂	Goldstream, Alaska
AMNH	VP	F:AM46886	♂	Upper Cleary Creek, Alaska
AMNH	VP	F:AM46888	♂	Goldstream, Alaska
AMNH	VP	F:AM46889	♂	Upper Cleary Creek, Alaska
AMNH	VP	F:AM46890	♂	El Dorado Creek, near Fairbanks, Alaska
AMNH	VP	F:AM46891	♂	Cripple Creek, Alaska
AMNH	VP	F:AM46892	♂	Engineer Creek, Alaska
AMNH	VP	F:AM46894	♂	Cripple Creek, Alaska
AMNH	VP	F:AM46895	♂	Ester Creek, Alaska
AMNH	VP	F:AM46908	♂	Ester Creek, Alaska
AMNH	VP	F:AM46911	♂	Upper Cleary Creek (?), Alaska
AMNH	VP	F:AM46914	♂	Upper Cleary Creek, Alaska
KU	VP	8348	♂	Fairbanks area, Alaska
NMC	VP	7700	♂	Last Chance Creek, Yukon Territory
NMC	VP	8145	♂	Bonanza Creek, Yukon Territory
NMC	VP	10458	♂	Hunker Creek, Yukon Territory
NMC	VP	10459	♂	Hunker Creek, Yukon Territory
NMC	VP	10460	♂	Hunker Creek, Yukon Territory
NMC	VP	11674	♂	Gold Bottom Creek, Yukon Territory
NMC	VP	11683	♂	Brewer Creek, Yukon Territory
NMC	VP	12098	♂	Aklavik (?), Northwest Territories
NMC	VP	13510	♂	Gold Run Creek, Yukon Territory
NMC	VP	17687	♀	Locality 11, Old Crow River, Yukon Territory
NMC	VP	17689	♂	Porcupine River, Yukon Territory
NMC	VP	20634	♂	Locality 11, Old Crow River, Yukon Territory
NMC	VP	24200	♂	Locality 11A, Old Crow River, Yukon Territory
NMC	VP	25219	♂	Bonanza Creek, Yukon Territory
NMC	VP	26026	♂	Quartz Creek, Yukon Territory
PPHM		1407-1	♂	Fairbanks Creek, Alaska
PPHM		1407-2	♂	Fairbanks Creek, Alaska
TTM		962-1	♂	Livengood, Alaska
TTM		962-2	♂	Head of Ester Creek, Alaska
USNM	P	2643	♂	Old Crow River, Yukon Territory
USNM	P	8584	♂	Kolyma River, Siberia
YPM	VP	5938	♂	Ketchum Creek, Alaska
YPM	VP	33873	♂	Upper Cleary Creek, Alaska

Note: Only those specimens which appeared to have diagnostic characters of both North American and Eurasian autochthons, and are from the Beringian area, are included in this table. Other probable hybrids are identified in table 72.

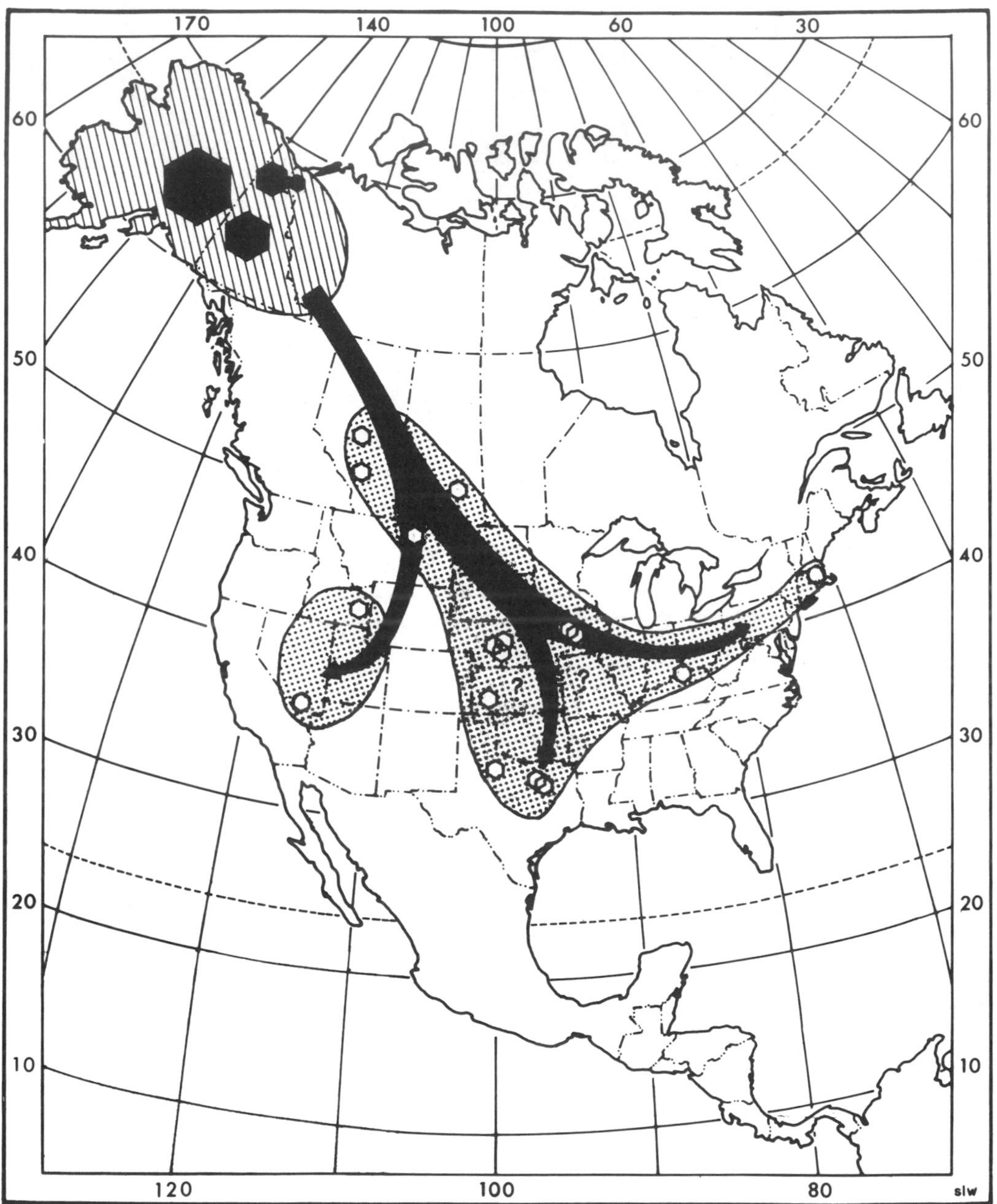

FIGURE 91 North American bison hybrid zones. Three hybrid zones are represented: (1) the eastern Beringian zone, which was probably the most strongly developed; (2) the periglacial arc—Great Plains zone, and (3) the Great Basin zone. Hexagons represent hybrid specimens, and arrows indicate dispersal salients of Eurasian bison entering mid-latitude North America.

glacial/stadial periods and thereupon came into contact with North American bison that had dispersed northward during the preceding interglacial/interstadial. The Eurasian *B. priscus* and North American *B. a. antiquus*, being ecological vicars, alternately reinvaded Beringia in phase with the climatic oscillations, and thereby allowed *B. "crassicornis"* to periodically recur. Probably, hybridization was most intense during the early phases of each stadial.

Eighteen specimens studied from midlatitude North America were identified as hybrids (table 72). These specimens are distributed in regions where southward dispersing *B. priscus* or *B. alaskensis* populations probably met and interbred with northward dispersing or resident *B. latifrons* or *B. a. antiquus* populations (fig. 91). This interbreeding took place, apparently, along the periglacial zone and along the axes of two salients of Eurasian autochthon dispersal which penetrated midlatitude North America. One of these salients extended southward across the Great Plains and one crossed the Rocky Mountains and entered the Great Basin. (Many additional specimens that appear to be hybrids are in the collections of the University of Nebraska's Division of Vertebrate Paleontology. I was not able to study these specimens, but they are unusually well represented in the Nebraska Collections, a fact that supports the midlatitude hybrid zone concept presented here.)

In summary, this discussion has attempted to present information that establishes hybridization as a logical explanation for the occurrence of certain specimens that do not conform

Table 72
Probable Hybrid Specimens

Institution and identification number		Sex	Provenience
AMNH	VP F:AM30052	♂	Clark County, Nevada
AMNH	VP F:AM39412	♂	American Falls Reservoir, southeast Idaho
ANSP	G 12990	♂	Big Bone Lick, Boone County, Kentucky
KU	VP 390	♂	Near Garden City, Kansas
KU	VP 7388	♂	Kansas (?)
MCZ	VP 4425	♂	Harvard, Massachusetts
PPHM	2797/1	♂	Montana
SM	60317	♂	Seale Pit, near Dallas, Texas
SM	61363	♀	Moore Pit, near Dallas, Texas
SMNH	P100(?)	♂	Ft. Qu'appelle, Saskatchewan
SMP	158-56-z	♂	Polk County, Iowa
SMP	159-56-z	♂	Polk County, Iowa
TMM	31041	♂	Stonewall County (?), Texas
UAI	627	♂	Dale Brothers Gravel Pit, Alberta
UCy	G A*	♂	Near Calgary, Alberta
UMo	VP 766	♂	Missouri (?)
UNSM	VP 30310	♂	Devils Gap, southwest of Gothenburg, Dawson County, Nebraska
UNSM	VP 30357	♂	Near Broken Bow, Custer County, Nebraska

*No identification number was found for this specimen.

phenotypically to species' or subspecies' norms. The argument for hybridization was based on the use of specific criteria (shared diagnostic traits) to identify hybrid specimens. Hybrid zones are identified in Beringia and midlatitude North America. Either the frequency and intensity of hybridization was insufficient to cause the breakdown or alteration of typical species' morphologies, or selection against hybrids was sufficiently strong to prevent this from happening.

UNBALANCED POLYMORPHISM

The term unbalanced polymorphism is used here to refer to samples in which abnormal characters are present. The abnormalities are considered the result of genetic drift and/or inbreeding, but excludes traumatically induced or developmental (e.g., malnutrition based) abnormalities. As with hybridization, however, proof that the unbalanced polymorphism discussed here is of genetic origin is difficult, if not impossible, to provide. Nonetheless, modern genetic theory recognizes that small populations of sexually reproducing animals can evidence phenotypic abnormalities (decanalized characters). The possibility of genetically based character abnormalities seems most simply to explain the occurrence of several unusual characters in bison.

An initial problem in working with this concept of unbalanced polymorphism is in deciding the limits of normal variation, beyond which a character may be considered abnormal. In many instances this decision is easy—if, after studying many specimens, a clearly unique or rare character appears, then it may reasonably be considered abnormal. When only a few specimens have been examined and expected variation is poorly defined, or when the particular characteristic is only marginally abnormal, a problem arises, however. The distribution, frequency, and intensity of abnormal characters must be considered in evaluating abnormalities. Frequent and patterned occurrences permit the most reliable and meaningful interpretations of abnormalities.

Most observed abnormalities of presumed genetic origin appeared in the skull. Horn cores, for example, exhibited abnormally excessive flattening of either the antero-posterior or dorso-ventral surfaces, roundness, reduction of the angle of posterior deflection, distal twisting, elongation, diminution, or subhorizontal emanation from the frontals. Many of the facial bones were malformed, exaggerated, or seemed to crowd (compete with other bones) during growth. Common tooth abnormalities included malformation, incomplete formation, overcrowding, disorientation, structural abnormalities, absence, or supernumerary teeth. Any of these abnormalities could, of course, have appeared in normal populations characterized by well-canalized mophologies, but the repeated appearance of many such abnormalities in a context suggestive of small and isolated populations leads to the conclusion that these specimens were, in fact, representative of small, isolated, inbreeding populations.

Peripheral Populations

B. a. antiquus from tropical America are consistently abnormal and collectively constitute the best-known example of presumed unbalanced polymorphism resulting from inbreeding within peripheral populations of North American bison (fig. 92). The holotype of *B. scaphoceras* (plate 10) from Nicaragua is diminutive and has a distorted shape. The "*B. occidentalis*" described and photographed by Hibbard (1955*b*) from locality 7, Barranca de Acatlan, Tequixquiac (IG 49–76), appears to be a *B. a. antiquus* with diminutive, posteriorly oriented, distorted, distally twisting horn cores. With but one exception, a suite of seven partial skulls, mostly horn cores from the Barranca de Caulapan near Valsequillo, Puebla, all show deviant characters: all deviant cores are flattened dorso-ventrally to a greater or lesser degree, two are especially diminutive, and one skull has exaggerated facial characters. A color transparency taken by Donald Brand of a *B. a. antiquus* (?) skull in a Morelia, Michoacan, museum shows that this specimen apparently had unusually shaped horns. These specimens probably represent small populations of *B. a. antiquus* which were probably typical of this taxon in tropical North America. Further north a medial section of what is probably a *B. a. antiquus* horn core (UALP no #) from near Tecoripa, Sonora, is abnormally flattened.

The relatively frequent occurrence of abnormal characters in some other regional samples suggests that these abnormalities too may have represented small isolated populations. *B. a. antiquus* from the Wacissa River, Florida (RO 5; UF VP 11861); Beaumont, California (AMNH VP 31070); and Contra Costa, California (UCMP 37615) are abnormal (fig. 92). Isolated teeth from numerous pre-Holocene localities, but especially Alaska and highland Mexico (table 73), show frequent malformations (plates 30 and 31). *B. b. bison* from the Malheur Valley, Oregon, show considerable dental abnormalities. The holotype of *B. b. oregonus* has an extra pair of premolars and 15 percent of all molar fossets in the Malheur Valley sample contain reentering enamel folds (table 73; plate 32). Females in the Malheur Valley sample are unusually large and robust. *B. b. bison* from Big Bone Lick, Kentucky, show abnormal doming of the fronto-parietal sutures. Specimen MCZ VP 2047 from Big Bone Lick shows unequal wear on the upper dentition which may be the result of malocclusion. The holotype of *B. b. athabascae* (NMC M 299), collected in 1892 at a time when the wood bison population was very much reduced, has extremely exaggerated lacrymal salients that completely separate the nasals and maxillae. These patterns suggest that the abnormal characters of tropical *B. a. antiquus* specimens and of similar abnormalities in other regions of North America may be the result of genetic drift and inbreeding in small, isolated populations.

The Latest Wisconsin-Early Holocene Bottleneck

Late Wisconsin and early Holocene bison populations from the western Great Plains show extremely unbalanced polymorphism. The Casper archeological site in Natrona County, Wyoming—a paleo-Indian bison kill site radiocarbon dated at about 10,000 BP—has yielded the best-known and most extreme example of a population evidencing excessive morphological abnormalities probably attributable to the direct or indirect effects of genetic drift

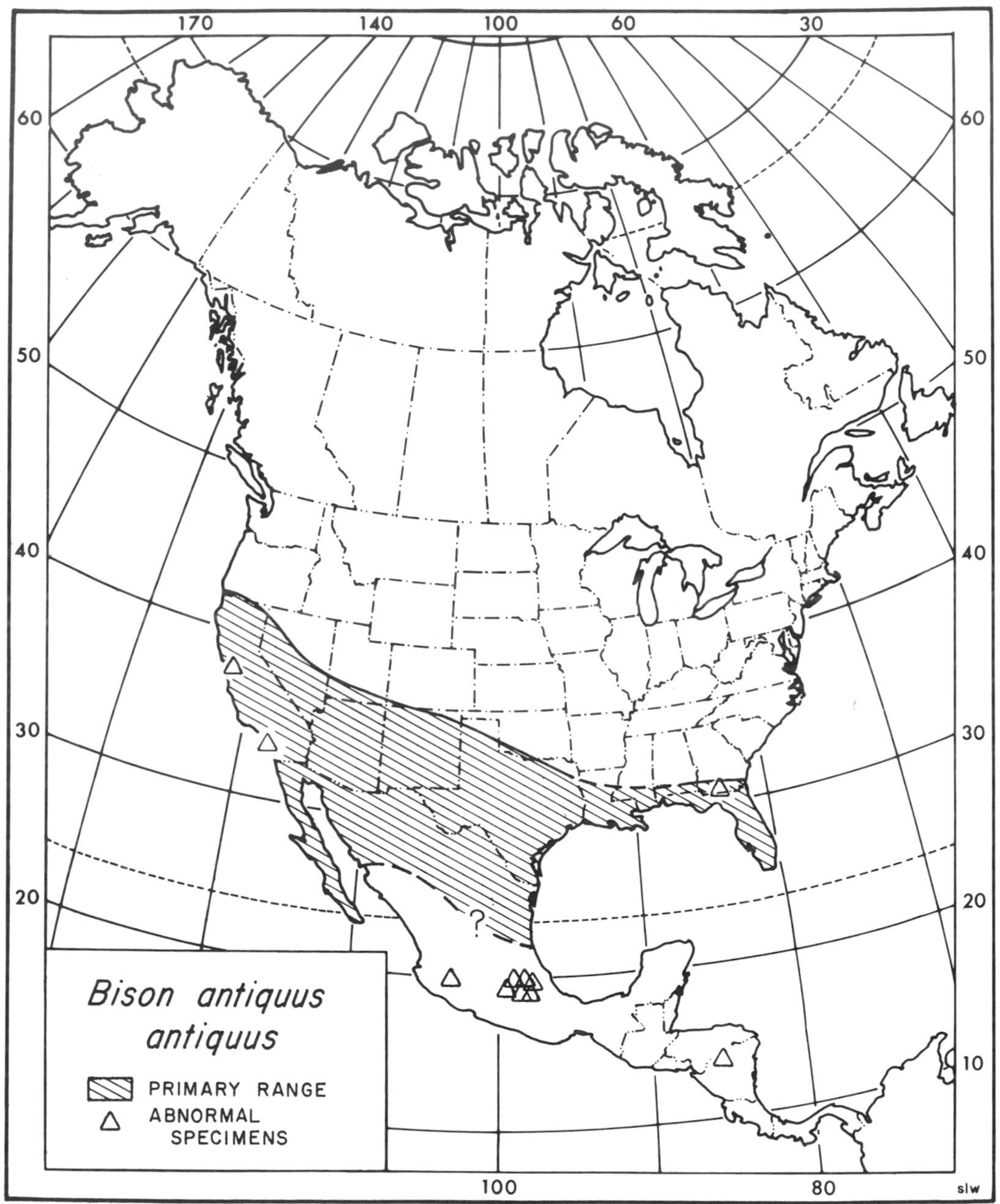

FIGURE 92 The distribution of abnormal *Bison antiquus antiquus* specimens, relative to the primary range of the taxon.

and inbreeding. Many of the abnormalities found in this deme have been described by Wilson (1974*b*), from whose study the following list of abnormalities is taken, supplemented by my own observations. Horn core deviations in the Casper sample consist of relative elongation, flattening, posterior orientation, and posterior rotation of the core tips. Dentition deviations include one case of shear mouth, unerupted premolars, overcrowding in the tooth row with associated disorientation of individual teeth, malformed and incompletely formed teeth, excessive development of cement, inadequate enamel formation, and malocclusion (plate 33). Three adult females show a single mode of nasal-maxillae deformation, although in differing degrees. In one specimen the maxillae are drawn inward toward the sagittal at the nasals, pinching the nasals into the shape of an hourglass (plate 34). In a second the pinching effect is present but less apparent. In the third the right maxilla has a scar-like mass at the nasal suture line. The extreme genetic debilitation probably rendered the Casper population (and others like it) more susceptible to somatic pathologies than would be true of a less inbred population (plate 33).

Bison from other archeological sites dating from the latest Wisconsin to middle Holocene (ca. 12,000–6,000 BP) reveal similar abnormalities (fig. 93), although these are less severe and frequent than in the Casper sample. Bison from the Clovis, New Mexico, area show a high frequency of dental abnormalities (table 73). Folsom, New Mexico, bison have elongate horns and deviant dentition. Rex Rodgers, Texas, bison have elongate horns with slight distal twist and upward orientation, deviant dentition, and crowding between the maxillae and nasals. Olsen-Chubbock, Colorado, bison skulls have horn

Table 73
Frequency of Selected Abnormalities in Molar Dentition

Sample	n	Abnormal fosset pattern (n)	Auxiliary labial styles (n)
Bison antiquus antiquus[a]	13	0	0
Bison bison bison:			
Central and northern Great Plains	48	5	0
Malheur Valley, Oregon	75	12	0
Bison antiquus occidentalis:			
Rex Rodgers archeological site, Texas	8	6	0
Specifically unidentified samples:			
Clovis area, New Mexico	12	6	1
Fairbanks area, Alaska	78	39	6
Folsom, New Mexico	10	4	0
Lago de Chapala, Jalisco	39	21	. . .

[a]The specimens providing this information are from various localities in California and Florida.

cores which emanate upward, are rotated rearward, and have distal twists. Hawken, Wyoming, bison show excessively developed maxillae along the nasal suture, and abnormal dentitions have been reported elsewhere for this population (Frison, Wilson, and Wilson, 1976). Detailed examination of other bison populations representative of the latest Wisconsin-early Holocene period might reveal similar high frequencies of abnormalities.

The sudden appearance of such a wide array and high frequency of morphological abnormalities over a limited area suggests that the formerly canalized *B. a. antiquus* population in the southwest and Great Plains regions suddenly became decanalized. In effect, the latest Wisconsin-early Holocene bison populations of these regions appear to have been fragmented into numerous small, relatively isolated populations, most of which probably developed unbalanced polymorphisms because of inbreeding and genetic drift. These bison populations were probably kept small and relatively isolated over a prolonged period of time by continuous and intense human hunting that reduced interdemic gene flow, especially during the period from about 11,000 to 9,000 BP, the time during which abnormalities are most frequent, intense, and varied. Most genetically based abnormalities recorded among bison did not become integrated into subsequent canalized populations, but some of the characters that appeared among latest Wisconsin-early Holocene *B. a. occidentalis* were subsequently integrated into later *B. a. occidentalis* and *B. bison* phenotypes. Most prominent among these characters that were retained are the rearward rotation and spiraling of horn cores.

Dental Abnormalities

A systematic survey of the frequencies of two abnormal dental characters was made in order to compare the abnormality rate in a population of known geologic age and normal gene flow with similar rates in other populations. The upper molars of nineteenth-century *B. b. bison* from Montana, Wyoming, and Kansas were examined for the presence or absence of two abnormal characters: reentrant folds in the enamel fosset margins and auxiliary styles on the labial side of the teeth. Dentition from peripheral populations (Fairbanks area, Alaska; Lago de Chapala, Jalisco; and Malheur Valley, Oregon); archeological kill sites of the early Holocene (Clovis, New Mexico; Folsom, New Mexico; Rex Rodgers, Texas); and canalized *B. a. antiquus* populations were examined for the same characters (table 73; plates 30 and 31).

The relatively low incidence of abnormalities in the *B. a. antiquus* sample and the higher frequencies of abnormalities in the other samples is noteworthy. The evidence suggests that reentrant folds in the enamel border of molar fossets and supernumerary styles can result from, and when in high frequencies can identify, isolated inbred populations. Crenulated fosset border patterns appear to be rather ubiquitous consequences of inbreeding, which implies that the genetic basis for the character is widespread throughout the genus—or at least in the shorter horned North American bison—and that it is rather easily released. The fact that such crenulated fosset borders are common in *Bos* may bespeak something of the antiquity of the genetic basis for the phenomenon. Auxiliary styles are less common than crenulated fosset borders, but

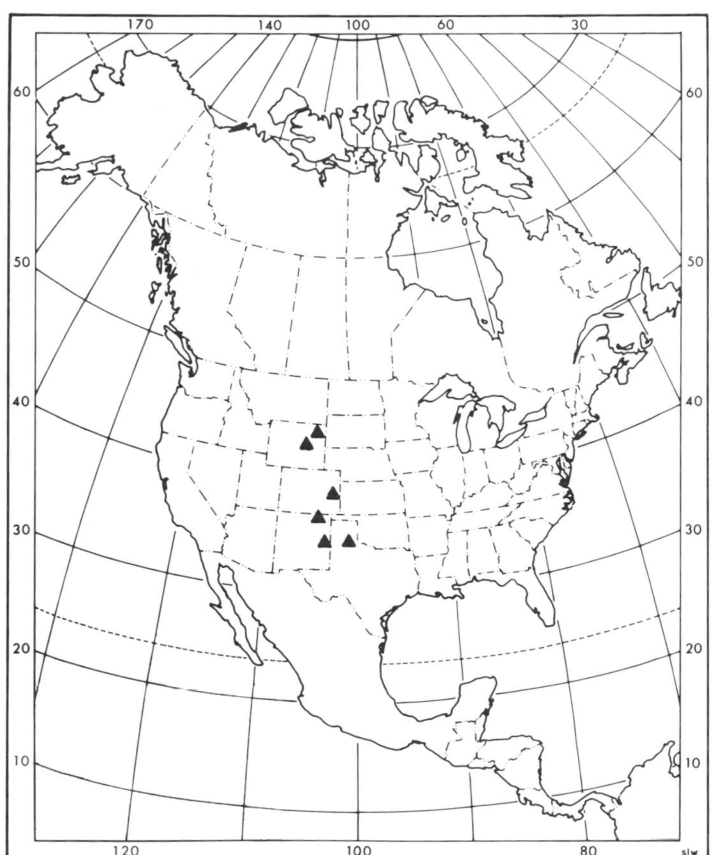

FIGURE 93 The distribution of abnormal *Bison antiquus occidentalis* during the latest Wisconsin to middle Holocene. The localities represented are, from north to south, Hawken (ca. 6,370 BP); Casper (ca. 9,945 BP); Olsen-Chubbock (ca. 10,150 BP); Folsom (ca. 10,000 BP); Clovis, New Mexico, area (probably spans entire period, with greater representation from earlier part of the period); and Rex Rodgers, Texas (ca. 9,391 BP). The abnormalities found in bison from these localities are probably the result of inbreeding and genetic drift, brought about by heavy human hunting.

nonetheless do occur with some frequency. This fact suggests that auxiliary styles, too, may be controlled by recessive genes widespread among bison and easily released in inbred populations.

Unusual Morphologies and Evolution

The distinction between the chance appearance of a character and active selection for a character must be kept in mind when evaluating both adaptations and unique characters. Normally, the appearance of a unique character in a phenotype as a result of genetic drift or inbreeding does not indicate that that character has been, or will be, adaptively favored, and such abnormal characters have little or no real evolutionary significance. Only after a character has become integrated into subsequent, canalized populations can it be considered adaptive.

CHAPTER 7

PATTERNS AND PROCESSES IN BISON EVOLUTION: A SYNTHETIC INTERPRETATION

The time, place, and environmental circumstances of the origin of *Bison* remain largely unknown. Scant and diffuse evidence suggests that the earliest species, *B. sivalensis*, came into existence during the late Pliocene in central or eastern Asia and was probably adapted to forest openings or forest edges.

Leptobos is often considered the most likely antecedent of *Bos* and *Bison* (Guthrie, 1978; Kurten, 1968; Sinclair, 1977; Skinner and Kaisen, 1947), an opinion with which I agree, although Guthrie (oral comm., September 2, 1978) has indicated that another as yet unidentified bovine from east Asia might be a more probable ancestor of *Bison*. *Bison* and *Bos* could have evolved independently from isolated *Leptobos* populations with but a few skeletal modifications, all of which could have resulted from genetic drift or chromosomal changes in small populations, and subsequent selection. *Leptobos, Bison*, and *Bos* horn cores are similar in many qualitative and quantitative respects. Broadening of the frontal-occipital regions and a more lateral orientation of horn cores in *Bison* and *Bos* relative to *Leptobos* constitute parallel and unidirectional morphological changes. Environmental events of the Pliocene probably created a highly varied mosaic of locally different habitats in eastern Asia (chapter 1). Small populations of *Leptobos* could conceivably have dispersed into many of these habitats and become differentially adapted to various habitat types. *Bison* and *Bos* could have come into existence in this way. *Leptobos*, therefore, probably preceded *Bison* and *Bos*, gave rise to each independently and perhaps simultaneously (but in different locations), and coexisted with them (at least temporally) until the middle Pleistocene.

Probison dehmi, from the Upper Siwalik Tatrot stage, has been considered the probable antecedent of *Bison* (Sahni and Khan, 1968; Khan, 1970). Flerow (1971) and M. Wilson (1972, 1974*b*, 1975) regard *Probison* as congeneric with *Bison*. In my opinion, however, *Probison* differs from *Bison* to too great an extent to consider them congeneric. The *Probison* horn core is the most immediately obvious basis of taxonomic differentiation—it has the wrong shape (flat oval) and the wrong orientation (curving downward and forward) for *Bison*. The orbits of *Probison* are nearly flush with the cranium and oriented markedly forward, and the face is proportionately too narrow to fit the long established (and stable) concept of *Bison*. For *Probison* to have been ancestral to *Bison*, several morphological reversals would have been necessary. The horn cores would have had to move from (1) a dorso-posterior orientation and supraorbital location of round, straight horns on primitive bovids; (2) downward to a more lateral position with flattened, downward and forward oriented horn cores in *Probison*; (3) then back to a more rounded core with upward and rearward orientation in *Bison*. The face would have been required to change from a relatively wider face in early bovids, to a narrow faced *Probison*, and back again to a wider face in *Bison*. It is not necessary that all known and probably related taxa constitute sequential steps in a common phyletic lineage. *Probison dehmi* appears to have been a short-lived genus that paralleled, rather than preceded, early *Bison*. I think that *Probison* and *Bison* would better be regarded as related but distinct genera sharing similar occipital, frontal, and dentition characters but possessing differing horn core, orbital, and facial characters.

Bison and the other large horned bovids of the late Pliocene-early Pleistocene probably inhabited relatively closed environments (forest openings, forest edges, woodlands, etc.) wherein populations were widely dispersed, densities were low, individual selection was intense, and competitive specialization was necessary. The success of the Bovini during the late Pliocene-early Pleistocene indicates that the competitive abilities of this tribe had evolved to a level equal to that of other contemporary large ungulates. The great number of genera with large horns known from the late Pliocene-early Pleistocene suggests that horns were important competitive organs and probably the principal organ of dominance and combat among the Bovini. The ruminant digestive system permitted efficient use of the varied forage available in the seral or ecotonal habitats. Each Bovini species probably evolved in a different part of the tribe's range, and only after relative genetic homeostasis and population increase had occurred among species did some of their ranges overlap.

B. sivalensis dispersed northward into Siberia no later than the middle Pleistocene (and probably earlier) in association with the Asian forests. If *B. sivalensis* did in fact disperse into Siberia during the early Pleistocene, the species was in a good position to enter North America whenever the Bering land bridge emerged, as it apparently did on several occasions during the early and middle Pleistocene (chapter 1). Several small populations of the species could have dispersed into North America along with members of the classic Blancan or Irvingtonian faunas (table 2), but could have been minor constituents of one or both of these faunas and left no fossil record. Palearctic autochthons that first appeared in North America as members of the Blancan fauna were predominantly forest adapted genera, with some possible steppe or savanna adapted genera (*Plesippus* and *Paracamelus*). These Palearctic genera which entered North America were, in great part, recently evolved; most are first known from the mid- or late Pliocene or early Pleistocene. The many Palearctic autochthons in the later Irvingtonian fauna were first known from the early or middle Pleistocene of Eurasia (Repenning, 1967). If *B. sivalensis* was adapted to forest openings or closed shrublands, and if the species did disperse northward and reach Siberia during the early or middle Pleistocene, there is no reason why the species could not have entered North America along with other genera of the Blancan or Irvingtonian faunas. Although no *B. sivalensis* fossils are known from North America, the morphological similarity between *B. sivalensis* horn cores and those of the North American autochthons suggests that the latter evolved from the former rather than from late Eurasian autochthons as is usually accepted (e.g., Guthrie, 1970, 1978; Schultz and Frankforter, 1946; Skinner and Kaisen, 1947; Wilson, 1974*a*). Also, the probable synchronous appearance of *B. latifrons* and *B. antiquus* in North America suggests that their separation occurred in North America from a common antecedent, not from the phyletic transition of one into the other.

THE ORIGINS OF *B. priscus* and *B. alaskensis*

B. priscus and *B. alaskensis* are the two Eurasian autochthons recognized here from the middle and late Pleistocene. *B. priscus* appears to have greater antiquity than *B. alaskensis*, but neither reached a level of great importance in the Palearctic or Beringian faunas until, probably, the Illinoian, when *Bison* became well established in the northern steppes and tundras (fig. 2). The phyletic relationship between the two is unclear, but the available evidence suggests that *B. priscus* evolved first as a savanna inhabitant in mid-latitude eastern Eurasia and thereafter gave rise to *B. alaskensis*, a forest opening or woodland inhabitant.

B. priscus is first known (previously as *B. paleosinensis*) from the early Pleistocene Nihowan fauna of northern China (Aigner, 1972; Teilhard de Chardin and Piveteau, 1930). The *B. paleosinensis* holotype is in most respects larger than the *B. sivalensis* holotype (table 15) and it shows several morphological advances that differentiate it from *B. sivalensis*: the cranium is proportionately broader and flatter, the parietals occupy more of the dorsal cranium, the orbits are more forward oriented as a result of the proportionately greater lateral expansion of the posterior cranium, and the horn cores are more laterally situated and directed. The horn cores also emanate subhorizontally from the frontal plane, they have a sinuous posterior profile, they grow spirally around a sinuous longitudinal axis, and they are apparently more strongly curved than in *B. sivalensis*. The diagnostic morphological characters of the middle and late Pleistocene Eurasian autochthons, then, had been acquired by the end of the early Pleistocene. The *B. paleosinensis* holotype was smaller than later *B. priscus*, but no idea of the actual range of variation among early populations of the species is possible with the resources at hand and the *B. paleosinensis* holotype might simply be a small specimen within the range of expected variation for *B. priscus*.

I suggest that *B. sivalensis* populations dispersed into the early Pleistocene steppe or savanna ecotone, which extended farthest east at about the latitude of northern China, or adapted *in situ* to the savannas that were then forming from the disintegrating forest. The new habitat provided an opportunity for the rapid evolution of savanna-adapted characteristics. The range of *B. priscus* expanded westward through the temperate transcontinental steppe/savanna zone into eastern Europe and north into Siberia by the early middle Pleistocene (Flerow, 1971; Kurten, 1968). *B. priscus* retained essentially (although not completely) a temperate range until the middle Pleistocene, at which time the principal range shifted to higher latitudes. This shift coincided with the Glacial Maximum of Eurasia and Beringia which was probably accompanied by environmental changes sufficiently new and intense to produce a significant change in the composition of the higher latitude fauna. *B. priscus* expanded into the high-latitude glacial environments, a mosaic of vegetation ranging from tundra and steppe to local shrub forests, and thereafter became a dominant member of the Palearctic high-latitude fauna.

The strong resemblance of *B. alaskensis* to *B. priscus* would imply that the former evolved from the latter rather than from the more different *B. sivalensis*. Also, the occurrence of shorter horned bison characteristics in *B. alaskensis*—occasional doming of frontals, depressed horn cores, and greater orbital protrusion—favors a *B. priscus* ancestry. The greatest problem in more fully understanding the relationship between these two species is the lack of clear size separation between them. This phenomenon can, however, be explained in part by the late Pleistocene Palearctic vegetation dynamics. If the Palearctic interglacial forests had remained intact as a discrete plant formation during the glacial periods, then *B. alaskensis* could have remained associated with the forest/woodland selection regime and evolved a more distinctive and differentiating morphology than appears to have in fact occurred. The same applies to *B. priscus*; if wooded steppe or savanna had remained intact, then greater species distinctiveness would probably have resulted. Intermediate-sized phenotypes resulted, however, as vegetation formations changed significantly between glacial and interglacial periods, and there consequently exists a near morphological continuum uniting the *B. priscus-B. alaskensis* modes.

THE ORIGINS OF *B. latifrons* and *B. antiquus*

B. latifrons and *B. antiquus* are inferred to have evolved directly from *B. sivalensis*, although neither *B. latifrons* nor *B. antiquus* fossils are known from Eurasia nor are *B. sivalensis* fossils known from North America. As mentioned above, the ecological potential existed for colonizing populations of forest/woodland adapted *B. sivalensis* to disperse from Siberia into North America along with other contemporary Eurasian autochthons that are known to have entered North America during the early and middle Pleistocene. The facts that *B. latifrons* and *B. antiquus* did overlap temporally for at least part of their histories, that they appear to have been essentially allopatric, and that *B. a. antiquus* probably originated and thereafter occurred south of the *B. latifrons* range suggest that *B. antiquus* came into existence at about the same time as did *B. latifrons*, apparently occupying and adapting to a savanna habitat much as *B. priscus* appears to have done in Eurasia whereas *B. latifrons* remained associated with a more wooded environment.

Proceeding with the assumption that *B. latifrons* and *B. antiquus* both evolved from *B. sivalensis*, the following evolutionary steps are suggested. Small populations of *B. sivalensis*—probably several over a long period of time, most of which became extinct—dispersed across Beringia, south through western lowland Canada, into midlatitude North America. Small dispersed populations evolved into two distinct phenotypes, both larger than the antecedent *B. sivalensis*. The smaller bodied, shorter horned species—*B. antiquus*—evolved in the more arid and open southwestern savanna biome and the larger bodied, larger horned species—*B. latifrons*—evolved in the more densely vegetated Great Plains and/or Great Basin region(s). The speciation of both *B. latifrons* and *B. antiquus* probably occurred during the Illinoian glacial period. The *B. latifrons* range and probably its overall population size were larger than those of *B. antiquus* because the Illinoian forest/woodland habitat was more widespread than the savanna habitat. The differences between the ranges and inferred population size of these two taxa help to explain why remains of *B. antiquus* are less common than those of *B. latifrons* from the early period of their existence.

It is important to note the difference between the prediction of this model and available evidence on the ages of *B. latifrons* and *B. antiquus*. The model presented here predicts a near simultaneous origin of both species during the Illinoian, even though positive evidence (i.e., identifiable horn cores) of *B. antiquus's* existence in the Illinoian is not available. The barrier effects of the zonal boreal forest and a taxon (*B. latifrons*) better adapted to the forests and woodlands north of the *B. antiquus* range suggest that *B. antiquus* probably evolved from ancestral populations already in North America south of the forest/woodland barrier. The small savanna region and the proportionately small *B. antiquus* populations inhabiting this region constitute plausible circumstances that explain the as yet unrecognized occurrence of *B. antiquus* in the Illinoian fossil record. Note must be made of the large number of nondiagnostic bison remains from the Palos Verde Sands formation of Los Angeles County, California. This formation has been assigned to the Illinoian (Miller, 1971). Until identifiable horn cores have been recovered from this formation, the possibility cannot be discounted that the bison remains already recovered represent *B. antiquus*.

BARRIERS, CORRIDORS, AND THE DIRECTIONS OF DISPERSALS

The directions and intensities of periodic bison dispersals were regulated to a greater or lesser extent by barriers and corridors. Physical barriers considered of importance here include water, ice, and mountains. The major water barrier influencing bison distribution was the Bering Strait, which prevented faunal exchange between Siberia and Alaska during interglacial and major interstadial periods. Continental glaciation during glacial stages covered large areas of land and constricted or eliminated the area available as a passage between eastern Beringia and midlatitude North America. The Rocky Mountains and the Sierra Nevadas were the major orographic barriers influencing bison distribution, but even these barriers could be penetrated in several places or circumvented by bison populations.

The most apparent biotic barriers to bison dispersal were dense forests, deserts, and true herbaceous tundra, all of which had low carrying capacities. Biotic barriers were undoubtedly more or less permeable by bison, so they probably functioned only to reduce the rate and magnitude of dispersal. The eastern deciduous forests, and especially the zonal boreal forest, were the major forests that influenced the southern dispersal and distribution of North American bison. The variability of the boreal forest as a filter is summarized in figure 94 and table 74.

The Bering land bridge, exposed during glacial stades, was the only Pleistocene corridor permitting the exchange of fauna between Eurasia and North America. The Yukon lowlands between the Brooks and Alaska ranges, or the arctic lowlands north of the Brooks Range, were corridors easily traversed by dispersing megaherbivore populations during the stadial periods, when

Table 74
Factors Influencing the Density of the Boreal Forest During Glacial/Interglacial Cycles

	Interglacial/ interstadial	**Transition to glacial/stadial**	**Glacial/ stadial**	**Transition to interglacial/interstadial**
	permafrost	permafrost	permafrost	permafrost
	winter drouth	winter drouth	winter drouth	winter drouth
Factors minimizing forest density	short summer	short summer		short summer
	fire	fire	fire	fire
	lithoseres			lithoseres
	hydroseres			hydroseres
	shallow soil			shallow soil
		migration		migration
	succession		succession	
		soil drainage	soil drainage	
		soil aeration	soil aeration	
		soil developed	soil developed	
Factors stimulating increased density			greatest insolation	
			greater evapotranspiration	
			permafrost less restrictive	

Note: This table attempts to identify in only a very general way environmental elements that either stimulated or retarded the development of the boreal forest during the four simplified phases of each glacial/interglacial cycle. The absence of any element from any column or row in the table does not imply its absence as an environmental element for the respective phase of the glacial/interglacial cycle; it is simply considered less significant than those elements listed.

vegetation was more open, nutritious, and accessible. The lowlands of western Canada east of the northern Rocky Mountains and the lowlands of midlatitude North America provided easily traversable terrain over which herbivores could disperse.

Perhaps the most basic question of bison dispersal is that of the direction of intercontinental exchange. Why did up to three species—*B. sivalensis* (?), *B. priscus*, and *B. alaskensis*—disperse into North America from Eurasia, when none (or perhaps only one, if *B. schoetensacki* is an Eurasian derivative of *B. a. antiquus*) dispersed from North America into Eurasia? The imbalance in the direction of Eurasian-North American faunal exchange during the Pleistocene, of which bison are representative, has been variously explained: more intense competition in the larger, faunally more diverse Palearctic than Nearctic; superior fitness of Palearctic genera; and a differential filtering effect of the Bering land bridge. The following model isolates critical processes that probably interacted to

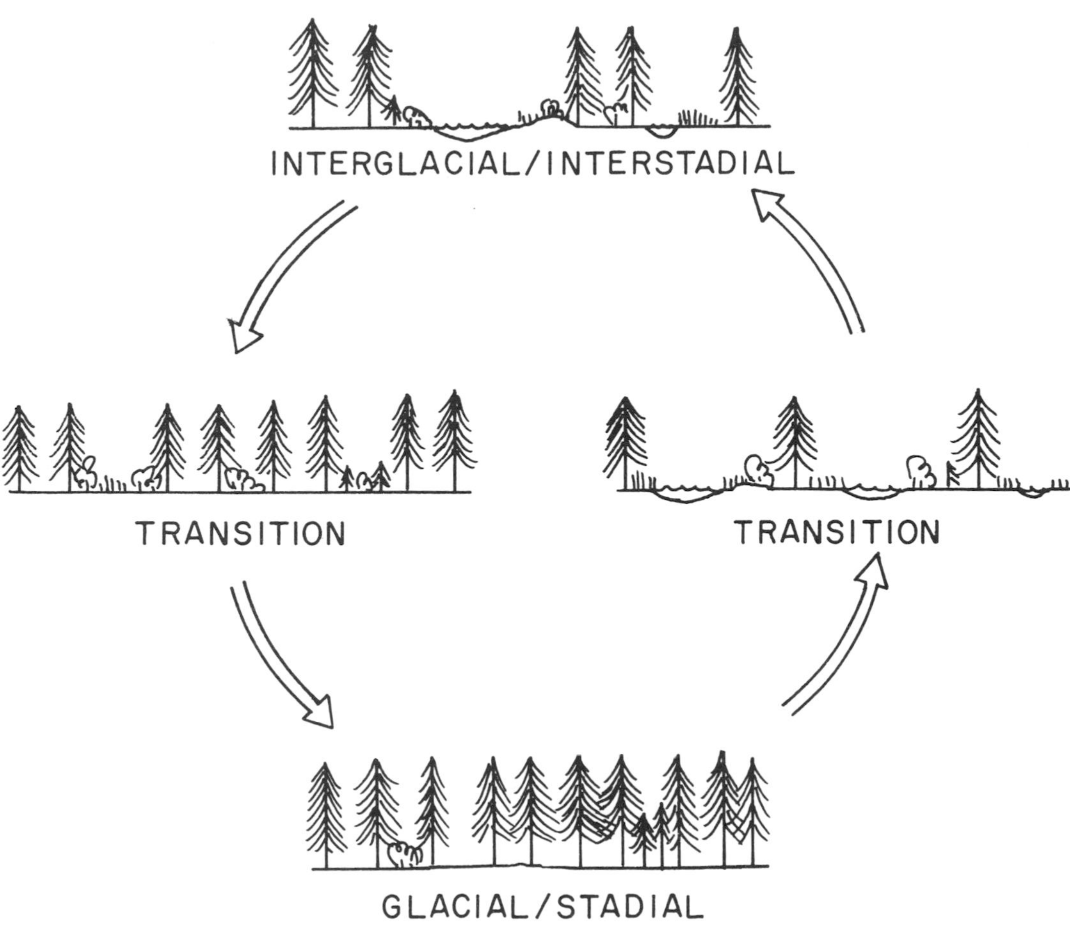

FIGURE 94 Variability of the boreal forest filter. The zonal boreal forest of North America may have fluctuated in its effectiveness as a filter or barrier to the dispersal of large mammals during the Quaternary. The abstraction presented here suggests that the boreal forest was probably most easily penetrated during the transition from a glacial/stadial period to an interglacial/interstadial period as vegetation spread across recently deglaciated terrain. The barrier was probably most effective during full glacial/stadial periods, when a greater proportion of the forest was probably in climax than at any other phase of the cycle. The barrier effect was moderately effective during interglacial/interstadial periods and during transitions from interglacial/interstadial to glacial/stadial periods (see table 74).

regulate the direction and magnitude of intercontinental bison dispersals.

During full glacial periods the Bering land bridge was normally open, but the Cordilleran and Laurentide glacial systems were extensive and they either constricted the north-south corridor separating them or formed an impenetrable ice barrier. South of the glaciers and the immediate periglacial zone, and lower on the mountains, was the glacial equivalent of a boreal forest. This forest was probably at its maximum density (nearest to total climax) during glacial stades (fig. 94; table 74; but see chapter 1). Populations of large North American herbivores were reduced in numbers because of reduced ice-free land area and consequently reduced carrying capacity (less steppe/savanna, more forest), whereas populations of large Siberian herbivores probably increased because of increased carrying capacity (more steppe/savanna, less forest) in that region. The situation resulted in Eurasian bison easily dispersing across the steppe tundra vegetated Bering land bridge into eastern Beringia. North American bison, however, were reduced in numbers and restrained from dispersing northward by relatively climax forests and a constricted north-south corridor or possible ice blockade. Successful North American emigrants (most likely *B. latifrons*, until it became extinct)—individualistic, few, and dispersal-oriented—would probably have been both quickly swamped by the more numerous Eurasian bison in Beringia and selected against by the savanna/steppe selection regime that dominated the region.

Full interglacial conditions were essentially the opposite of the full glacial conditions. The Bering land bridge was submerged. The glaciers had essentially disappeared. The interglacial boreal forest was more northerly than during glacials and was probably characterized by more open vegetation than was the glacial boreal forest (fig. 94). It was, therefore, somewhat more penetrable by megaherbivores than was the full glacial forest. Midlatitude North American bison populations increased because of increased habitable area and savanna, less forest), whereas the Siberian bison populations decreased because of reduced carrying capacity (more forest, less steppe/savanna). Midlatitude North American bison populations were, therefore, more capable of dispersing northward during interglacials than glacials. Relict Eurasian and hybrid bison probably still existed in eastern Beringia during interglacials, so the magnitude of actual dispersal northward by North American autochthons was less than would have been the case if these relicts had not been present. The Eurasian relicts were probably *B. alaskensis*, adapted to the more forested or woodland habitat of interglacial Alaska and were, as well, nearly the competitive equivalent of *B. latifrons*, the more northerly Pleistocene North American autochthon and the more likely North American species to disperse northward. Northward dispersal of Noth American bison was, therefore, minimal, even during interglacials.

Most northward dispersals probably occurred during glacial-interglacial transitions. As ice masses withdrew, the boreal forest vegetation shifted northward across recently glaciated terrain. North American bison populations increased with increased area and carrying capacity, whereas eastern Beringian bison populations decreased because of reduced carrying capacity as steppe/tundra changed to forest/woodland. The transition from a more r- to K-type selection regime in Beringia during glacial to interglacial periods resulted in expulsion (i.e., net forced emigration) of much of the region's bison population from eastern Beringia toward midlatitude North America. Environmental changes favored the net dispersal of Eurasian bison southward toward midlatitude North America at a time when the midlatitude environment was capable of accommodating a larger overall bison population.

The transition from interglacial to glacial conditions produced environmental pulsations with different effects. The Bering land bridge was probably opened early in the transition period, relative to the cumulative effects of continental glaciation (Hopkins, 1967*a*). This means that the Bering corridor existed before the maximum glacial inundation of land area and before extensive midlatitude habitat change had occurred. The carrying capacity and bison population of eastern Beringia probably increased with the transition from more closed wooded habitat to more open herbaceous habitat. Beringia probably acquired increased bison populations from *in situ* forest/woodland populations, from populations dispersing northward from midlatitude North America, and from Siberia. Dispersals from Siberia were probably more important, because the habitat transition taking place in Beringia would have increasingly favored steppe/savanna adapted taxa over forest/woodland adapted taxa. The most likely exception to this generalized pattern would have been the possible northward dispersal of *B. a. antiquus* during the late Wisconsin stade, a process made possible by the prior extinction of *B. latifrons*.

The foregoing considerations suggest a general pattern of bison dispersals as regulated by environmental oscillations. According to this model, *B. priscus*, being adapted to steppe/savanna habitats, entered eastern Beringia during transitions from interglacial/interstadial to glacial/stadial conditions. The first populations to reach Beringia might have hybridized with bison already in Beringia (*B. latifrons*; *B. antiquus*) but subsequent eastward dispersal of *B. priscus* would continue and this would result in a strong *B. priscus* influence on the morphology of the Beringian bison. Subsequent dispersal southward into midlatitude North America was minimized by the lower carrying capacity of the northern Rocky Mountains, which had to be crossed or bypassed through regional tundra; a constricted zone of suitable habitat between the Cordilleran and Laurentide ice sheets; and the boreal forest, generally impenetrable by the shorter horned, socially cohesive savanna mode K-selected *B. priscus*, especially if and when this species came into direct competition with established regional populations of *B. latifrons*, such as would have occurred in midlatitude North America. *B. priscus* was, therefore, largely a full glacial/stadial inhabitant of arctic Eurasia and Beringia, which occasionally dispersed southward as far as the midlatitude periglacial zone and, rarely, even further south.

B. alaskensis populations arose in eastern Beringia during the transition from glacial-to-interglacial/

stadial-to-interstadial conditions and probably remained in this region throughout the interglacial (Sangamon?) period, although in low overall numbers. Habitat succession from more open to more closed conditions contributed to the dispersal of *B. alaskensis* populations southward through generally forested or parkland habitat. Some *B. alaskensis* dispersals extended as far south as the highlands of tropical Mexico. *B. alaskensis* was capable of dispersing further southward than *B. priscus* because it was adapted to forest/woodland habitat through which it would necessarily have had to pass in order to reach southern North America, and it could effectively compete with *B. latifrons* in that forest/woodland habitat. The Rocky Mountain barrier kept most *B. alaskensis* to the east, although some populations did penetrate this barrier and reach the Great Basin and Pacific Coast. The relatively more dense eastern forest kept *B. alaskensis* from much or all of eastern North America.

The *B. latifrons* population and range probably increased during the Sangamon, primarily because an increased area of suitable habitat was available. *B. latifrons* dispersals probably reached a maximum during the transition from the Sangamon to the Wisconsin as both the area of suitable habitat and the openness of habitat decreased and competition for available resources intensified. The major direction of dispersal was probably primarily to the north because the main *B. latifrons* range was in or near the Great Plains and Great Basin, north of and adjacent to the main *B. antiquus* range.

The *B. antiquus* population and range probably increased during the Sangamon because of increased area of savanna. *B. antiquus* population density was probably higher than that of *B. latifrons* during interglacials, but the total population was probably smaller than that of *B. latifrons*, considering the greater probable expanse of forest/woodland habitat than savanna habitat during the Sangamon. Pressure for *B. antiquus* dispersal would have increased during the Sangamon-Wisconsin transition, but the opportunity for dispersal was minimized or prevented by either water (Pacific Ocean, Gulf of Mexico), closed vegetation barriers on all sides, or competition from a better adapted forest/woodland congener. An exception to this pattern possibly occurred during the late Wisconsin after *B. latifrons* had become, or was very near becoming, extinct. Apparently, some *B. antiquus* populations did penetrate the boreal forest barrier, which may have been more open or possibly nonexistent during part of this time (the maximum North American glaciation) because of regional climatic conditions attributable to the continental ice sheet (Wright, 1971). The absence or weakness of the forest barrier, especially in the central Great Plains (Hoffmann and Jones, 1970), would have facilitated the northward dispersal of *B. antiquus* at this time. The coalescence of the Cordilleran and Laurentide ice sheets would have prevented dispersal, but if these ice sheets were not fused prior to the onset of the late Wisconsin stade, populations of *B. antiquus* could have dispersed north into Beringia prior to the end of the Wisconsin.

These range dynamics can logically explain the distribution pattern of species and their hybrids documented in chapters 2 and 6, especially when the behavioral and ecological adaptations hypothesized in chapter 5 are considered.

EXTINCTION OF *B. latifrons*

Radiocarbon and stratigraphic evidence indicates that *B. latifrons* was extant during the mid-Wisconsin interstade (table 20), but no convincing evidence exists that *B. latifrons* survived beyond this time. Fuller and Bayrock (1965) described a large horn core from an early Holocene terrace near Edmonton, Alberta, which they referred to *B. latifrons*. This specimen (UA1 627) appears, however, to be a hybrid and to have been redeposited from an older matrix; it is more abraded than the other, smaller horn cores from this terrace which I examined. The *B. latifrons* from Rancho La Brea come from accumulations that span the period from about 40,000 BP to about 12,000 BP. The large astragalus from the Holocene of Avery Island, Louisiana, and referred to *B. latifrons* by Dillehay (1974), is not diagnostic.

The extinction of *B. latifrons* probably took place as the late Wisconsin stadial environment formed. This environmental change would have imposed severe competitive stress on the species, not only by reducing the area of suitable habitat but in creating environmental circumstances that accelerated forest succession. The late Wisconsin was the most severe glacial phase of the North American Quaternary and, presumably, more effective moisture was available to the midlatitude vegetation in the form of precipitation, runoff, and reduced evaporation than was true in previous stades, with the immediate consequence of fewer seres of shorter duration being available at any one time. These environmental circumstances brought about lowered population levels for much of the continent's forest/woodland megaherbivores. The *B. latifrons* population may have been reduced below the threshold at which the species' behavior and competitive ability could maintain a viable population under existing environmental circumstances, with extinction being the consequence. Although evidence does not exist that humans hunted *B. latifrons*, evidence is mounting that humans were in North America during and after the middle Wisconsin interstade about 30,000–22,000 BP (Bada, Schroeder, and Carter, 1974; Bryan, 1978; Hopkins, 1967*a*). A modest amount of hunting pressure might, therefore, have contributed to the extinction of *B. latifrons*, but, if so, this was probably not the main cause of extinction.

LATE WISCONSIN FAUNAL REBOUND AND EXTINCTION

A major theme in this interpretation of bison evolution has been the regulatory effect of environmental change on the distribution and numbers of bison. A model was proposed which correlates species population size and distribution with available area and the carrying capacity of that area. Distribution and numbers of midlatitude bison are thought to have been greatest during interglacials/interstadials and least during glacials/stadials, whereas the opposite was true for eastern Beringia. This, insofar as most of the late Pleistocene is concerned, is a conclusion based on a mixture of empirical data and ecological theory. Additional empirical evidence bearing directly on the problem, however, is available for the late Wisconsin and Holocene.

An increase in population size and range of *B. antiquus* in midlatitude North America would be expected after the late Wisconsin maximum, as the forest/woodland-dominated glacial period vegetation changed to a more open savanna/steppe interglacial vegetation, which resulted in more available habitat and increased carrying capacity for megaherbivores. The radiocarbon record is one possible measure of megaherbivore increase; occurrences of late Wisconsin-Holocene megafauna for the period 30,000 BP to the present are presented in figures 95 and 96 and in Appendix 2. All but two genera (*Bison* and *Ovibos*) represented in these data became extinct in North America during the late Wisconsin-early Holocene period.

The radiocarbon dates in figure 95 represent both archeologic and paleontologic sites, a grouping that imposes several biases: (1) the greater likelihood of bone preservation in archeologic than paleontologic sites because of the greater initial quantity of bones (usually more than one individual represented) and greater potential for preservation (frequently on moist or steeply sloping sites, where deposition can take place relatively rapidly); (2) the selective killing of certain species by human hunters, which amplifies the representation of these game species in the archeological record while deemphasizing the representation of nongame species; and (3) the differential dating of faunal remains—archeologists use dating more frequently than paleontologists, certain genera appear to be of more interest than others to both archeologists and paleontologists, certain periods of the prehistoric period have been more consistently dated than others, and more radiocarbon dates are available from some regions of North America than others. Acknowledging these biases, the following assumptions are nonetheless made: (1) the radiocarbon record of the late Quaternary megafauna is a representative sampling of potentially datable sites; (2) human hunters, being biological organisms, existed in numbers in proportion to the overall carrying capacity of their range, so they too should have increased in population as the Wisconsin maximum waned; (3) the archeologic record should contain human sampling of the existing megafauna in approximate proportion to the relative composition and economic value of that megafauna, because economic specialization should be proportional to the supply of resources available; and (4) the number of radiocarbon dates for each period of time should be inversely proportional to the age of the period if factors of time alone (i.e., assuming a constant rate of decay of bone over time) were responsible for different frequencies of radiocarbon measurements over time.

Given the preceding assumptions the following conclusions can be reached. (1) The frequency of dates per time period is not consistently inversely proportional to the age of each period, so factors other than time are responsible for the differences in class frequencies (figs. 95 and 96; the correlation coefficient between years BP and radiocarbon date frequency for the last 15,000 years = .086). (2) There occurred a decrease in the megafaunal populations of midlatitude North America after ca. 22,000 BP. This coincides relatively well with the onset of the late Wisconsin maximum stade, which peaked between ca. 22,000 and 18,000 BP. (3) The megafaunal population remained relatively low throughout the late Wisconsin stadial maximum, ca. 22,000 to 16,000 BP. (4) The megafaunal population rebounded after ca. 16,000 BP., but was possibly truncated by a subsequent and widespread readvance of glaciers, with a consequent reduction in carrying capacity, ca. 14,500–14,000 BP (Mercer, 1972). (5) The final late Wisconsin rebound began about 14,000 BP. After that time the human factor becomes potentially important in amplifying the relative presence of the various genera, and beginning at about 12,000 BP the human factor clearly becomes important. The resulting exaggeration in the rebound curve for the late Wisconsin need be of no great consequence, however, if the assumption is valid that man and megafauna increased in rough proportion to each other. The important point is that there apparently was a rebound, as indicated by the differential frequency of dates per time period during the late Wisconsin, ca. 15,000–11,000 BP.

Bison and most of the genera that became extinct show an abrupt decline in frequency after about 11,000 BP. *Mammut*, however, may actually have increased for a brief period following the decline of the other genera but, by about 9,000 BP, at a time when *Bison* was again increasing, *Mammut* began a rapid population decline that led to extinction. By the end of the first quarter of the Holocene, only *Bison* and *Ovibos* among the megaherbivores here considered, appear to have been extant in North America. The decline of the North American megafauna between about 11,500–9,000 BP is here attributed primarily to the effect of human hunting (cf. Martin, 1967). Habitat for each of these genera was increasing and their numbers should have increased as well. But they did not increase in numbers—they decreased and most became extinct. The pattern of population increase and decrease, as revealed by the radiocarbon record, can be interpreted in the following way.

The late Pleistocene population peak for these genera was probably between 12,000 and 11,000 BP. Environmental changes after about 14,000 BP resulted in increased land area and carrying capacity for these genera, so a population increase during this period is to be expected. Human competition, however, probably increased at a greater rate than did the absolute numbers of late Pleistocene megafauna, due to both increases in the human population and the development of more efficent hunting methods. The North American megafauna was simply unable to adapt physiologically, behaviorally, or ecologically to the modified set of adverse selective factors operating on them, especially the hunting pressure imposed by humans—an ecological process new to North American fauna in which relative intelligence and its synergistic potentials was of greater importance in determining fitness and competitiveness than size and strength. Apparently the thresh-

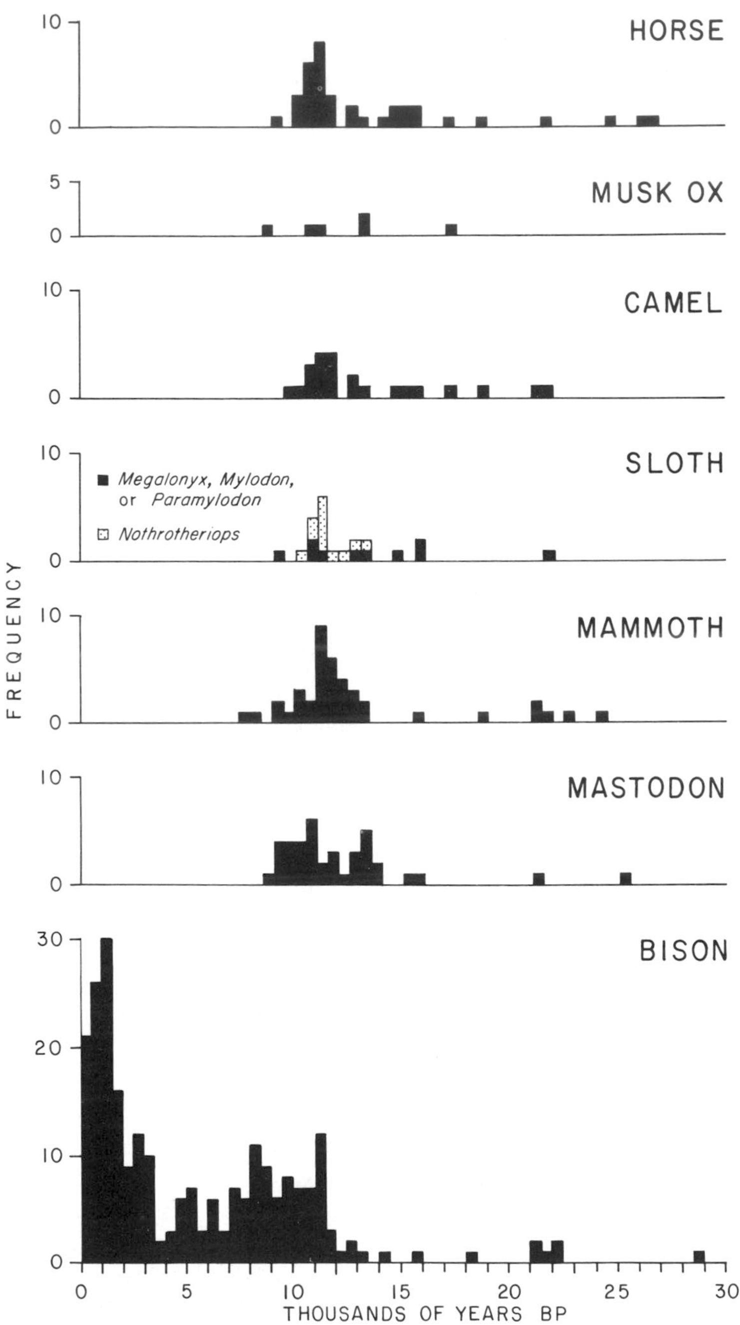

FIGURE 95 Radiocarbon dated occurrences of selected midlatitude megafauna.

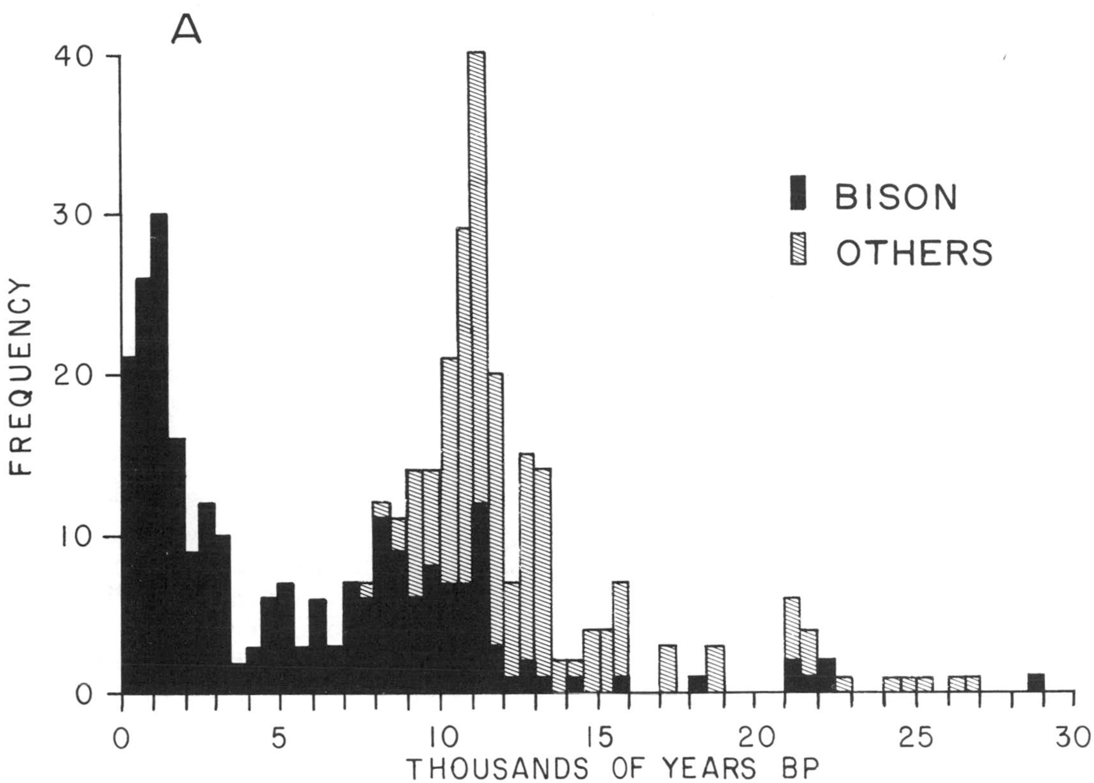

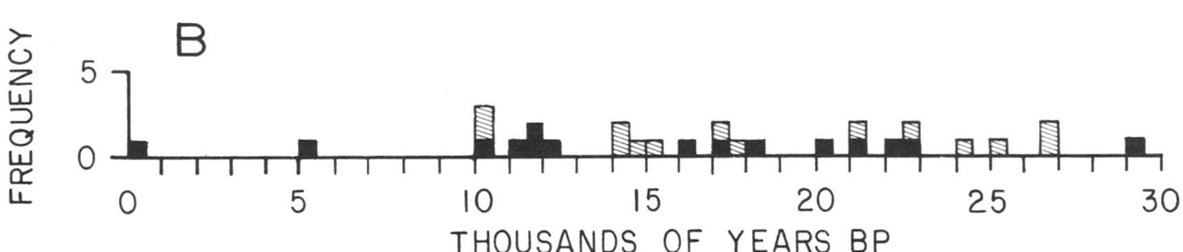

FIGURE 96 Radiocarbon dated transition from a complex to a simple megafauna, 30,000 BP to present. Bottom: All eastern Beringian radiocarbon dates. Many genara of the Beringian megafauna became extinct by ca. 10,000 BP. The two holocene dates are for bison which probably dispersed into Alaska from midlatitude North America. Top: All midlatitude dates plotted in figure 95. Notice the decreasing frequency of dates and the proportionate increase in bison representation after 11,000 BP. Periods with lowered frequencies of dates, at least for the period after 21,000 BP, can be correlated with environmental conditions which probably reduced the continent's carrying capacity: 21,000 to about 16,000 was the maximum late Wisconsin glaciation: 14,500 to 14,000 was a brief period of reglaciation; and 8,000 to 5,000 was a period of relative dryness on the southern Great Plains especially.

old of sustained yield for the continent's megafauna was crossed between 12,000 and 11,000 BP, and population levels of most genera declined sharply thereafter. The relatively high number of conventional radiocarbon dates between 11,000 and 9,000 BP (fig. 95) is considered a reflection of greater hunting pressure on bison resulting from the demise or extinction of alternate game species.

Less information is available for the eastern Beringian region than for midlatitude North America, but accumulating evidence indicates that human hunting of the region's megafauna was taking place throughout the late Wisconsin (Dixon, 1975; Müller-Beck, 1967; Péwé, 1975). The pressures imposed on the region's megafauna by human hunting and environmental changes, from habitats more suitable to habitats less suitable for sustaining viable populations of large herbivores, probably resulted in the extinction of much of the region's late Pleistocene megafauna by the onset of the Holocene (Dixon, 1975; R. Guthrie, 1968; Péwé, 1975). The extinction of the eastern Beringian megafauna during the late Wisconsin-early Holocene constitutes the final disappearance of Eurasian *Bison* from North America.

In summary, the late Wisconsin megafaunal populations oscillated according to available land area and suitable habitat, and the carrying capacity of that habitat. A general decrease in megafaunal populations in midlatitude North America followed the middle Wisconsin interstade, whereas the megafaunal population of eastern Beringia probably increased during this time. A period of climatic amelioration (between ca. 16,000 and 14,000 BP) was accompanied by an increase in midlatitude megafaunal populations, but was possibly followed by a decrease in populations immediately thereafter. The final midlatitude megafaunal rebound of the late Wisconsin commenced after ca. 14,000 BP. About 11,000 BP there occurred a population crash involving most large bodied genera in midlatitude North America. A similar crash occurred in *Mammut* ca. 9,000 BP. These population crashes, which occured at a time when populations should have actually been increasing, considering the nature of habitat changes which were then taking place (chapter 1), are interpreted as a result of excessive hunting pressure by man. The two genera that survived this hunting pressure and habitat change (*Bison* and *Ovibos*) apparently did so because they were better adapted to occupy expansive, open, and physiognomically simple habitats, and to utilize them to a greater extent than man.

THE SURVIVAL OF LATE PLEISTOCENE BISON

Bison barely survived the late Pleistocene-early Holocene extinctions. The point has already been made that a crash in bison numbers occurred after about 11,000 BP and that bison did not rebound until after about 9,000 BP. The high rate of unbalanced polymorphism (chapter 6) which existed in bison during the 11,000 to 9,000 BP period is empirical evidence that supports this theorized reduction in numbers. How and why, then, did bison survive?

Much of the explanation rests with the chance coincidence of (1) location; (2) reproductive potential; (3) direction of environmental change; and

(4) adaptive potential. The principal range of late Wisconsin *B. a. antiquus* was southwestern North America and, perhaps, the Great Basin (fig. 97). This principal range habitat became steadily drier during the transition from glacial to interglacial environments but, simultaneously, the central North American grasslands were developing where glacial forests, woodlands, and savannas had previously existed. The structural and compositional simplification of habitat which occurred during the transition from late Wisconsin to Holocene should have resulted in increased populations of most large herbivore species and, simultaneously, a more distinct partitioning of the habitat due to greater specialization in feeding behavior among these species. Human hunting also increased during this period. As a result, the potential for population increase afforded by vegetation changes was nullified by the proportionately more adverse selective factors operating against most of the genera in the form of human hunting. (The counter argument that loss of habitat was responsible for most megafaunal extinctions is weakened by two facts. Similar habitat changes took place earlier during the Pleistocene but were not characterized by catastrophic megafaunal simplification. Also, most if not all large bodied late Wisconsin mammals were euryecious and, practically speaking, many habitats they occupied at that time remained throughout the Holocene.)

The transition from woody to herbaceous vegetation on the Great Plains resulted in an abundance of resources which could be utilized by fully grazing species. The formation of this grassland was synchronous with the desiccation of the savanna, and in a region generally contiguous to what was then the main bison range. Basic herding tendencies and the capability to forage largely, if necessary, by grazing were established parts of the *B. a. antiquus* behavior and ecology. The development of grasslands provided impetus to bison population growth in two ways—it segregated bison from previous competitors that might have been unable to adapt to the grassland environment and it provided less diverse but more fully utilizable food resources than were available in the savanna habitat. Hunting pressure against bison, however, not only continued but probably intensified as other genera became extinct. As a bison population increased and herding tendencies strengthened, because of either adaptively increased social tendencies or simply greater opportunity to aggregate in the wake of decreased competition and habitat diversity, humans apparently evolved specialized hunting methods that were able to more efficently exploit the newly organized resource. The period from about 11,000 to 9,500 BP, therefore, was a time of both expanding suitable habitat for bison and increased hunting pressure on bison imposed by man during which time the bison range was minimal and bison populations appear (on the basis of unbalanced polymorphisms) to have been kept small and relatively isolated by the human hunting (fig. 98). It was during this period that the widespread effects of genetic drift and inbreeding brought about the appearance of *B. a. occidentalis*, a highly variable taxon in apparent disequilibrium with its environment. Populations of *B. a. antiquus* continued to exist, however, until the middle Holocene.

After about 9,500 BP bison populations appear to have rebounded and expanded their range (fig. 99). Extinction was probably avoided, ultimately, because bison were able to adapt to the open grasslands more completely than was man. This point is simple—bison depended primarily on the grasses, forbs, and water of the grasslands for survival. Man, an omnivore, required more shelter and regularly utilized a greater variety of diet items than the open grassland provided in abundance. Bison populations on the open plains, away from the plains margins where human hunters regularly and effectively cropped bison, were able to increase in numbers rapidly. Indeed, one of the most important factors in the survival of the bison might have been the relatively high reproductive potential and relatively short maturation period which probably characterized late Wisconsin-early Holocene bison vis-à-vis those genera that became extinct.

During and following the early Holocene bison rebound, environmental changes were occurring which resulted in a major expansion and shift of the bison range. The onset of the Atlantic climatic episode ca. 8,500 BP initiated the desiccation of the southern plains and a consequent reduction in the region's carrying capacity. At the same time the grasslands and the bison range were expanding northward and eastward (figs. 100 and 101). The boreal forest to the north was relatively permeable because of retarded succession over the recently glaciated terrain. Human hunting of bison continued during this period, probably strongest on the northern, eastern, and western periphery of the range, but the net effect of hunting was probably greater in the southern and central plains where it, together with reduced carrying capacity, virtually eliminated bison from the region (Dillehay, 1974; McDonald, 1974, 1976). One salient of this early Holocene range change extended eastward into the midwestern prairie peninsula, whereas another extended north along the western lowland corridor toward Alaska.

A gradual southward expansion of the bison range began after 6,000 BP but accelerated after about 5,000–4,000 BP, coinciding with the dispersal of the newly evolved late Holocene grassland bison, *B. b. bison*. The principal environmental factors that contributed to these range dynamics were a slight cooling trend and correlated decrease in the aridity of the southern plains. The boreal forest was also probably becoming less penetrable as succession proceeded. The cumulative effects of these habitat changes on bison were to reduce dispersal northward and eastward, but to increase dispersal southward and (less so) westward. The primary and generally historic *B. bison* range was established by about 3,000–2,000 BP (figs. 102 and 103).

HOLOCENE DWARFING OF BISON

Bison diminished in body size during the period from about 11,000 BP until about 5,000–4,000 BP, after which time the reduction generally terminated for most characters (table 75). The terminal Wisconsin-early and middle Holocene may be considered the transitional phase in the evolution of *B. antiquus* to *B. bison*, the period during which late Wisconsin savanna bison were adapting

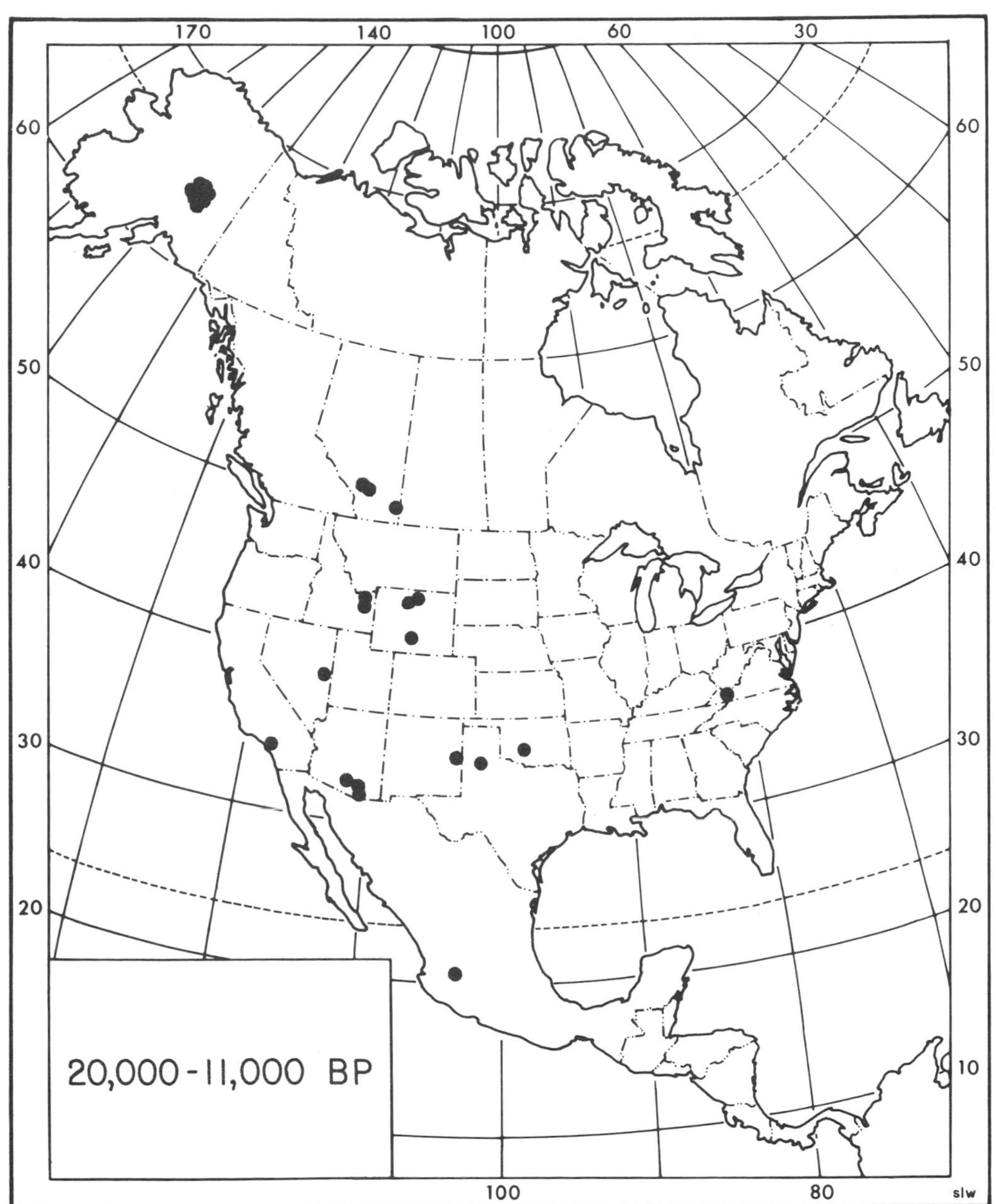

FIGURE 97 Radiocarbon dated bison distribution: 20,000–11,000 BP. Although southwestern North American was the primary bison range during the late Wisconsin period, populations were nonetheless widely distributed throughout nonglaciated North America. *Bison antiquus* was the dominant, and perhaps only, taxon in midlatitude and tropical North America; its range possibly extended northward into Beringia. Eurasian bison were present in Beringia during at least part of the late Wisconsin.

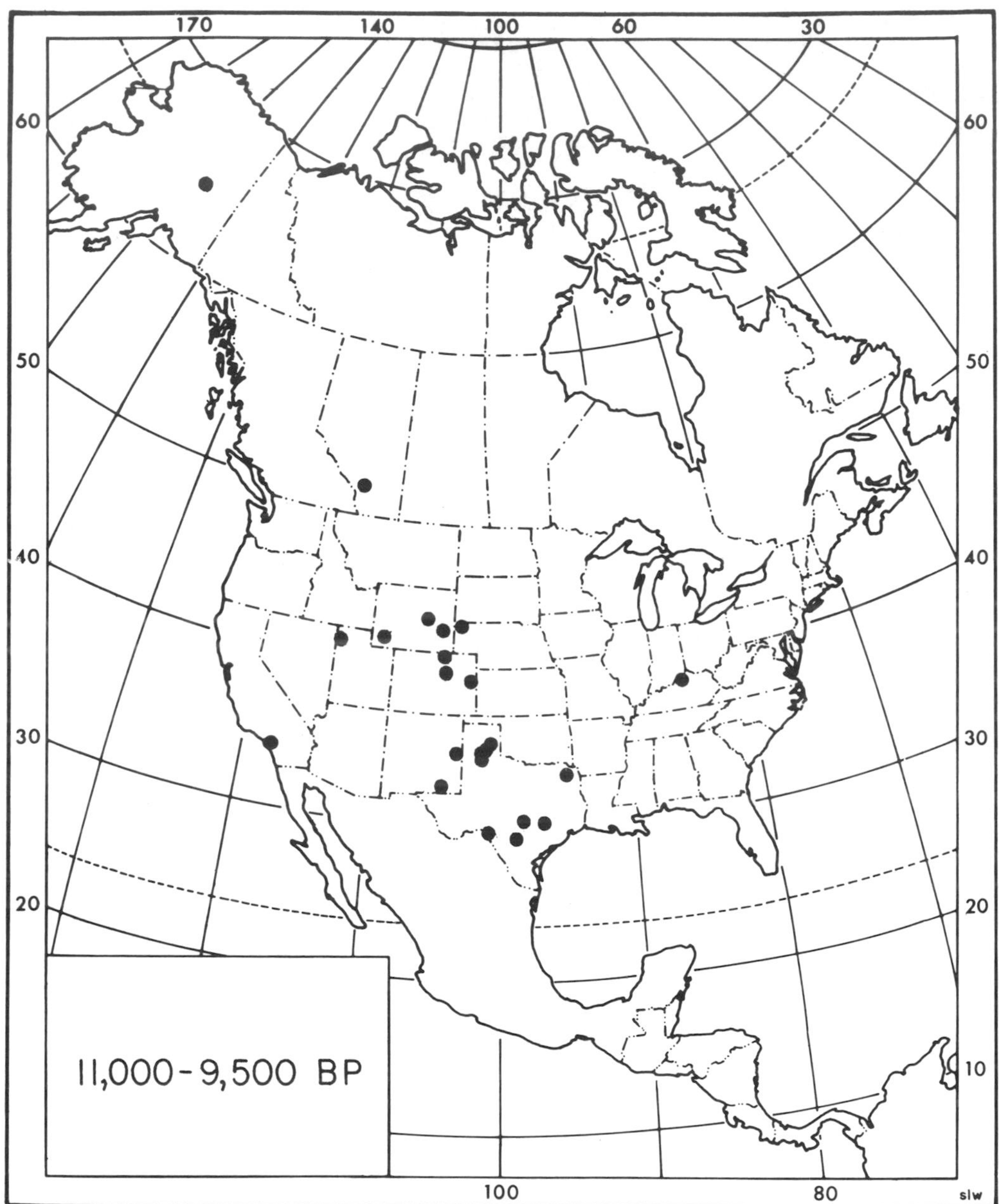

FIGURE 98 Radiocarbon dated bison distribution: 11,000–9,500 BP. The primary range is now shifting northeastward onto the Great Plains as the central grassland develops. The most intense, varied and widespread instances of abnormal specimens also date from this period, suggesting that the bison population probably came nearer the threshold of extinction at this time than at any other time prior to the nineteenth century. Bison became extinct in Alaska during this period.

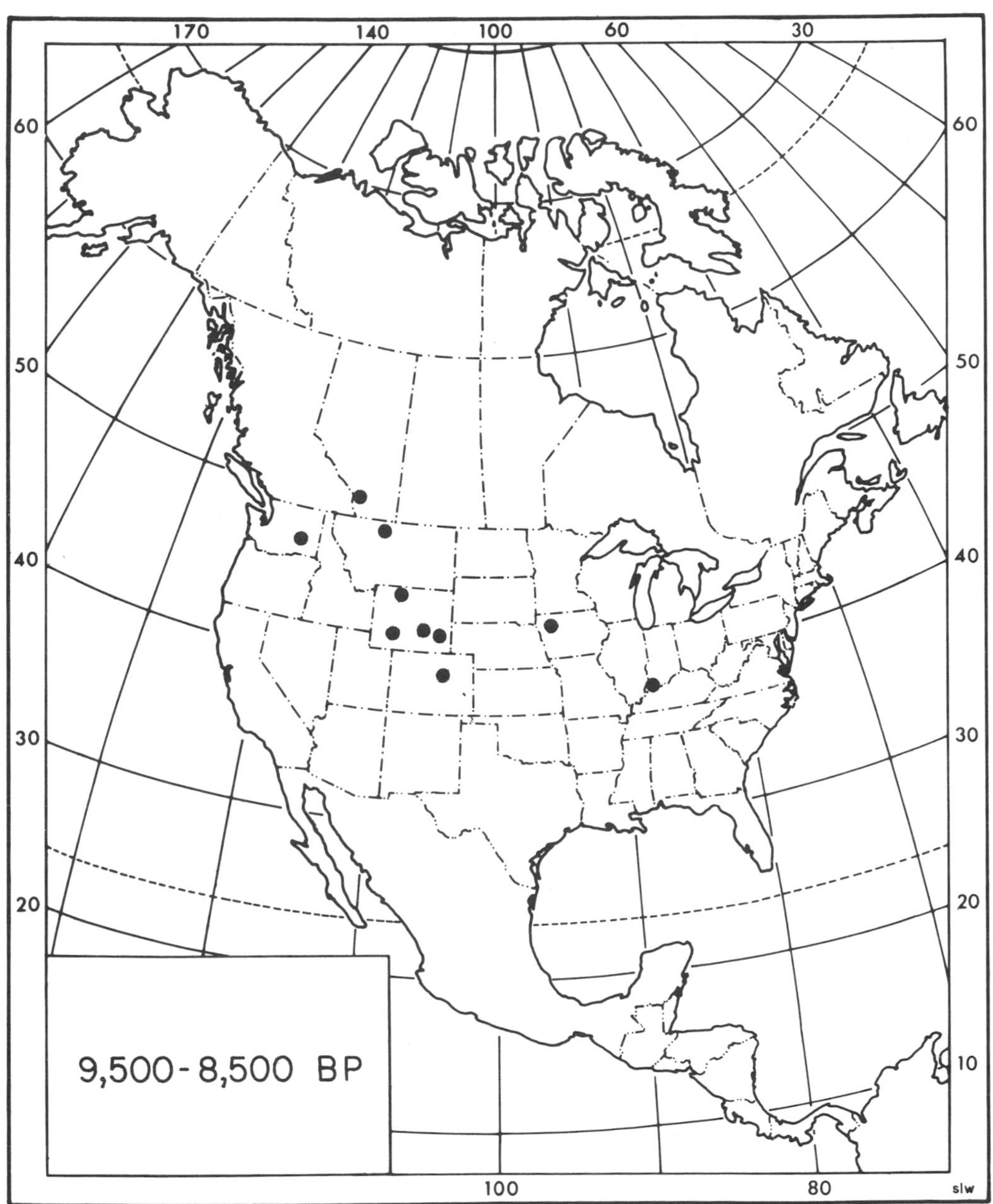

FIGURE 99 Radiocarbon dated bison distribution: 9,500–8,500 BP. The primary bison range has now shifted onto the northern Great Plains and adjacent prairie margins. A slight decrease in abnormal specimens from this period suggests that gene flow is becoming reestablished and the species' population might be increasing modestly, probably away from the plains margins where human hunting would have been generally heaviest.

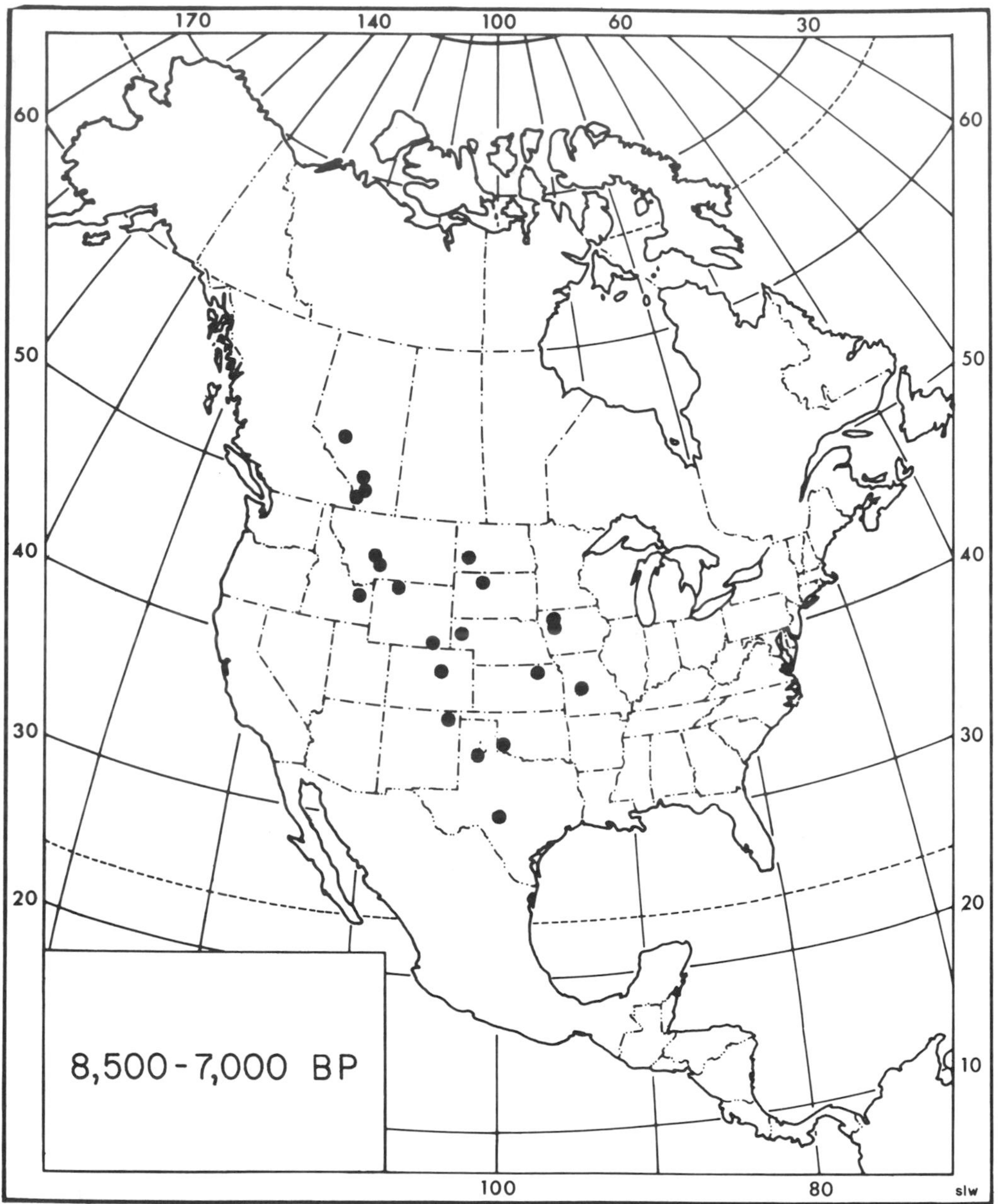

FIGURE 100 Radiocarbon dated bison distribution: 8,500–7,000 BP. The primary bison range remained on the northern plains during this period, but some bison were also on the southern and central plains. Most southern bison were gone by, or soon after, 8,000 BP as the Atlantic climatic episode got under way and rendered the southern plains less productive.

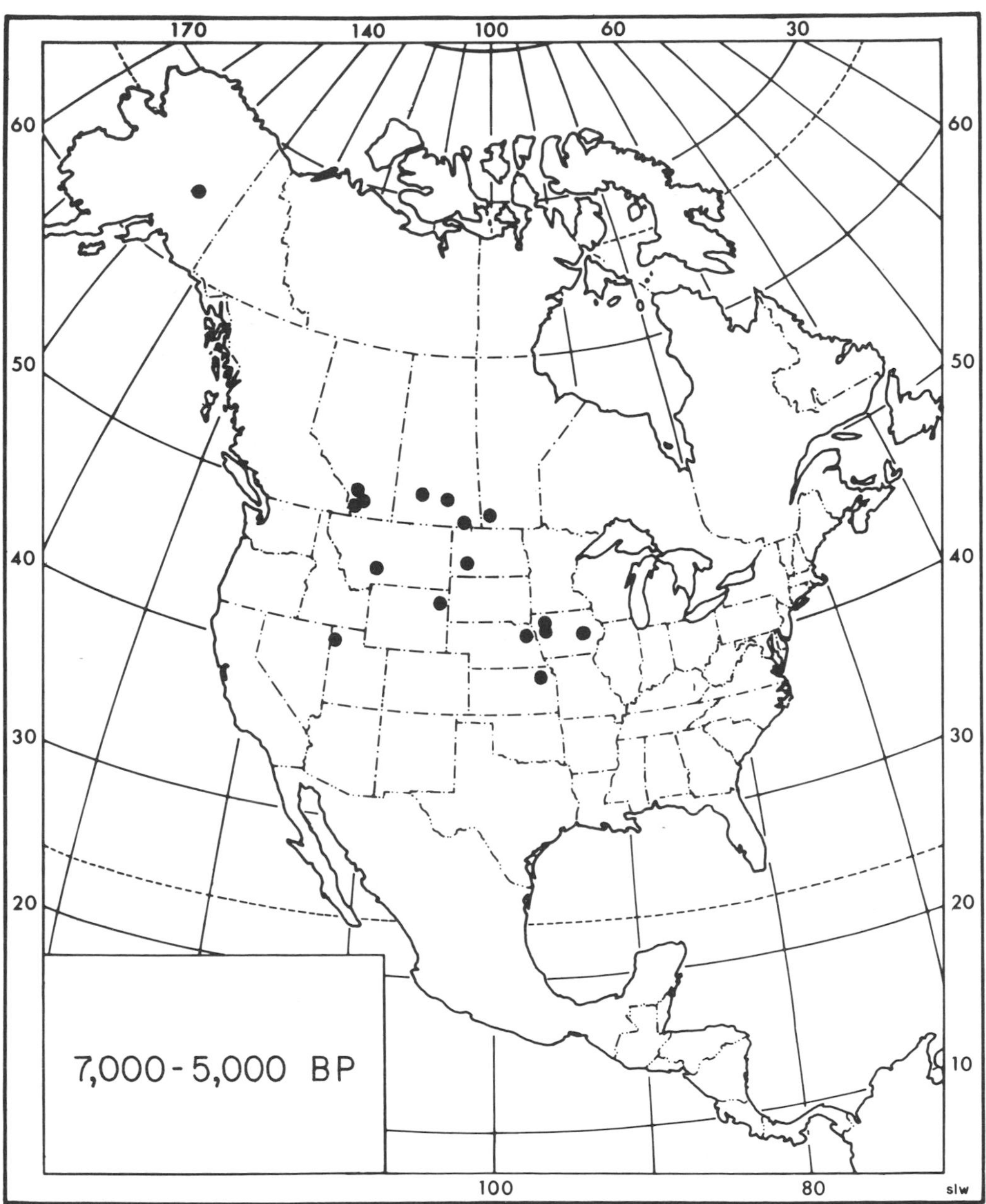

FIGURE 101 Radiocarbon dated bison distribution: 7,000–5,000 BP. The primary bison range remained anchored on the northern plains until after about 5,000 BP. During this period, human predation continued the selective pressure upon bison and brought the regional population to the verge of a new adaptive homeostasis with the Holocene selective regime. Also during this period, bison again appear in Alaska, some populations apparently having dispersed northward through the boreal forest filter.

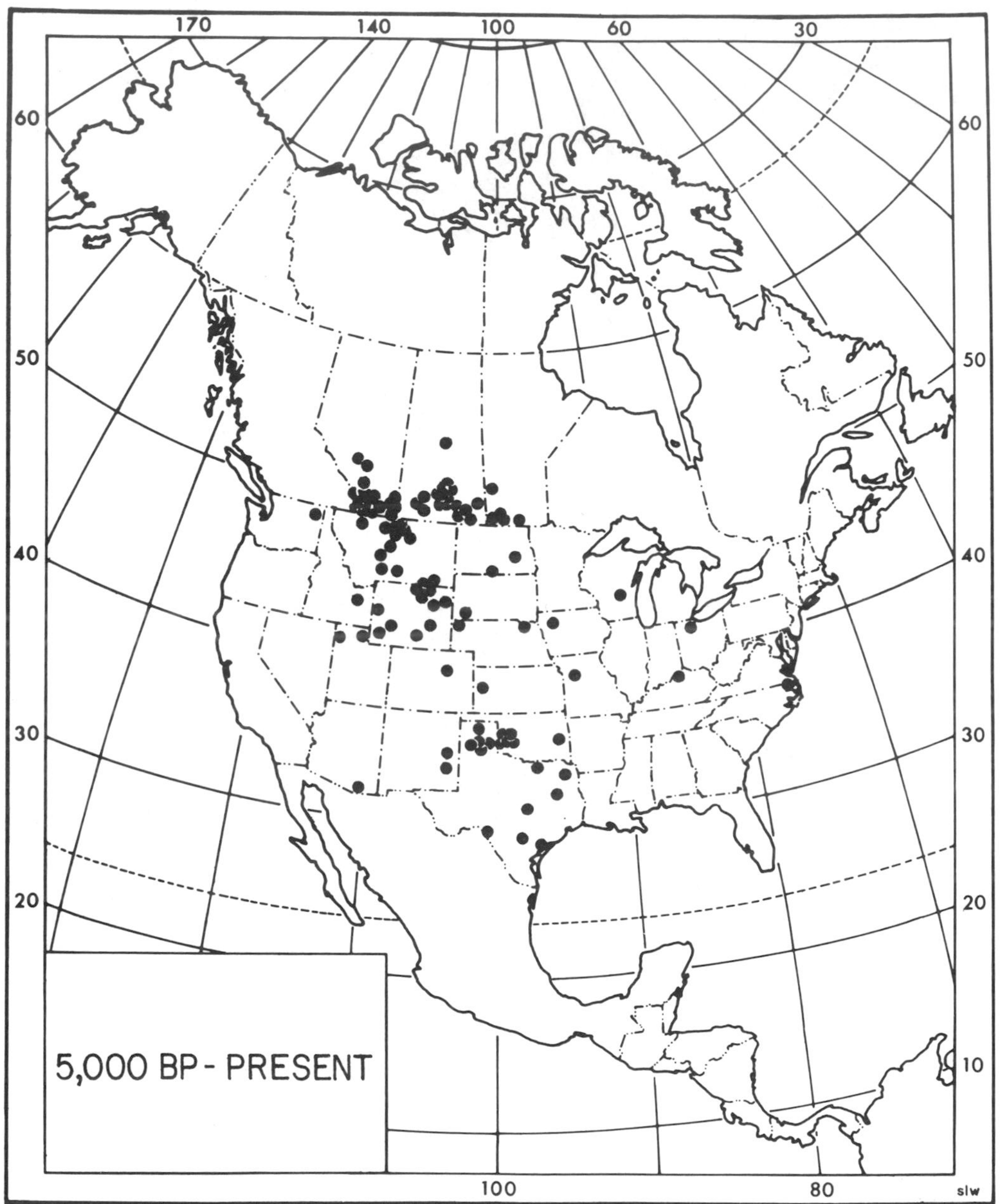

FIGURE 102 Radiocarbon dated bison distribution: 5,000–present. The bison population started expanding its range southward after about 5,000 BP as the central and southern plains habitat improved as a result of regional climatic changes, and as a new phenotype, *Bison bison*, appeared. The southward expansion apparently accelerated around 3,000 BP and, by about 2,000 BP, the species maximum range had probably been achieved.

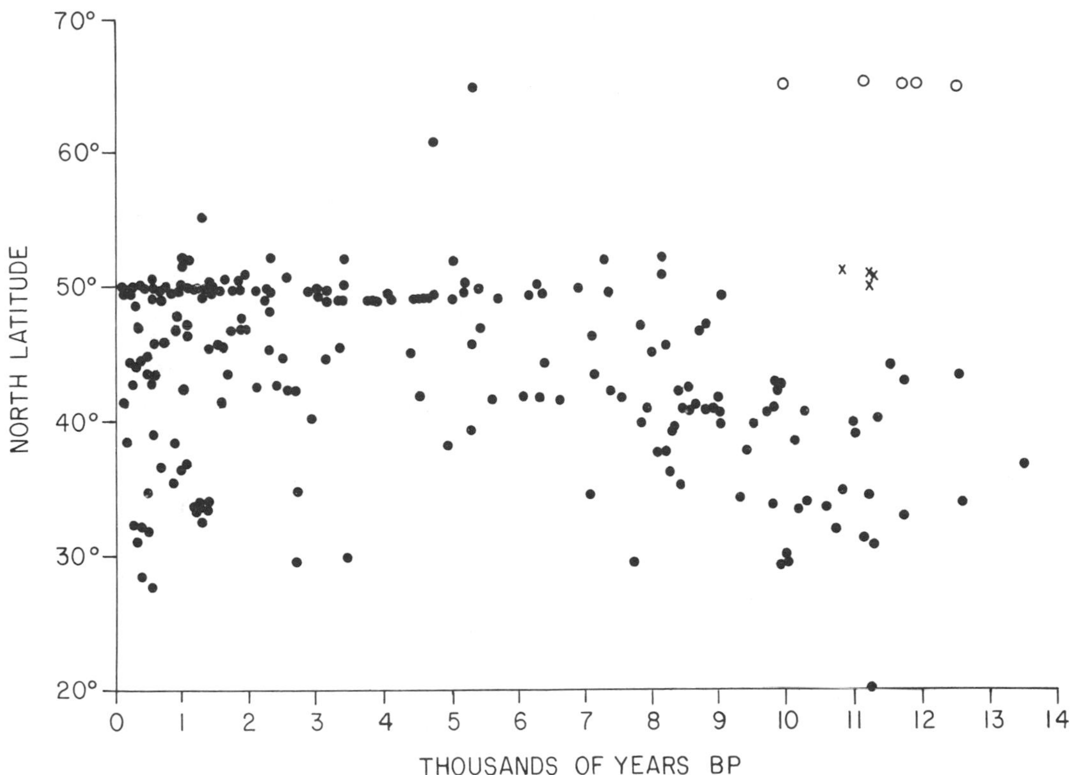

FIGURE 103 Radiocarbon dated bison distribution: years BP vs. latitude. Shown here are dated occurrences of midlatitude and tropical bison (dots), periglacial bison (x's), and eastern Beringian bison (circles). Notice especially (a) the northward shift in the midlatitude bison range which occurred between about 10,000 and 7,000 BP; (b) the virtual absence of bison on the southern Great Plains during the period from about 8,000 to 3,500 BP; and (c) the range expansion which occurred after about 5,000 BP. The Beringian and periglacial populations are assumed to have become extinct during the latest Wisconsin–earliest Holocene.

to the new Holocene environment and its reordered selection regime.

The progressive diminution of bison has been documented for horn cores and metapodials (Bedord, 1974; Hillerud, 1970; Hughes, 1978; McDonald, 1978*a*; M. Wilson, 1975). Not all characters, however, diminished in size at the same rate (table 75). In fact, the size of some characters in some populations actually increased during the early part of this transitional period, soon after disequilibrium was established.

Usual explanations given for the net size reduction in Holocene bison include density-dependent stress or resource limitations, low nutritional levels, or climatic conditions (e.g., Geist, 1971*a*; Guthrie, 1970, 1978; Wilson, 1974*b*). Of these explanations the first is nonadaptive, whereas the second and third are environmentally deterministic but not adaptive. The dwarfing was, of course, general—but it also included qualitative structural changes and was, ultimately, a genetically based phenomenon.

Possibly the simplest and most meaningful explanation for Holocene size reduction in North American bison lies in the selective advantage of accelerated maturation and increased biotic potential. Holocene bison were continuously hunted by Indians (Davis and Wilson, 1978; Frison, 1975), and the bison population was, as a result, probably always kept below the habitat's carrying capacity (plates 35 and 36). This constituted a relatively r-type selection regime in which progenic individuals in continuously rebounding populations were favored simply because smaller but earlier maturing phenotypes could outreproduce the larger, probably more slowly maturing phenotypes.

Accompanying the selection for progenic individuals imposed by human hunting was the reduction or elimination of selective forces favoring larger individuals. Competition for food, water, and space resources from other large herbivores was nearly eliminated and fewer large nonhuman predators preyed on bison. Simultaneously, flight rather than seclusion from threatening situations became more important to bison inhabitating an open grassland habitat; small animals accelerate faster and are more agile than larger animals.

The combined selective factors, therefore, favored largeness less than smallness and, through the early and middle Holocene, improved adaptations evolved. By ca. 5,000–4,000 BP, a sufficently successful combination of characters, behavior, and ecology had evolved to allow a new taxon, *B. bison*, to reach adaptive homeostasis with its environment and to undergo an increase in numbers and expand across the post-Atlantic central grasslands. The selective environment has remained sufficiently stable since the end of the Sub-Boreal climatic episode (table 4) to maintain general phenotypic equilibrium throughout the *B. b. bison* population. *B. b. athabascae*, however, may have undergone, or may be undergoing, selection favoring a larger phenotype in a parkland environment.

THE IDENTITY OF *B. antiquus occidentalis*

Wilson (1974*b*) has recently synonymized *B. antiquus* and *B. bison*. This is less desirable than keeping them as distinct species because both were adapted to different selection regimes at different times. Considering these species synonymous ignores the adaptive aspect of species' identity and differentiation. *B. a. occidentalis* was the transition taxon bridging the gap between the relatively K-selected *B. a. antiquus* and the relatively r-selected *B. bison*. Even though *B. a. occidentalis* did depart from *B. a. antiquus* morphologically, and probably behaviorally and ecologically, when subjected to the new selective forces of the times, it is still better to retain this transitional taxon as a subspecies of *B. antiquus* than *B. bison*. Several factors support this position. *B. a. antiquus* and *B. a. occidentalis* were apparently contemporaneous during much if not all of the transition period (ca. 11,000–5,000 BP). *B. a. occidentalis* apparently represented regional populations of the species; populations that first appeared on the western

Table 75
Character Dwarfing During the Holocene

Sample	Radiocarbon age (years BP)	Horn core length, dorsal curve (SM 3) ♂♂	Horn core length, dorsal curve (SM 3) ♀♀	Width of occipital condyles (SM 9) ♂♂	Width of occipital condyles (SM 9) ♀♀	Least width of frontals (SM 14) ♂♂	Least width of frontals (SM 14) ♀♀	Length of metatarsal ♂♂	Length of metatarsal ♀♀
Rancho la Brea, California	> 12,000	267.0 mm	208.0 mm	144.7 mm	136.4 mm	323.2 mm	264.0 mm	277.3 mm	271.3 mm
Lindenmeier, Colorado	10,990*	292.0 mm	....	148.0 mm	....	330.0 mm	....	280.8 mm	....
Lubbock Lake, Texas (Clovis/Folsom)	10,155*	....	....	....	....	....	....	276.9 mm	....
Olsen-Chubbock, Colorado	10,150	240.0 mm	197.0 mm	148.0 mm	....	309.6 mm	243.0 mm	265.7 mm	262.2 mm
Folsom, New Mexico	(10,000)	267.0 mm	166.0 mm	143.6 mm	....	....	....	280.2 mm	267.8 mm
Bonfire, Texas (Bed 2)	10,083*	....	....	....	....	....	....	....	269.5 mm
Casper, Wyoming	9,945*	381.0 mm	205.0 mm	147.0 mm	124.0 mm	310.0 mm	250.0 mm	....	....
Hudson-Meng, Nebraska	9,820	....	....	....	....	....	....	274.6 mm	255.1 mm
Plainview, Texas	9,800	303.0 mm	214.0 mm	143.0 mm	....	329.0 mm	....	268.0 mm	263.7 mm
Rex Rodgers, Texas	9,391	351.0 mm	215.0 mm	151.0 mm	....	327.0 mm	221.0 mm	270.0 mm	....
Wasden, Idaho	(8,000)	....	....	....	....	....	....	273.5 mm	263.9 mm
James Allen, Wyoming	7,900	....	....	....	....	....	....	....	266.0 mm
Zap, North Dakota	7,840	330.0 mm	....	125.0 mm	....	270.0 mm	....	....	....
Duffield, Alberta	7,750*	269.1 mm	188.6 mm	131.3 mm	117.0 mm	288.2 mm	232.6 mm	....	....
Hawken, Wyoming	6,370*	262.1 mm	190.0 mm	140.0 mm	124.0 mm	300.5 mm	232.0 mm	263.1 mm	252.4 mm
Hughes Bog, Iowa	5,640	315.0 mm	....	130.0 mm	....	293.0 mm	....	....	....
Zap, North Dakota	5,440	253.0 mm	....	145.0 mm	....	280.0 mm	....	....	....
Scoggin, Wyoming	4,540	....	....	126.0 mm	....	295.0 mm	....	....	....
Southern plains Archaic sites, Texas	(2, 500 – 3,500)	....	141.0 mm	....	....	....	....	249.3 mm	242.5 mm
Sitter, Texas	(2,500 – 3,500)	....	....	....	....	....	....	....	245.0 mm
Bonfire, Texas (Bed 3)	2,795	....	106.4 mm	....	....	....	....	255.6 mm	248. 6 mm
Buffalo Creek, Wyoming	2,530	208.0 mm	129.0 mm	128.0 mm	113.0 mm	270.0 mm	227.0 mm	256.0 mm	247.1 mm
Glenrock, Wyoming	245*	202.0 mm	....	131.0 mm	....	281.0 mm	....	254.2 mm	244.6 mm
Rocky Ford, South Dakota	(200)	202.0 mm	123.0 mm	129.8 mm	114.5 mm	275.7 mm	212.1 mm	263.0 mm	248.0 mm
Montana, 1880s	(100)	190.5 mm	125.3 mm	128.4 mm	119.0 mm	264.6 mm	220.3 mm	254.6 mm	247.5 mm
B. b. athabascae	(50±)	246.7 mm	....	192.5 mm	....	295.7 mm	....	275.6 mm	264.0 mm

Note: Figures in parentheses are estimates or historic.

*Designates the arithmetic mean of two or more acceptable radiocarbon dates.

plains when and where the combined pressures of the new selection regime were most intense and thereafter spread across the plains and into the Great Basin, whereas *B. a. antiquus* populations shifted northward, surviving in decreasing proportions as the transition period proceeded (fig. 104). *B. a. occidentalis* was not adapted to the new selection regime, the taxon was not canalized and the morphologic response to the combination of circumstances was not unidirectional as some populations apparently experienced even some increase in character size (tables 73 and 75). Character size in *B. a. occidentalis* was clearly nearer that of *B. a. antiquus* than of *B. bison.* This combination of factors suggests that *B. a. occidentalis* should be considered a highly variable subspecies of *B. antiquus* (with which it apparently coexisted), not of *B. bison* (which did not then exist), possessing a distinct range and responding multidirectionally to a newly ordered selection regime.

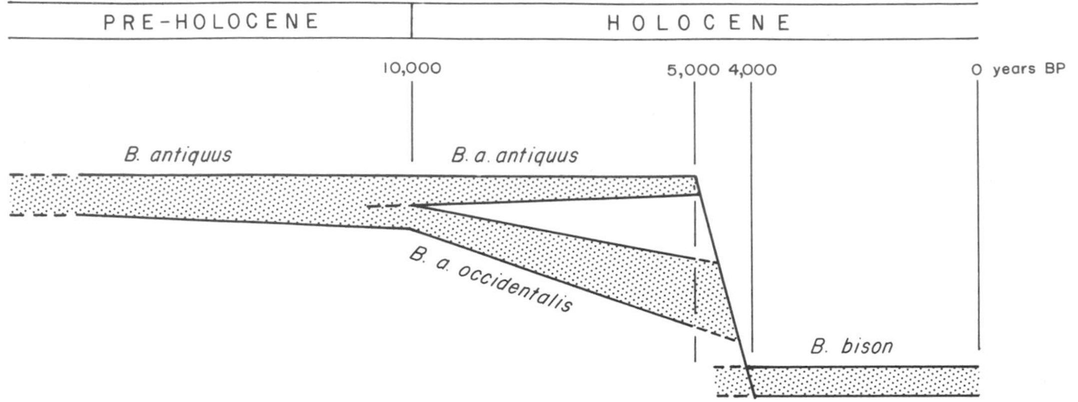

FIGURE 104 This model illustrates several important facets of the evolution of *Bison bison* from *Bison antiquus*. Until about 11,000 BP, *Bison antiquus* was a monotypic, well canalized species, but at that time a sudden increase in abnormal specimens occurred. Some populations (*B. a. antiquus*) remained unchanged throughout the early and perhaps middle Holocene, but the highly variable and still uncanalized populations (*B. a. occidentalis*) became more numerous as the Holocene progressed. During the millenium between 5,000 and 4,000 BP, while the primary bison range was still largely anchored on the northern plains, adaptive homeostasis was reached and *Bison bison*, a smaller, then-and-thereafter canalized taxon adapted to the selective regime then prevailing in North America, appeared.

Bison bison

Bison bison represents the current stage of North American bison evolution. More information is available on morphological variation, population dynamics, behavior, ecology, and range dynamics for *B. bison* than any other species. This provides an opportunity to evaluate these characteristics in greater detail than is possible for other species. Such a detailed examination, however, is beyond the scope of the present study. Here only three specific aspects of *B. bison* range and phenotypic variation of evolutionary significance will be briefly discussed. These are (1) differences between *B. b. athabascae* and *B. b. bison*; (2) the retention of an r-selection regime throughout the species existence; and (3) range oscillations and population distribution within the species range.

B. b. athabascae occupied the boreal parklands and woodlands at the northern end of the primary *B. bison* range. *B. b. athabascae* is larger than *B. b. bison* because of the selective forces operating in the forest opening/woodland environment, although the *B. b. athabascae* selection regime has more of an r-selection character than was true for pre-Holocene North American bison. The greater body size of *B. b. athabascae* could have been reached by either, or both, of two methods: (1) suspended equilibrium, in which case *B. b. athabascae* retained large body size as it evolved characters from *B. antiquus* adapted to the new r-type forest/woodland selection regime; or (2) more recent adaptive differentiation, in which case *B. b. athabascae* has increased in body size from *B. b. bison* as populations of the latter dispersed from the grasslands and adapted to the forest/woodland environment. Elements of both alternatives could have operated in the origin and continuation of *B. b. athabascae*, but the second has probably been more important over time in influencing the nature of *B. b. athabascae*. The *B. b. athabascae* gene pool has probably been regularly supplemented by gene flow from *B. b. bison* populations to the south. Periodic, and perhaps continuous, contact between grassland and parkland bison probably occurred for as long as *B. bison* has existed, or at least until Euroamerican instigated hunting of bison accelerated the decline of bison numbers (plate 37) and isolated many regional populations. Moodie and Ray (1976) have documented annual movements of *B. b. bison* into the boreal parkland during the historic period, this after hunting pressure had increased. The seasonal movement was probably even greater before the Euroamerican impact, and gene flow between both subspecies probably regularly occurred as individuals periodically joined and remained with herds dominated by members of other subspecies.

The differentiation of *B. b. athabascae* from *B. b. bison* is somewhat analogous to the speciation of *B. alaskensis* from *B. priscus*, although the time over which *B. b. athabascae* has differentiated is much shorter, and the area and the bison population involved are much smaller than were probably true for *B. alaskensis*. *B. b. athabascae*, therefore, has been in a tension situation between the selection for larger body size by intraspecific resource competition and hierarchal dominance organization in small, presumably socially organized (although numerically unstable) herds, and the selection for more rapid maturation imposed by regular hunting by man. *B. b. athabascae*, then, probably existed in a forest/woodland mode r-selection regime in which human hunting probably kept population levels below the habitat's carrying capacity. Many aspects of the distribution, behavior, and ecology of *B. b. athabascae* are described by Allen (1900), Fuller (1950, 1960), Roe (1951), Seton (1885–1886), and Soper (1941).

B. b. bison also remained under a relatively r-type selection regime after adapting to the late Holocene grassland environment. This was because adverse environmental factors were sufficiently strong to keep the overall population below the carrying capacity of its environment. Drought, fire, winter storms, and other environmental phenomena all probably functioned to periodically reduce regional populations, but human hunting was probably the single most important and widespread process that kept the overall population level below its potential (plate 36). As with paleo-Indians, however, neo-Indians appear to have hunted primarily and most consistently along, or from, the margins of the main bison range, although they did, of course, hunt in the interior of the bison range as well, both before and after their acquisition of the horse and displacement toward or onto the plains by the Euroamerican advance. This hunting, in effect, minimized or prevented the peripheral expansion of the bison range. Hunting in or outward from the ecotonal periphery of the Great Plains grassland may have been the most influential factor separating the *B. b. athabascae* and *B. b. bison* gene pools, to the extent that they were, in fact, separated.

Range and population oscillations did occur for *B. b. bison*; the Great Plains were not completely or uniformly occupied by populations of *B. b. bison*. Within the main range regional droughts caused displacements in the distribution of bison, and severe snow and ice storms destroyed or reduced local populations (Allen, 1876, 1877; Cole, 1954; McHugh, 1972; Meagher, 1971; Roe, 1951; see also Bryson, 1974). Alternatively, improved conditions brought regional population increases, such as apparently occurred in the northwestern Great Plains during the "Little Ice Age" of the sixteenth to nineteenth centuries (Reher, 1978). Unrestricted hunting by Indians often dislodged bison populations from an area, whereas contested hunting in intertribal buffer zones (i.e., regions that experienced intergroup conflict over hunting rights) appears to have provided relative security and temporary refuges for some bison populations. Within the main bison range there was, then, a regular ebb and flow of bison populations locally (see, e.g., Davis and Wilson, 1978; Roe, 1951), resulting in a mosaic of areas with and without bison at any given time.

Periodic range expansions beyond the Great Plains also occurred. Westward expansion from the northern and central plains consisted of two phases, occupation and penetration of the Rocky Mountain barrier and dispersal across the Great Basin. The Rocky Mountain valleys and parks were probably regularly occupied by bison, either small resident populations or immigrants from the western Great Plains. The carrying capacity of these habitats was relatively low because of their small area, and occupation was largely seasonal, especially at higher altitudes (Christman, 1971; Meagher, 1973; Reeves, 1978*a*). Penetration of the Rocky Mountain barrier was, therefore, minimized by environmental circumstances and probably accompanied by selection for larger phenotypes (chapter 5).

Populations of *B. b. bison* had filtered through the Rocky Mountain barrier and become well established in the northern Great Basin steppes and brushlands by about 2,500 BP (Schroedl, 1973). Between that time and the early nineteenth century, this regional population was seriously reduced, probably by a combination of drought (such as that of the late thirteenth century) and human hunting (Jennings, 1957; Jennings and Norbeck, 1955; Schroedl, 1973). Euroamerican accounts indicate that bison were present but only in small numbers in the Great Basin in the midnineteenth century, and were soon thereafter hunted to extinction (Hornaday, 1889; Roe, 1951; Schroedl, 1973).

The great southwestern salient of bison probably reached its maximum extent by about 2,000 BP, although more information, especially from Mexico, is needed before this can be stated with certainty. The southwestern bison range apparently expanded during relatively cool and moist periods between about A.D. 1100–1460 (Agenbroad and Haynes, 1975; P. Johnson, 1974). Drier climate probably brought about a reduction in the regional population, but human hunting must have contributed to the reduction also. Euroamerican explorers in the southwest during the sixteenth to eighteenth centuries encountered evidence that bison either then existed, or had recently existed, in the area from Zacatecas to northern Arizona. Bison were, however, relatively rare throughout the region south and west of the Rio Grande even in the early sixteenth century, and by the late eighteenth century most, if not all, had disappeared from the region. The last population of any significance probably existed in northeastern Arizona (Brand, 1967, oral comms., 1972–1976; Gunnerson, 1974; Reed, 1955; Roe, 1951).

Bison probably dispersed into the humid, generally forested eastern United States regularly, just as they dispersed into the Rocky Mountains, although the human barrier in the east was probably more difficult to penetrate. Little evidence is available to determine how early or how rapidly dispersal eastward occurred, but bison (identified as *B. bison*) are known from the Atlantic Coast of North Carolina by at least 2,610 BP (Painter, 1978), and occurred from Florida to New York and Ontario at or near the time of European contact (i.e., sixteenth century). Accounts from the sixteenth to the eighteenth centuries indicate that bison were widely distributed throughout the east during this time, although in relatively dispersed and small populations. The displacement of eastern Indians westward and the westward advance of Euroamerican settlement resulted in the elimination of bison from most of the humid east by the early nineteenth century (see, e.g., Allen, 1876; Dunbar, 1962; Goodwin, 1936; Griffin and Wray, 1945; Hays, 1871; Hornaday, 1889; Jakle, 1968; Painter, 1978; Roe, 1951; Rostlund, 1960; Sherman, 1954).

The main bison range was steadily reduced from the early seventeenth through the middle nineteenth centuries by both Euroamerican and Indian hunting (Allen, 1876; Dary, 1974; Hornaday, 1889; Newcomb, 1950; Roe, 1951). The major range reduction and near extinction of *B. b. bison*, however, occurred between the late 1860s and early 1880s when millions of bison were killed, primarily by Anglo-American hunters supplying the meat and robe trade (plate 37). Very few individual bison were saved from this slaughter, but from that small sample several thousands now exist, although mainly as quasi-domesticates (plate 38). The overall effect of this second Holocene bottleneck and subsequently established new selection regime has not yet been systematically assessed, but it does appear that some morphological changes have occurred (e.g., the southern Great Plains phenotype may have been completely lost), and probably behavioral changes as well. The bison range, of course, has been fragmented and gene flow seriously reduced and artificially regulated. Lastly, new selective factors have replaced those of the pre-nineteenth century late Holocene; general and semiselective hunting have been replaced by selective removal or retention of individual bison, and idiosyncratic human preferences control the distribution, composition, and density, and in many ways the behavior and ecology, of *B. bison*.

CHAPTER 8

CONCLUSIONS AND PERSPECTIVE

The preceding chapters have presented information and developed models that attempt to fulfill the objectives identified in the preface. In addition, the information and models developed earlier can usefully be compared with existing theories of bison evolution to better ascertain their reliability. Lastly, the accumulated, organized, and evaluated information, and the models derived from this, provide some insight into informational deficiencies and needs. A useful way to close this book might, therefore, be to summarize the previous chapters vis-à-vis the stated objectives, to comment on the findings of this study vis-à-vis other current theories, and to indicate some areas in which more information seems to be especially needed.

The revised classification presented here recognized five species of North American bison. Three of these—*B. latifrons, B. antiquus,* and *B. bison*—are considered to be North American autochthons. *B. priscus* and *B. alaskensis* are considered to be Eurasian autochthons that periodically dispersed into North America through Beringia. The North American and Eurasian lineages probably both developed from *B. sivalensis*. Two subspecies are recognized for *B. antiquus* (*B. a. antiquus* and *B. a. occidentalis*) and *B. bison* (*B. b. bison* and *B. b. athabascae*). Each species included in this classification is considered to have been adapted to a distinct selection regime. *B. a. occidentalis* is considered the transitional phase leading from *B. antiquus* to *B. bison*.

Measures of variation are given for selected skeletal characters of both sexes of each taxon. These measures provide insight into the absolute, relative, and statistical nature of continuous variation, sexual dimorphism, allometry, and variance. Clinal gradients for some characters are also provided. The data presented in chapter 3 indicate that not all characters vary identically between sexes, among taxa, or in space, and that not all characters are equally plastic and equally responsive to evolutionary pressures.

Interspecific differences among selected characters and character complexes are identified and described in chapter 4, and their adaptive significances (i.e., functions) are considered. The horns, fronto-facial region, superior molars, occipital region, limbs, and overall body size were evaluated for various adaptive qualities of size, shape, placement, and articulation. Skull characters were shown to have the greatest interspecific differences, a fact that underscores the socially and ecologically adaptive functions of the head and its elements.

Bison occupying forest openings or woodlands are larger than those occupying open grasslands. This fundamental difference forms the basis for the adaptation models presented in chapter 5. Generally, pre-Holocene forest opening/woodland and savanna/wooded steppe bison were both adapted to relatively K-type selection regimes in which selection favored large body size, longevity, and competitiveness, but the forces of selection peculiar to each habitat type differentially emphasized certain characters. Holocene bison are adapted to a relatively r-type selection regime that favors high biotic potential as a result of the recurrent significant catastrophic reduction of their populations. Even within the general r-type selection regime of the Holocene, however, the differential selection of forest/woodland and grassland environments is evident in both North American and Eurasian bison, with the forest/woodland taxa (*B. b. athabascae* and *B. b. bonasus*) being larger than the savanna/grassland taxa (*B. b. bison* and *B. b. caucasicus*).

Hybridization, genetic drift, and inbreeding were considered possible mechanisms by which certain problematic phenotypes could be explained. Hybridization is assumed for most bison in Canada's Wood Buffalo National Park since the late 1920s, and morphological characteristics of post-1928 adults indicate some of the phenotypic responses to be expected in hybrids. It was not possible to identify hybrid *B. latifrons* X *B. antiquus* or *B. priscus* X *B. alaskensis* (but see the East Asia/Beringia data in fig. 24) because their qualitative diagnostic characters are identical. Some specimens studied shared Eurasian and North American characters, however, and these were considered hybrid specimens. *B. priscus* X *B. antiquus* and *B. latifrons* X *B. alaskensis* hybrids would probably not be competitively disadvantaged vis-à-vis pure individuals of either taxon because both species of each pair were essentially vicarious. This helps to explain the relatively high frequency of hybrid specimens. *B. priscus* X *B. latifrons* or *B. antiquus* X *B. alaskensis* hybrids would, however, probably be competitively disadvantaged (as would *B. antiquus* X *B. latifrons*). Hybridization appears to have occurred in well-defined zones, strongest in Beringia and less frequently in periglacial and central North America.

Genetic drift and inbreeding are probably responsible for several types of phenotypic abnormalities which appear in bison, including numerous types of dental abnormalities, facial abnormalities, deviant horn cores, and, less frequently, occipital and limb abnormalities. Some inbred populations were probably also more susceptible to pathologies than were less inbred populations. Some of the dental pathologies of the Casper bison and the rapid tooth wear evident in some early Holocene southern plains populations might be attributable to the debilitating consequences of inbreeding. Two general situations were probably especially conducive to genetic drift and inbreeding: isolated populations on the periphery of the main bison range (e.g., the tropical American bison) and bottlenecking (e.g., the very late Wisconsin-early Holocene bison).

An attempt to integrate bison chronology and phylogenesis with environmental patterns and regulatory processes of the Quaternary resulted in the synthetic model of bison evolution presented in chapter 7. Drawing on environmental information presented in chapter 1 and systematic information presented in chapters 2–6, the origin, dispersal, distribution, differentiation, and extinction of taxa was described as a systematic process occurring through time and space, given detail and direction by the dynamic Quaternary environment.

The most important points made in this study insofar as the current theories of bison evolution are concerned are that (1) *B. latifrons* survived, and apparently in significant numbers, into the mid- and possibly late Wisconsin; (2) *B. latifrons* and *B. antiquus* were apparently synchronous; and (3) *B. bison* evolved from *B. antiquus*, not a Beringian taxon. Other important relevant elements of this study include findings that (4) *B. latifrons* and *B. antiquus* were largely if not completely allopatric, with *B. antiquus* located south of *B. latifrons*; (5) Eurasian taxa dispersed into North America and these taxa were not, therefore, phases of North American bison phylogenesis; (6) some early Holocene bison actually experienced an increase in character size, including horn size; (7) the mainstream of Holocene bison shifted from a generally southwestern location northward to a northern plains location during the early to middle Holocene, and then expanded southward to occupy its present range by about 3,000–2,000 BP; and (8) forest/woodland taxa are typically larger than savanna/grassland taxa.

Five of the six theories identified in the preface are based, ultimately, on the notion that North American bison horn core size diminished regularly (either step-wise or continuously) over time, following the establishment of the "original" North American species, *B. latifrons*. The Horn Core Theory (Allen, 1876) appears invalid, however, because *B. latifrons* and *B. antiquus* were contemporaries (which they should not have been if the larger had evolved into the smaller), were allochthonous (which implies adaptive differentiation, not sequential occupance of the same general range), and their horn cores are differentiated by a decided morphological gap (which should not be the case if they were phyletically sequential). Inserting the two Eurasian taxa into the horn core gradation observed for North American bison requires that *B. latifrons* evolve into a taxon like Eurasian taxa, and this or these in turn evolve into a subsequent taxa showing typical North American characters (i.e., *B. antiquus*). This is an unlikely and unnecessarily complex arrangement, and is otherwise incompatible with the findings of this study. If the Horn Core Theory is invalid, so too is the Orthogenetic Theory (Schultz and Frankforter, 1946), and the Wave Theory (Skinner and Kaisen, 1947), Synthetic Theory (Guthrie, 1970), and Cline Theory (Wilson, 1974*a*) are seriously weakened.

The Wave Theory is probably accurate in that it recognizes multiple immigrations into Alaska from Siberia, but there were probably more than two such immigrations (immigration probably took place with each exposure of the Bering land bridge); the bison immigrating were probably not typical North American autochthons and, with the exception of the earliest immigrants, they probably did not give rise to typical North American autochthons; and they were probably not strictly allopatric until the Holocene.

The Synthetic Theory contains three essential elements: North American bison species each evolved from a Beringian taxon, each successive species evolved horns smaller than the previous species, and each North American species, except for the last (*B. bison*), evolved horns larger than the parent species (*B. priscus*) as a result of higher quality forage available at lower latitudes. The first two elements are variants of the Horn Core and Wave theories, discussed above. The third elements introduces a new consideration—the idea that phenotypic development is dependent on the quality of forage. The quality of forage must certainly be recognized as an important element in phenotypic development, but I doubt that this parameter alone is sufficient to generate the degree of phenotypic difference which exists among North American bison. Two problems with the nutrition-based explanation for differences among bison are the failure of subsequently evolving taxa (1) to attain the size of previous taxa (when bison chronology is viewed according to the Synthetic Theory model); or (2) to duplicate the pre-Holocene spatial pattern of larger versus smaller horned bison. Another problem is that this view subordinates, excessively I think, the genotype's role in phenotypic development. The Dispersal Theory (Geist, 1971*a*) is based on the same notion that forage quality, coupled with social changes, associated with the spread of peripheral populations into an unoccupied range is responsible for phenotypic development. Here, too, the same criticism applies—too much emphasis is placed on forage quality and too little on genotype influence. Beyond this fundamental objection, I have differed quite strongly with the specifics of the Dispersal Theory in my interpretation of bison adaptation to different environments. Although both the Synthetic and Dispersal theories attempt to establish an environmental basis for observed differences among bison taxa, both rely on a deterministic role of environment, rather than treating the environment as a source of stimuli influencing adaptive differentiation among bison.

The Cline Theory is weakened by the general synchroneity and allopatry of *B. antiquus* and *B. latifrons* during the pre-Holocene, and *B. b. bison* and *B. b. athabascae* during the late Holocene (both of which are exactly opposite of the pattern predicted by the Cline Theory); the northward shift of early Holocene bison (which is exactly the opposite predicted by the Cline Theory); and the invalidity of the Horn Core Theory.

The specific objectives and tests of current theories have been accomplished with varying degrees of completeness and confidence. Much additional information is needed to either disprove or support the conclusions reached here, as well as to press ahead with new investigations into bison evolution. The following seem to me to be especially critical needs:

1. More evidence is needed on the origin of the genus *Bison* in Asia. I have here noted the close association between *Leptobos* and *Bison*, and acknowledged Gutherie's information that current work on bovids from China has yielded a more likely ancestor for *Bison* than *Leptobos*. Also, *B. sivalensis* must actually be regarded as a tentative taxon only. A "primitive" species is not to be unexpected at or soon after the origin of a new genus, and *B. sivalensis* fills this position. Both specmens identified herein have nonetheless been identified by others as females and they could, in fact, be female *B. priscus*, although they are too large and too male-like, vis-à-vis the early, small *B. priscus* males, to convince me that they are females. We have here the possibility that early males might simply resemble later females. Additional information bearing on the time and place of origin, the phyletic pathway, and the morphology, behavior, and ecology of early bison is sorely needed.

2. Systematically derived information on the morphology and chronology of Eurasian bison is needed. Information describing morphological variation and the geochronology of Eurasian bison is essential to an understanding of the *B. priscus-B. alaskensis* problem and to an assessment of the role of Eurasian bison in the origin and evolution of bison in North America. Information on Eurasian bison morphology is available, but in limited quantities only. Good chronology, ideally based on absolute measurements, is also available only in limited quantities. These data are needed for the entire Eurasian bison range, but especially for the Siberian area insofar as North American-Eurasian bison relations are concerned.

3. The temporal and spatial relations between *B. latifrons* and *B. antiquus* need to be better documented. Radiocarbon

measurements are gradually providing reliable resolution of the time of disappearance of these two species, but the time and place of their origin is still very poorly known. Careful attention to the interpretation of the age of new-found specimens, preferably employing some absolute dating technique, would be extremely helpful.

4. Many additional radiocarbon measurements for late Pleistocene megafauna are needed. Radiocarbon dates have provided important chronological control for the examination and assessment of late Pleistocene bison specimens and populations, and their relations to other faunal elements. Available radiocarbon dates have permitted a reasonably clear picture of the rate and direction of Holocene bison evolution, but much greater temporal resolution would be possible if additional dates were available. This is especially true for bison occurrences on the margins of the main bison range, and for regions in the Great Plains not yet well represented by adequate dates.

5. Certain geographic regions need additional attention. Mexico is probably the region for which additional information on the presence of bison is most needed, but more information is also needed from the Great Basin and the southwestern United States for the period before 12,000 BP. Good chronology for eastern Beringia is also needed to help clarify what is, in fact, still a relatively confusing history of bison in that region.

6. Additional biometric analyses of populations are needed. Much more information on the nature of variation among populations, especially on other than (but certainly not excluding) skull and metapodial characters, is necessary to better understand all aspects of morphological variation. The measures of variation presented in chapter 3 need to be supplemented, perhaps with the use of additional quantitative techniques. New data for females of taxa represented here by only small samples would be very useful. Studies of the nature of ontogeny could be especially useful, as would investigations of evolutionary differences in nonskeletal characters (see, e.g., Smiley, 1978; M. Wilson, 1975).

7. Further empirical work is needed on the physiology, ethology, and ecology of bison and other ungulates. This serves the immediate function of providing a better understanding of modern populations, but it also provides baselines for making inferences about past behavior and ecology and to influence predictions and management decisions for future behavior and ecology. Bison evolution is continuing, and the more completely we understand the present needs and responses of modern bison, the better will we be able to evaluate its future needs and evolutionary directions in this human dominated and directed world.

APPENDIX 1

INSTITUTIONAL AND DEPARTMENTAL ABBREVIATIONS USED IN TEXT

A. Institutional Abbreviations

AMNH — The American Museum of Natural History, New York, New York

ANSP — The Academy of Natural Sciences of Philadelphia Philadelphia,Pennsylvania

BBL — Big Bone Lick Museum, Big Bone Lick State Park, Big Bone Lick, Kentucky

BM — The British Museum (Natural History), London, England.

BeMNH — James Ford Bell Museum of Natural History, University of Minnesota, Minneapolis, Minnesota

CAS — California Academy of Science, Golden Gate Park, San Francisco, California

CM	Carnegie Museum of Natural History, Pittsburgh, Pennsylvania
CMNH	Cincinnati Museum of Natural History, Cincinnati, Ohio
CoMNH	Denver (formerly Colorado) Museum of Natural History, Denver, Colorado
CSC	Chadron State College, Chadron, Nebraska
DM	Drumheller and District Museum, Dinosaur and Fossil Museum, Drumheller, Alberta
DMNH	Dayton Museum of Natural History, Dayton, Ohio
EC	Earlham College, Richmond, Indiana
FC	"Fauna Cedazo," private collection of Oswaldo Mooser, Aguascalientes, Aguascalientes
FMNH	Field Museum of Natural History, Chicago, Illinois
IG	Instituto de Geologia, Universidad Nacional de Mexico, Mexico, D.F.
INAH(DP)	Departamento de Prehistoria, Instituto Nacional de Antropologia e Historia, Mexico, D.F.
ISM	Illinois State Museum, Springfield, Illinois
IMNH	Idaho Museum of Natural History, Idaho State University, Pocatello, Idaho
KU	Museum of Natural History, University of Kansas, Lawrence, Kansas
LACM	Natural History Museum of Los Angeles County, Los Angeles, California
MCZ	Museum of Comparative Zoology, Harvard University, Cambridge, Massachusetts
MMMN	Manitoba Museum of Man and Nature, Winnipeg, Manitoba
MNA	Museum of Northern Arizona, Flagstaff, Arizona
MNHN	Museo Nacional de Historia Natural, Mexico, D.F.
MPM	Milwaukee Public Museum, Milwaukee, Wisconsin
MRG	Museo Regional de Guadalajara, Guadalajara, Jalisco
MU	Department of Biology, Midwestern University, Wichita Falls, Texas
NMC	National Museums of Canada, Ottawa, Ontario
NDSU	University of North Dakota Zoology Museum, Grand Forks, North Dakota
OHS	Ohio Historical Society, Columbus, Ohio
OSU	Orton Museum of Geology, Ohio State University, Columbus, Ohio
PMA	Provincial Museum of Alberta, Edmonton, Alberta
PPHM	Panhandle-Plains Historical Museum, Canyon, Texas
RO	Private collection of Richard Ohmes, Chaires, Florida
SM	Schuler Museum of Paleontology, Southern Methodist University, Dallas, Texas

SMNH	Museum of Natural History, Regina, Saskatchewan
SMP	Sanford Museum and Planetarium, Cherokee, Iowa
SPSM	The Science Museum of Minnesota, St. Paul, Minnesota
SU	Stanford University, Stanford, California
SUI	Museum of Natural History, The University of Iowa, Iowa City, Iowa
TMM	Texas Memorial Museum, University of Texas at Austin, Austin, Texas
TTU	The Museum, Texas Tech University, Lubbock, Texas
UAl	University of Alberta, Edmonton, Alberta
UASM	Arizona State Museum, University of Arizona, Tucson, Arizona
UALP	Laboratory of Geochronology, University of Arizona, Tucson, Arizona
UCy	University of Calgary, Calgary, Alberta
UCM	University of Colorado Museum, University of Colorado, Boulder, Colorado
UCMP	Museum of Paleontology, University of California, Berkeley, California
UF	Florida State Museum, University of Florida, Gainesville, Florida
UM	University of Montana, Missoula, Montana
UMMP	Museum of Paleontology, University of Michigan, Ann Arbor, Michigan
UMo	University of Missouri, Columbia, Missouri
UNSM	University of Nebraska State Museum, University of Nebraska, Lincoln, Nebraska
UP	University of Pennsylvania, Philadelphia, Pennsylvania
UPM	Museo dell' Istituto di Anatomia Comparata, University of Pavia, Pavia, Italy
USGS	United States Geological Survey, Denver, Colorado
USNM	National Museum of Natural History, Smithsonian Institution, Washington, D.C.
UTEP	Museum of Arid Land Biology, University of Texas at El Paso, El Paso, Texas
UU	University of Utah, Salt Lake City, Utah
UW	Laboratory of Anthropology, University of Wyoming, Laramie, Wyoming
WM	Woolaroc Museum, Bartlesville, Oklahoma
WPBSM	Science Museum and Planetarium of Palm Beach County, Inc., West Palm Beach, Florida
WTSU	West Texas State University, Canyon, Texas
YPM	Peabody Museum of Natural History, Yale University, New Haven, Connecticut

B. Departmental Abbreviations

A	Anthropology/Archeology
G	Geology/Geological Sciences
M	Mammalogy
P	Paleobiology
VP	Vertebrate Paleontology
Z	Zoology

APPENDIX 2

RADIOCARBON MEASUREMENTS

Radiocarbon measurements were collected primarily from published radiocarbon laboratory lists, and secondarily from other published or unpublished sources. The laboratory number is given for dates that have been published in radiocarbon laboratory lists. Deevey, Flint, and Rouse (1967) have indexed radiocarbon measurements published prior to 1966. Each volume of *Radiocarbon* lists laboratory numbers of measurements published in that volume. The Geological Survey of Canada measurements were published in *Radiocarbon* until 1972, since which time they have been published annually as a *Paper* of the Geological Survey of Canada. Sources have been provided at the end of the appendix for those measurements which are unpublished or have been published in other than radiocarbon laboratory lists.

All acceptable radiocarbon measurements dating *Bison* and other selected late Pleistocene megafauna are listed here. Several dates associated with megafauna were rejected as unacceptable, usually because (1) their validity was questioned or rejected in the source; (2) they did not clearly and directly date megafauna; or (3) they were rendered suspect by other information. I have occasionally accepted a date questioned in its source(s) when other information supported the validity of the date. The radiocarbon measurements listed here are arranged alphabetically by country, state or province, and locality name. The fauna represented by each measurement is identified by the following code: A = antilocaprids; B = *Bison*; C = camelids; E = *Equus*; M = *Mammut*; Mu = *Mammuthus*; Mx = *Megalonyx*; My = *Mylodon*; N = *Nothrotheriops*; O = *Ovibos*; P = *Paramylodon*; S = undifferentiated sloth; Sy = *Symbos*/*Bootherium*; and T = *Tapirus*.

All radiocarbon measurements presented here are based on the Libby half-life of 5,568 ± 30 years and have not been calibrated to true calendar years because their age range considerably exceeds tree ring calibrations presently available. When calibrations to true calendar years have been extended over the entire range of dates available, then all measurements presented here can be calibrated to true calendar years and evaluated accordingly.

Locality	C^{14} Years BP	Fauna present	Laboratory number(s)
Canada			
Alberta			
Bayrock site, Oldman River	9,050 ± 220	B	S-68
Cactus Flower (EbOp-16)[a]			
Occupation level VIII	4,130 ± 85	B	not given[a]
Occupation level VI	3,755 ± 130*	B	not given[a]
Occupation level IV	3,625 ± 90*	B	not given[a]
Occupation level II	3,475 ± 20	B	not given[a]
Occupation level I	2,770 ± 95	B	not given[a]
Calgary, Bow River floodplain	11,300 ± 290	B	RL-757[b]
Castle River	6,340 ± 140	B	GSC-705
Castle River	6,125 ± 160*	B	GSC-447, GSC-490
Cochrane, middle terrace	10,760 ± 160	B, E	GSC-612
Cochrane, middle terrace	11,235 ± 165*	B, E	GSC-613, GSC-989
Del Bonita	4,270 ± 135	B	Gx-1770[c]
DgPh-2	130 ± 85	B	S-844
DgPk-75	630 ± 135	B	S-723
DhPh-13	1,150 ± 120	B	S-829
Duffield	8,150 ± 100	B	S-106
Duffield	7,350 ± 100	B	S-107
East Battle Creek	7,300 ± 150	B	GaK-2334
EbOp-44			
Occupation level II	2,010 ± 55	B	S-1213
Occupation level I	210 ± 110	B	S-1214
Flach (FjPh-101)	3,260 ± 320	B	S-874
Gallelli gravel pit, Calgary	8,145 ± 320	B	Gx-2104[c]
Head-Smashed-In			
Levels 10 (bottom, north) & 16c (south)	5,530 ± 200*	B	GSC-803, RL-334[d]

Locality	C^{14} Years BP	Fauna present	Laboratory number(s)
Levels 10 (top, north) & 16a (south)	4,565 ± 110*	B	GaK-1476, RL-333[d]
Level 8 (north)	2,865 ± 105*	B	GaK-1474, RL-332[d]
Level 11 (south)	1,925 ± 80	B	Gx-1253[d]
Levels 2c (north) & 10 (south)	1,780 ± 115*	B	GaK-1475, Gx-1252, RL-330[d]
Levels 7b, 8a, & 9 (south)	1,075 ± 115*	B	GSC-983, RL-256, RL-257[d]
Level 5 (south)	700 ± 170	B	GSC-992[d]
"Island bluff"	0 ± 140	B	GSC-704
Kajewski	3,100 ± 80	B	GaK-1272
"Lindoe bluff"	11,200 ± 200	B, E	GSC-805
Manyfingers (DhPj-31)			
Unidentified level	1,380 ± 70	B	S-722
Levels 6 & 7	1,090 ± 90*	B	S-865, S-866
Levels 4 & 6	555 ± 90*	B	S-864, S-947
Mona Lisa	8,080 ± 150	B	GSC-1209[e]
Old Women's Buffalo Jump			
Level 25	1,840 ± 70	B	S-91
Level 17	1,650 ± 60	B	S-90
Level 13	1,060 ± 80*	B	S-87, S-89
Point Beazer (DhPh-3)	430 ± 90	B	S-828
Ramillies (EcOr-35)			
Feature A1	965 ± 65	B	S-1015
Feature A2	660 ± 115	B	S-1016
Ross Creek	1,180 ± 140	B	GSC-1296
Saamis (EaOp-6)			
Area C	435 ± 125	B	S-824
Area B	210 ± 80	B	S-827
See-Everywhere (EcOr-34)	160 ± 60	B	S-1014
Shaw Burial (EdOn-7)	1,390 ± 90	B	S-1017
Travers	6,880 ± 280	B	Gx-1769[f]
Waldron Ranch	2,300 ± 140	B	M-1724
Waterton Lakes National Park	565 ± 130	B	Gx-(?)[c]
Weed Creek	2,765 ± 90	B	S-804
Windle	2,530 ± 140	B	GSC-1744
British Columbia			
Babine Lake	34,000 ± 690	Mu	GSC-1754
Chilliwack	22,700 ± 320	Mu	GSC-2232
Peace River	7,670 ± 170	Mu	I-2244
Manitoba			
Carberry	1,260 ± 130	B	GSC-990
Grandview	8,620 ± 190	O	I-1623[g]
Harris #2	210 ± 50	B	S-519
Paddon	2,995 ± 105	B	S-588
Stendall			
Level 3	965 ± 70	B (?)	S-786
Level 2	755 ± 60	B (?)	S-785
Swan River	2,320 ± 130	B	GSC-1219
Ontario			
Ferguson	8,910 ± 150	M	GSC-614
Perry Farm	11,930 ± 185*	M	GSC-211, S-172
Port Talbot	>29,500	M	L-440
Rodney	11,700 ± 475*	M	S-29, S-30
Thamesville	11,380 ± 170	M	GSC-611

Locality	C^{14} Years BP	Fauna present	Laboratory number(s)
Saskatchewan			
Avonlea	1,500 ± 100	B	S-45
Carruthers (FbNs-3)	3,050 ± 80	B (?)	S-742
Dunn (DgNf-1)	5,000 ± 120	B (?)	S-168
Eagle Creek	2,365 ± 70	B	S-1074
Eagle Creek	850 ± 60	B	S-1073
Frenchman's Flat	1,035 ± 60	B	S-266
Garratt, level 6	1,365 ± 65*	B	S-406, S-408
Gull Lake			
Level 31A	1,740 ± 60	B	S-255
Level 26	1,290 ± 60	B	S-254
Level 24B	1,220 ± 80	B	S-149
Level 21	1,165 ± 80	B	S-150
Harder	3,393 ± 115*	B	S-490, S-668
Katepwa Beach (EaMv-1)	4,780 ± 195	B	S-1067
Kyle	12,000 ± 200	Mu	S-246
Long Creek			
Level 9	5,000 ± 125	B	S-54
Level 8, upper	4,635 ± 115*	B	S-52, S-53
Level 7	4,620 ± 170	B	S-50
Level 5, upper	3,370 ± 115	B	S-63a
Level 4, upper	2,230 ± 100	B	S-49a
Melhagen	1,960 ± 90	B	S-491
Moon Lake	5,000 ± 90	B	S-403
Mortlach Midden			
7' level	3,400 ± 200	B	S-2
3' level	1,580 ± 159	B	S-22
Oxbow Dam	5,200 ± 130	B	S-44
Riddell	15,340 ± 500	E	S-1305
Riddell	4,560 ± 115	B	S-1306
Rousell	1,185 ± 70	B	S-670
Russell	6,320 ± 140	B	GSC-280
South Saskatchewan River	585 ± 205	B	I-1393
Tschetter	1,005 ± 75	B	S-669
Walter Felt, level 4	400 ± 40	B	S-280
Wiseton	10,600 ± 140	Mu	S-232
Yukon Territory			
Gold Run Creek	>39,000	B	I-5405[h]
Gold Run Creek	32,250 ± 1750	Mu	I-4226[h]
Gold Run Creek	22,000 ± 1400	B	I-3570[h]
Near Gold Run Creek	14,870 ± (?)	E	I-3569[h]
Mexico			
Jalisco			
Jocotepec	11,275 ± 160*	B, C, Mu	I-6162, I-6163[i]
Mexico			
Ciudad de los Deportes	18,700 ± 450	C, E, Mu	UCLA-111[j]
San Bartolo Atepehuacan	9,670 ± 400	Mu	M-776
Santa Isabel Iztapan	>16,000	E, Mu	C-204
Puebla			
Barranca de Caulapan			
Base of alluvial fill	>35,000		W-1898[k]
Near base of alluvium	30,600 ± 1000		W-2189[k]
Upper limit of alluvium	9,150 ± 500		W-1896[k]

Locality	C^{14} Years BP	Fauna present	Laboratory number(s)
United States			
Alaska			
Alaska	>39,000	B	I-2246[l]
Baldwin Peninsula	$26{,}900^{+2400}_{-3400}$	Mu	AU-90
Chester Creek	470 ± 90	B	SI-852
Cleary Creek	11,735 ± 130	B	ST-1631[m]
Cripple Creek	>39,000	B	SI-840
Cripple Creek	29,295 ± 2440	B	SI-842
Cripple Creek	21,065 ± 1365	B	SI-839
Dome Creek	32,700 ± 980	Mu	ST-1632[m]
Dome Creek	31,400 ± 2040	B	ST-1721[m]
Dome Creek	>28,000	B	L-127
Dome Creek	17,695 ± 445	Sy	SI-851
Dry Creek			
Component I	11,120 ± 250	B	SI-2880[n]
Component II	10,015 ± 225*	B, E, Mu (?)	SI-1561, SI-2329[n,o]
Ester Creek	>35,500	Mu	RL-402
Fairbanks area	>40,000	Sy	SI-291[m]
Fairbanks area	31,980 ± 4490	B	SI-843
Fairbanks area	22,540 ± 900	B, Sy	SI-292[m]
Fairbanks area	21,300 ± 1300	Mu	L-601
Fairbanks area	16,400 ± 2000	B	M-38
Fairbanks Creek	24,140 ± 2200	O	SI-455
Fairbanks Creek	20,445 ± 885	B	SI-837
Fairbanks Creek	17,210 ± 500	O	SI-454
Fairbanks Creek	17,170 ± 840	B	SI-838
Fairbanks Creek	15,380 ± 300	Mu	SI-453
Fairbanks Creek	11,980 ± 135	B	ST-1633
Goldstream area	5,340 ± 110	B	SI-845
Little Eldorado Creek	>40,000	Sy	SI-291
Little Eldorado Creek	>35,000	B	SI-844
Lost Chicken Creek	26,760 ± 300	E	SI-355
Manley Hot Springs	18,000 ± 200	B	SI-841
Trail Creek, Cave 9	14,410 ± 315*	B, E	K-1210, K-1327
Upper Cleary Creek	25,090 ± 1070	Sy	SI-850
Upper Cleary Creek	12,460 ± 320	B	SI-290
Arizona			
Glen Canyon	24,600 ± 1400	E	A-526
Hurley	21,210 ± 770	Mu	A-988
Lehner Ranch	11,200[p]*	B, E, Mu, T	A-40a, A-40b, A-42, A-378, K-554, M-811
Murray Springs			
Level F_1	11,106 ± 180*	B, E, Mu	A-805, SMU-18, SMU-27, SMU-41, SMU-42
Levels G_2-H contact	340 ± 370	B	Tx-1174
Mauv Caves	11,115 ± 165*	N	A-1212, A-1213[q]
Rampart Cave			
99 cm level	32,560 ± 730	N	A-1210[q]
Rat layer	26,300 ± 760	E	A-1791
67 cm level	13,140 ± 320	N	A-1207[q]
61-46 cm level	12,245 ± 350*	E, N	A-1070, L-473c[q]
50 cm level (intrusive)	11,140 ± 250	N	A-1453

Locality	C^{14} Years BP	Fauna present	Laboratory number(s)
5–0 cm level	10,962 ± 250*	E, N	A-1041, A-1392[q]
Surface	10,800 ± 205*	N	A-1066, A-1067, A-1068, I-442[q]
Ventana Cave, level 5B	11,300 ± 1200	A, B, E, N, T	A-203
California			
Costeau Pit	42,000 ± 4400	A, B, C, E, Mu, P	UCLA-1324
Costeau Pit	>40,000	A, B, C, E, Mu, P	not given[r]
La Mirada	10,690 ± 360	B, C, E, M, Mx	not given[r]
McKittrick	38,000 ± 2500	A, B, C, E, M, Mu, Mx, My, N, T	UCLA-728
Rancho La Brea	37,000 ± 2660	B, C, E, M, Mx, N, T	UCLA-773c
Rancho La Brea	12,650 ± 160	B, C, E, M, Mx, N, T	UCLA-1292b
Colorado			
Dent	11,200 ± 500	Mu	I-622
Frazier	9,600 ± 130	B	SMU-31[s]
Jurgens	9,070 ± 90	B	SI-3726[s]
Kassler	10,200 ± 350	Mu	W-401
Lamb Springs			
108″–96″ depth	13,140 ± 1000	C, Mu	M-1464
42″–40″ depth	8,370 ± 295*	B	M-1463, SI-45
Lindenmeier	10,990 ± 320*	B, C	I(UW)-141, I-622[t]
Merino	2,945 ± 1475	B	UGa-661
Olsen-Chubbock	10,150 ± 500	B	A-744
Florida			
Palm Beach County	21,150 ± 400	B, C, M, Mu	I-4910
Idaho			
American Falls Reservoir	>32,000	B	W-358
American Falls Reservoir	31,300 ± 2300	B, C, E, Mu, Mx, S	WSU-1424
American Falls Reservoir	21,500 ± 700	B, C, E, Mu, Mx, S	WSU-1423
Bison Cave/Birch Creek	2,600	B	not given[u]
Bison Cave/Birch Creek	100	B	not given[u]
Jaguar Cave	11,580 ± 250	B, C, E	Gx-395[v]
Wasden			
Bottom level	12,550 ± 175*	B, Mu	WSU-1259, WSU-1281
Level 16	7,100 ± 350	B	M-1853
Wilson Butte Cave			
Stratum E	15,000 ± 800	C, E	M-1410
Stratum C	14,500 ± 500	C, E, S	M-1409
Illinois			
Urbana	9,190 ± 200	M	ISGS-17c
Indiana			
Cromwell	12,630 ± 1000	M	M-139
Elkhart	9,320 ± 400	M	M-694
Evansville	9,400 ± 250	B, E, Mx, T	W-418
Muncie	9,755 ± 300	M	W-325
Rochester	12,000 ± 450	M	I-586[j]

Locality	C^{14} Years BP	Fauna present	Laboratory number(s)
Iowa			
Burkholder-Mether	13,520 ± 135	M	Wis-712
Cherokee Sewer			
Paleo-Indian	8,535 ± 200*	B	not given[w]
Lower Archaic	7,395 ± 80*	B	Wis-882, Wis-886, Wis-888, Wis-891[w]
Upper Archaic	6,080 ± 70	B	Wis-889
Hughes Peat Bog	5,640 ± 540	B	ISGS-56
Quimby (Simonsen)	8,430 ± 520	B	I(UW)-79
Turin, 11′8″ level	6,080 ± 300	B	M-1071
Witrock, feature 2, 20″ level	550 ± 80	B	Wis-34b
Coffey, level B	5,263 ± 70*	B	Wis-624, Wis-629
Herl	1,067 ± 110*	B	SI-132, SI-133, SI-134
Sutter	7,825 ± 240*	B	SM-1420, SM-1421, SM-1423[x]
Wilson Ford	31,000 ± 6000	B	M-997
Kentucky			
Big Bone Lick	17,200 ± 600	Sy	W-1617
Big Bone Lick	10,600 ± 250	B, E, M, Mu, My	W-1358
Big Bone Lick	<225*	B	W-908, W-1357
Welsh	12,950 ± 550	Mu	I-2982
Louisiana			
Tunica Bayou	12,740 ± 300	M	W-944[j]
Massachusetts			
Harvard	21,200 ± 1000	B	W-544
Michigan			
Bailer Farm	24,000 ± 4000	Mu	M-2145
Colville Farm	13,200 ± 600	Sy	M-639
Genesse County	11,400 ± 400	Mu	M-1361
Jackson County	12,200 ± 700	Mu	M-507
Lenawee County	9,568 ± 1000	M	M-282
Pontiac	11,900 ± 350	M	not given[j]
Prillwitz	8,260 ± 300	Mu	M-1400
Rappuhn	10,565 ± 400*	M	M-1746, M-1782
Scotts	11,100 ± 400	Sy	M-1402
Smith Farm	10,700 ± 400	M	M-1254
Missouri			
Boney Spring	15,552 ± 285*	E, M, P, T	A-1079, I-3922, M-2211, Tx-1477, Tx-1478
Grundel	25,100 ± 2200	M	I-1559
Guthrey, feature 2	600 ± 80	B	Wis-81
Rodgers Shelter			
Level 19	8,200	B	not given[y]
Level 17	8,100	B	not given[y]
Montana			
Antonsen			
Area C	1,605 ± 350	B	I-7027[z]
Area A	<180	B	I-7849[z]
Avocet	1,130 ± 110	B	GaK-6268[aa]
Boarding School	360 ± 150	B	M-1066
Cascade	8,800 ± 300	B	W-594
Clarkston Valley	600 ± 250	B	W-918

Locality	C^{14} Years BP	Fauna present	Laboratory number(s)
Donovan	1,900 ± 100	B	GaK-6272[aa]
Fresno, Besant level	1,700 ± 120	B	GaK-6266[aa]
Keaster	1,945 ± 250	B	W-1366
Kremlin	1,100 ± 100	B	GaK-6270[aa]
Kremlin	970 ± 90	B	GaK-6269[aa]
Lindsay	11,925 ± 350	Mu	S-918
MacHaffie	7,100 ± 300	B	L-578a[f]
Myers-Hindman			
Level 1	8,195 ± 210*	B	GaK-2627 GaK-2634[bb]
Level 3	5,315 ± 185*	B	GaK-1490 GaK-2632[bb]
Level 4	3,340 ± 110*	B	GaK-2629 GaK-2630[bb]
Level 5	2,300 ± 120	B	GaK-2628[bb]
Level 6	1,470 ± 70	B	GaK-2633[bb]
Level 7	790 ± 90	B	GaK-2631[bb]
Pishkun, Cascade County	550 ± 200	B	LJ-8
Powers-Yonkee	4,450 ± 125	B	I-410
Rinehardt Kill	945 ± 120	B	Gx-147
Sun River Canyon	11,500 ± 300	Mu	W-1753
Nebraska			
Allen	8,274 ± (?)	B	C-108a
Cedar Canyon (25SX101)	2,147 ± 150	B	C-469
Hudson-Meng	9,820 ± 160	B	SMU-224[cc]
Knox County, #1	2,660 ± 110	B	I-1308
Logan Creek, level B	6,633 ± 300	B	M-837
Nevada			
Astor Pass	17,150 ± 600	C, E	L-364
Fishbone Cave	11,200 ± 250	C, E	L-245
Gypsum Cave	11,690 ± 250	N	LJ-452
Gypsum Cave	11,360 ± 260	N	A-1202
Gypsum Cave	10,455 ± 340	C, E, N	C-221
Las Vegas Wash	>23,800	B, C, E	C-914
Smith Creek Cave	11,020 ± 175*	B, C	Tx-1420, Tx-1421, Tx-1637, Tx-1638
Tule Springs	>35,000	B, C, E, Mu	UCLA-513
Tule Springs	12,645 ± 205*	A, C, E, Mu	UCLA-507, UCLA-509, UCLA-521, UCLA-522, UCLA-604[dd]
New Mexico			
Aden Crater	11,080 ± 200	N	Y-1163b
Blackwater Draw			
McCullum Ranch	15,750 ± 760	B, C, E, Mu, S	A-375
Clovis level	11,280 ± 420*	B, C, E, Mu	A-481, A-490, A-491
Folsom level	10,350 ± 415*	B, Mu	A-380, A-386, A-488, A-489
Cody/Firstview level	8,620 ± 350*	B	A-512, (?)[ee]
Dry Cave			
Harris' Pocket	14,470 ± 250	E	I-3365
Unidentified location	11,880 ± 250	C	I-5987
Test Trench II	10,730 ± 150	B, C, E	I-6200

Locality	C^{14} Years BP	Fauna present	Laboratory number(s)
Garnsey			
P — 3/55	490 ± 110	B	Gx-5099[ff]
P — 2/27	390 ± 130	B	Gx-5097[ff]
Pigeon Cliffs	8,280 ± 1000	B	W-636
18 Mile Bend			
Middle Pecos III	850 ± 200	B	M-742
Mesita Negrita Phase	820 ± 110	B	M-1466
New York			
Byron	10,450 ± 400	M	W-1038
Kings Ferry	11,410 ± 410	M	Y-460
Malloy	12,100 ± 400	Mu	I-838
Sheridan	9,200 ± 500	M	M-490
North Carolina			
Currituck	2,610 ± (?)	B	UGa-(?)[gg]
North Dakota			
Boundary	1,540 ± 160	B	I-499
Zap	7,840 ± 250	B	I-2536[c]
Zap	5,440 ± 200	B	W-1537
32BA1	1,860 ± 150	B	I-497
Ohio			
Akron	15,315 ± 625	M	OWU-190
Akron	13,695 ± 460	M	OWU-321
Akron	13,025 ± 550*	M	M-1970, M-1971
Ansonia	10,230 ± 150	M	I-5843
Clyde	150 ± 100	B	M-1518
Johnstown	10,190 ± 190	M	OWU-141
Novelty	10,654 ± 188	M	OWU-126
Orleton Farms	9,600 ± 500	M	M-66
Pontius Farm	13,180 ± 520	M	OWU-220
Ross County	13,070 ± 340*	M	OWU-260 A, B & C
Oklahoma			
Barton	11,990 ± 170	Mu	A-582
Domebo	11,210 ± 550*	B, Mu	SI-172, SI-175
Goff Shelter	910 ± 55	B	Wis-261
Edwards II (34Bk44)	1,070 ± 70*	B	Tx-109, Tx-807, Tx-808[hh,ii]
McLemore	720 ± 85*	B	C-1245, R-892-1 & 2[ii]
Mouse	1,000 ± 150	B	M-1091[ii]
Perry Ranch (34Jk81)	7,030 ± 190	B	Tx-2190
34Gr12	2,770 ± 150	B	GaK-694[jj]
34Gr51	555 ± 115*	B	GaK-695, GaK-696[jj]
Oregon			
LaGrand	11,030 ± 800	My	SI-331
Pennsylvania			
Cheming River	13,320 ± 200	M	Y-2619
South Dakota			
Conata Basin	660 ± 150	B	M-947
Walth Bay			
Level B	8,030 ± 1100	B	RL-308
Level A	7,010 ± 210	B	RL-309
Texas			
Alibates 28	1,350 ± 70	B	not given[kk]
Barton Road	3,450 ± 150	B	not given[ll]
Ben Franklin			
"Pleistocene"	10,345 ± 415*	B, E, M, Mu	SM-532, SM-533[mm]

Locality	C^{14} Years BP	Fauna present	Laboratory number(s)
"Recent"	1,375 ± 220*	B	SM-598, SM-599, SM-600[mm]
Bonfire			
Level 2	10,083 ± 205*	B	Tx-153, Tx-657, Tx-658
Level 3	2,795 ± 130*	B	Tx-46, Tx-47, Tx-106, Tx-131
Cave Without a Name	10,900 ± 190	B, E, M, O/Sy	not given[ll]
Clear Creek	28,840 ± 4740	B	SM-534[j]
Dye Creek	1,350 ± 150	B	M-845[nn]
Felton Cave	7,770 ± 130	B	not given[ll]
Hinojosa	580 ± 50	B	Tx-2207
Kincaid Rock Shelter, level 5	9,890 ± 130*	B	Tx-17, Tx-18, Tx-19, Tx-20[oo]
Kyle	530 ± 145*	B	not given[ll]
Levi Rock Shelter, level 2	10,000 ± 175	B, E	0-1106
Palo Duro Creek	10,800 ± 835	B	SM-936[pp]
Pickett Ruin	1,240 ± 70	B	not given[kk]
Plainview	9,800 ± 500	B	L-303
Rex Rodgers	9,391 ± 83	B	SMU-274[qq]
Roper	1,300 ± 70	B	not given[kk]
Sanford Ruin	1,250 ± 90	B	not given[kk]
Seagoville	22,130 ± 600	B	W-1719
Sims Bayou	>23,000	B, E	not given[rr]
Spring Canyon	1,400 ± 90	B	not given[kk]
Tortuga Flat	410 ± 60	B	Tx-1515
Toyah	12,140 ± 140	Mu	I-7088
Lubbock Lake			
Clovis level	11,725 ± 165*	B, C, E, Mu	I(TTC)-246, SMU-263[ss,tt]
Folsom level	10,155 ± 295*	A, B	C-558, L-283G, SMU-292[tt]
Late Paleo-Indian	7,967 ± 80	B	SMU-262[tt]
Utah			
Bear River #1	1,065 ± 120	B	Gx-359
Danger Cave,			
Level 2	9,789 ± 630	B	C-611[uu]
Level 3	6,600	B	not given[uu]
Level 4	5,400	B	not given[uu]
Level 5	3,000	B	not given[uu]
Summit County	>40,000	B, C, E, Mu, P	not given[vv]
Virginia			
Saltville	13,460 ± 420	B, E, M(?), Mu, Mx, Sy	SI-641[ww]
Washington			
Chewelah	2,300 ± 120	B	WSU-1453
Lind Coulee	8,700 ± 400	B	C-827
Seattle	12,300 ± 200	S	UW-8
Wisconsin			
Jefferson	9,065 ± 90	Mu	Wis-704
Overhead	440 ± 65	B	Wis-573
Schimelpfenig Bog	9,630 ± 110	M	Wis-267
Schimelpfenig Bog	9,480 ± 100	M	Wis-265
Stiles	9,065 ± 90	Mu	Wis-704

Locality	C^{14} Years BP	Fauna present	Laboratory number(s)
Wyoming			
Big Goose Creek	490 ± 110*	B	M-1859, M-1860[xx]
Brewster			
Folsom level	10,375 ± 700	B	I-472
Agate Basin level	9,670 ± 450*	B	M-1131, O-1252
Buffalo Creek (Mavrakis-Benson-Roberts)	2,530 ± 60*	B	I-644, RL-160[xx]
Casper	9,955 ± 260*	B, C	RL-125, RL-208[dd,xx]
Colby	11,200 ± 220	B(?), Mu	RL-392[xx]
Essex Mountain	755 ± 90	B	I-6320[yy]
Finley	9,028 ± 118	B	SHU-250
Glenrock	245 ± 100*	B	M-2349, M-2350[xx]
Hawken	6,370 ± 155*	B	RL-185, RL-437[xx]
Hell Gap			
Eden level	8,970 ± 135*	B	A-753A, A-753C
Alberta	8,590 ± 350	B	A-707
Frederick level	8,600 ± 300	B	A-501
Scottsbluff level	8,600 ± 600	B	I-245
Horner (Sage Creek)			
Locality 1	8,885 ± 120*	B	UCLA-697A & B
Locality 2	7,880 ± 1300	B	SI-74
James Allen	7,900 ± 300	B	M-304
Lance Creek	2,450 ± 75	B	A-364
Natural Trap Cave	11,845 ± 600	B, C, E, Mu	not given[zz]
North Walker Pit	9,050 ± 220	B	RL-454[c]
Pine Spring, level 1	9,695 ± 195	B	Gx-354
Piney Creek	355 ± 100*	B	M-1747, M-1748[xx]
Ruby	1,735 ± 135*	B	Gx-1157, M-2348[xx]
Scoggin	4,540 ± 110	B	RL-174[xx]
Union Pacific	11,280 ± 350	B, Mu	I-449
Vore			
17′ level	370 ± 140	B	RL-349[xx]
9′ level	200 ± 90	B	RL-173[xx]
Wardell	1,080 ± 100*	B	RL-103, RL-111[aaa]
48SH312	2,910 ± 140	B	RL-162[xx]

Note: Alford (1973), Haynes (1967), and Hester (1960), among others, have earlier presented lists of radiocarbon dates relevant to megafaunal geochronology.

Sources other than radiocarbon laboratory lists: [a]Brumley, 1978; I have used only those dates reported by Brumley which progressively increased with increasing depth. [b]Wilson, 1978. [c]M. Wilson, 1975. [d]Reeves, 1978*b*. [e]Wilson, 1974*c*. [f]Reeves, 1973. [g]Harington, 1970. [h]Harington and Clulow, 1973. [i]Cabrera Castro, 1972. [j]Martin, 1967. [k]Kelley, Spiker, and Rubin, 1978; These dates essentially bracket the period of accumulation of alluvium in which several bison specimens have been found. [l]Guthrie, 1970. [m]Péwé, 1975. [n]Powers and Hamilton, 1978. [o]Thorson and Hamilton, 1977. [p]This is the age accepted by Saunders, 1977. [q]Long and Martin, 1974. [r]Miller, 1971. [s]Wheat, 1979. [t]Wilmsen, 1974. [u]Butler, Gildersleeve, and Sommers, 1971. [v]Anderson, 1974. [w]Anderson and Shutler, 1978. [x]Katz, 1973. [y]McMillan, 1976. [z]Davis and Zeier, 1978. [aa]Keyser, 1977. [bb]Lahren, 1971. [cc]L. Agenbroad, written comm., January 28, 1977. [dd]Frison et al., 1978; Mehringer, 1967. [ee]Agogino, Patterson, and Patterson, 1976. [ff]Speth and Parry, 1978. [gg]Painter, 1978. [hh]Eighmy, 1970. [ii]Lintz, 1974. [jj]Leonhardy, 1966. [kk]Duffield, 1970. [ll]Lundelius, 1967. [mm]Slaughter and Hoover, 1963. [nn]Dalquest and Hibbard, 1965. [oo]Haynes, 1967. [pp]Schultz and Cheatum, 1970. [qq]J. Hughes, written comm., March 6, 1978. [rr]Slaughter and McClure, 1965. [ss]C. Johnson, 1974. [tt]E. Johnson, 1974. [uu]Jennings, 1957. [vv]Miller, 1976. [ww]Bottoms, 1969; Ray, Cooper, and Benninghoff, 1967. [xx]Frison, Wilson, and Wilson, 1974. [yy]Ahlbrandt, 1974. [zz]Martin, Gilbert, and Adams, 1977. [aaa]Frison, 1973.

*Denotes the arithmetic average of two or more radiocarbon dates.

REFERENCES CITED

Agenbroad, L. D. 1978. The Hudson-Meng site: An Alberta bison kill in the Nebraska high plains. In *Bison procurement and utilization: A symposium*, ed. L. B. Davis and M. Wilson, pp. 151–174. Memoir 14, *Plains Anthropol.*

Agenbroad, L. D., and Haynes, C. V. 1975. *Bison bison* remains at Murray Springs, Arizona. *The Kiva* 40:309–313.

Agogino, G. A., Patterson, D. K., and Patterson, D. E. 1976. Blackwater Draw locality no. 1, south bank: Report for the summer of 1974. *Plains Anthropol.* 21:213–223.

Ahlbrandt, T. S. 1974. Dune stratigraphy, archaeology, and the chronology of the Killpecker dune field. In *Applied geology and archaeology: The Holocene history of Wyoming*, ed. M. Wilson, pp. 51–60. Geol. Survey Wyoming, Rept. Invest. no. 10.

Aigner, J. S. 1972. Relative dating of north Chinese faunal and cultural complexes. *Arctic Anthropol.* 9:36–79.

Alford, J. J. 1973. The American bison: An ice age survivor. *Proc. Ass. Am. Geog.* 5:1–6.

Allen, J. A. 1876. *The American bisons, living and extinct.* Memoirs Mus. Comp. Zool., vol. 4, no. 10. Cambridge: Harvard Univ. Press.

———. 1877. Northern range of the bison. *Am. Nat.* 11:624.

———. 1900. Note on the wood bison. *Bull. Am. Mus. Nat. Hist.* 13:63–67.

Anderson, B. A. 1977. Overview of bison remains from the Plum Creek area, Lake Meredith Recreation Area, Texas. Typed MS, Southwest Cultural Resources Center, Santa Fe.

Anderson, D. C., and Shutler, R., Jr. 1978. The Cherokee sewer site (13CK405): A summary and assessment. In *Bison procurement and utilization: A symposium,* ed. L. B. Davis and M. Wilson, pp. 132–139. Memoir 14, *Plains Anthropol.*

Anderson, E. 1974. A survey of the late Pleistocene and Holocene mammal fauna of Wyoming. In *Applied geology and archaeology: The Holocene history of Wyoming,* ed. M. Wilson, pp. 79–87. Geol. Survey Wyoming Rept. Invest., no. 10.

Axelrod, D. I. 1975. Evolution and biogeography of Madrean-Tethyan sclerophyll vegetation. *Annals Missouri Bot. Gard.* 62:280–334.

Bada, J. L., Schroeder, R. A., and Carter, G. F. 1974. New evidence for the antiquity of man in North America deduced from aspartic acid racemization. *Science* 184:791–793.

Bailey, V. 1932. Buffalo of the Malheur Valley, Oregon. *Proc. Biol. Soc. Wash.* 45:47–48.

Banfield, A. W. F., and Novakowski, N. S. 1960. The survival of the wood bison (*Bison bison athabascae* Rhoads) in the Northwest Territories. *Nat. Mus. Can. Nat. Hist. Papers* 8:1–6.

Basrur, P. K., and Moon, Y. S. 1967. Chromosomes of cattle, bison, and their hybrid, the cattalo. *Am. J. Vet. Res.* 28:1319–1325.

Bayrock, L. A., and Hillerud, J. M. 1964. New data on *Bison bison athabascae* Rhoads. *J. Mammal.* 45:630–632.

Bedord, J. N. 1974. Morphological variation in bison metacarpals and metatarsals. In *The Casper site: A Hell Gap bison kill on the high plains,* ed. G. C. Frison, pp. 199–240. New York: Academic Press.

Bell, R. H. V. 1971. A grazing ecosystem in the Serengeti. *Sci. Am.* 225:86–93.

Benson, L. V. 1978. Fluctuation in the level of pluvial Lake Lahontan during the last 40,000 years. *Quat. Res.* 9:300–318.

Berggren, W. A., and Van Couvering, J. A. 1974. The late Neogene. *Palaeogeog. Palaeoclim. Palaeoecol.* 16:1–216.

Bernabo, J. C., and Webb, T., III. 1977. Changing patterns in the Holocene pollen record of northeastern North America: A mapped summary. *Quat. Res.* 8:64–96.

Berti, A. A. 1975. Pollen and seed analysis of the Titusville section (mid-Wisconsin), Titusville, Pennsylvania. *Can. J. Earth Sci.* 12:1675–1684.

Bhambhani, R., and Kuspira, J. 1969. The somatic karyotypes of American bison and domestic cattle. *Can. J. Genet. Cytol.* 11:243–249.

Black, R. F. 1969. Climatically significant fossil periglacial phenomena in northcentral United States. *Biul. Peryglac.* 20:225–238.

Blackwelder, B. W., Pilkey, O. H., and Howard, J. D. 1979. Late Wisconsin sea levels on the southeast U.S. Atlantic shelf based on in-place shoreline indicators. *Science* 204:618–620.

Blake, W. P. 1898a. Remains of a species of Bos in the Quaternary of Arizona. *Am. Geol.* 22:65–72.

———. 1898b. Bison latifrons and Bos arizonica. *Am. Geol.* 22:247–248.

Bohlken, H. 1967. Beitrag zur Systematik der rezenten Formen der Gattung *Bison* H. Smith, 1827. *Z. Zool. Syst. Evol.* 5:54–110.

Bojanus, L. H. 1827. De uro nostrate eiusque sceleto commentatio. Scripsit et *Bovis primigenii* sceleto auxit. *Nova Acta Leopoldina* 13:413–467.

Bolton, H. E. 1949. *Coronado: Knight of Pueblos and plains.* New York: Whittlesey House.

Borowski, S., Krasinski, Z., and Milkowski, L. 1967. Food and role of the European bison in forest ecosystems. *Acta Theriol.* 12:367–376.

Bottoms, E. 1969. Notes on the geology, Pleistocene paleontology, and archaeology of Saltville, Virginia. *The Chesopiean* 7 (nos. 4–5):80–89.

Brand, D. D. 1967. Unpublished letter to the editor of *The American West.* Austin, Texas. March 27, 1967.

Bright, R. C. 1966. Pollen and seed stratigraphy of Swan Lake, southeastern Idaho: Its relation to regional vegetational history and to Lake Bonneville history. *Tebiwa* 9:1–47.

Brophy, J. A. 1965. A possible *Bison (Superbison) crassicornis* of mid-Hypsithermal age from Mercer County, North Dakota. *Proc. N. Dak. Acad. Sci.* 19:214–223.

Brumley, J. H. 1978. McKean complex subsistence and hunting strategies in the southern Alberta plains. In *Bison procurement and utilization: A symposium*, ed. L. B. Davis and M. Wilson, pp. 175–193. Memoir 14, *Plains Anthropol.*

Bryan, A. L. ed. 1978. *Early man in America from a circum-Pacific perspective.* Occasional Papers Dept. Anthropol. Univ. Alberta, no. 1.

Bryson, R. A. 1974. A perspective on climatic change. *Science* 184:753–760.

Bryson, R. A., Baerreis, D. A., and Wendland, W. M. 1970. The character of late-glacial and post-glacial climatic changes. In *Pleistocene and recent environments of the central Great Plains*, ed. W. Dort, Jr. and J. K. Jones, Jr., pp. 53–74. Dept. Geol. Univ. Kan. Spec. Publ. 3. Lawrence: Univ. Press Kan.

Bryson, R. A., and Wendland, W. M. 1967. Tentative climatic patterns for some late glacial and post-glacial episodes in central North America. In *Land, life and water*, ed. W. J. Mayer-Oakes, pp. 271–302. Winnipeg: Univ. Manit. Press.

Butler, B. R. 1972a. The Holocene in the desert west and its cultural significance. In *Great Basin cultural ecology: A symposium*, ed. D. D. Fowler, pp. 5–12. Desert Research Institute Publications in the Social Sciences no. 8. Reno.

———. 1972b. The Holocene or post-glacial ecological crisis on the eastern Snake River Plain. *Tebiwa* 15:49–63.

———. 1976. The evolution of the modern sagebrush-grass steppe biome on the eastern Snake River Plain. In *Holocene environmental change in the Great Basin*, ed. R. Elston, pp.

4–39. Nevada Archaeological Survey Research Paper no. 6. Reno.

Butler, B. R., Gildersleeve, H., and Sommers, J. 1971. The Wasden site bison: Sources of morphological variation. In *Aboriginal man and environments on the plateau of northwest America*, ed. A. H. Stryd and R. A. Smith, pp. 126–152. Calgary: Students' Press.

Cabrera Castro, R. 1972. Fauna fósil Pleistocénica in Jocotepec, Jalisco. *Bol. Inst. Nac. Antrop. Hist.*, ser. 2 3:37–44.

Canby, T. Y. 1979. The search for the first Americans. *Nat. Geog. Mag.* 156:330–363.

Catesby, M. 1754. *The natural history of Carolina, Florida, and the Bahama Islands.* Revised by George Edwards. London: C. Marsh.

Chandler, A. C. 1916. *A study of the skull and dentition of Bison antiquus Leidy, with special reference to material from the Pacific coast.* Univ. Calif. Publ., Bull. Dept. Geol., vol. 9, no. 11.

Christman, G. M. 1971. The mountain bison. *Am. West* 8:44–47.

Cole, J. E. 1954. Buffalo (*Bison bison*) killed by fire. *J. Mammal.* 35:453–454.

Colinvaux, P. A. 1967. Quaternary vegetational history of arctic Alaska. In *The Bering land bridge*, ed. D. M. Hopkins, pp. 207–231. Stanford: Stanford University Press.

Cook, H. J. 1928. A new fossil bison from Texas. *Proc. Colo. Mus. Nat. Hist.* 8:34–36.

Cope. E. D. 1895. Extinct Bovidae, Canidae, and Felidae from the Pleistocene of the plains. *J. Acad. Nat. Sci. Phila.*, ser. 2, 9:453–459.

Craig, A. J. 1969. Vegetational history of the Shenandoah Valley, Virginia. In *United States contributions to Quaternary research*, ed. S. A. Schumm and W. C. Bradley, pp. 283–296. Geol. Soc. Am. Spec. Paper no. 123.

Crandall, L. S. 1964. *The management of wild mammals in captivity.* Chicago: Univ. Chicago Press.

Creager, J. S., and McManus, D. A. 1967. Geology of the floor of Bering and Chukchi seas—American studies. In *The Bering land bridge*, ed. D. M. Hopkins, pp. 7–31. Stanford: Stanford Univ. Press.

Curray, J. R. 1965. Late Quaternary history, continental shelves of the United States. In *The Quaternary of the United States*, ed. H. E. Wright, Jr. and D. G. Frey, pp. 723–735. Princeton: Princeton Univ. Press.

Cuvier, G. 1825. *Recherches sur les ossemens fossiles.* 3rd ed. Paris: Dufouret d'Ocagne.

———. 1827. *The animal kingdom, arranged in conformity with its organization.* London: Geo. B. Whittaker.

Dalimier, P. 1955. *Les buffles du Congo Belge.* Brussels: Institut des Parcs Nationaux du Congo Belge.

Dalquest, W. W. 1957. First record of Bison alleni from a late Pleistocene deposit in Texas. *Texas J. Sci.* 9:346–354.

———. 1959. Two unusual subfossil Bison specimens from Texas. *J. Mammal.* 40:567–571.

———. 1961a. A record of the giant Bison (Bison latifrons) from Cooke County, Texas. *Texas J. Sci.* 13:41–44.

———. 1961b. Two species of Bison contemporaneous in early recent deposits in Texas. *Southwest. Nat.* 6:73–78.

——. 1962. The Good Creek formation, Pleistocene of Texas, and its fauna. *J. Paleont.* 36:568–582.

Dalquest, W. W. and Hibbard, C. W. 1965. 1,350-year-old vertebrate remains from Montague County, Texas. *Southwest. Nat.* 10:315–316.

Dary, D. A. 1974. *The buffalo book: the full saga of the American animal.* Chicago: Swallow Press.

Davis, A. M. 1977. The prairie-deciduous forest ecotone in the upper Middle West. *Ann. Ass. Am. Geog.* 67:204–213.

Davis, L. B., and Wilson, M. eds. 1978. *Bison procurement and utilization: A symposium.* Memoir 14, *Plains Anthropol.*

Davis, L. B., and Zeier, C. D. 1978. Multi-phase late period Bison procurement at the Antonsen site, southwestern Montana. In *Bison procurement and utilization: A symposium*, ed. L. B. Davis and M. Wilson, pp. 222–235. Memoir 14, *Plains Anthropol.*

Davis, M. B. 1967. Late-glacial climate in northern United States: A comparison of New England and the Great Lakes region. In *Quaternary Paleoecology*, ed. E. J. Cushing and H. E. Wright, Jr., pp. 11–43. New Haven: Yale Univ. Press.

——. 1969. Climatic changes in southern Connecticut recorded by pollen deposition at Rogers Lake. *Ecology.* 50:409–422.

Deevey, E. S., Flint, R. F., and Rouse, I. eds. 1967. *Radiocarbon measurements: Comprehensive index, 1950–1965.* New Haven: *Am. J. Sci.*

Delcourt, P. A., and Delcourt, H. R. 1977. The Tunica Hills, Louisiana-Mississippi: Late glacial locality for spruce and deciduous forest species. *Quat. Res.* 7:218–237.

Delpech, F., Le Tensorer, J.-M., Pineda, R., and Prat, F. 1978. Un nouveau gisement du Pléistocène moyen: Camp-de-Peyre à Sauveterre-la-Lémance (Lot-et-Garonne). *C. R. Acad. Sci. Paris* 286 (ser. D):1101–1103.

de Terra, H., Romero, J., and Stewart, T. D. 1949. *Tepexpan Man.* New York: Viking Fund Publ. Anthropol., no. 11.

Dillehay, T. D. 1974. Late Quaternary bison population changes on the southern plains. *Plains Anthropol.* 19:180–196.

Dillon, L. S. 1956. Wisconsin climate and life zones in North America. *Science* 123:167–176.

Dixon, E. J. 1975. The Gallagher flint station, an early man site on the north slope, arctic Alaska, and its role in relation to the Bering land bridge. *Arctic Anthropol.* 12:68–75.

Dodson, M. M. 1975. Quantum evolution and the fold catastrophe. *Evol. Theory* 1:107–118.

Donn, W. L., Farrand, W. R., and Ewing, M. 1962. Pleistocene ice volumes and sea-level lowering. *J. Geol.* 70:206–214.

Downs, T. 1958. Fossil vertebrates from Lago de Chapala, Jalisco, Mexico. *Papers, 20th International Geological Congress*, session 7:75–77.

Dubost, G. 1979. The size of African forest artiodactyls as determined by the vegetation structure. *African J. Ecol.* 17:1-18.

Duffield, L. F. 1970. Some Panhandle Aspect sites in Texas: Their vertebrates and paleoecology. Ph.D. Dissertation, Univ. Wisc. Madison.

Dunbar, G. S. 1962. *Some notes on*

bison in early Virginia. Virginia Place-Name Society, Occasional Papers no. 1, January 30. Charlottesville.

Eighmy, J. 1970. Edwards II: Report of an excavation in western Oklahoma. *Plains Anthropol.* 15:255–281.

Einarsson, T., Hopkins, D. M., and Doell, R. R. 1967. The stratigraphy of Tjörnes, northern Iceland, and the history of the Bering land bridge. In *The Bering land bridge*, ed. D. M. Hopkins, pp. 312–325. Stanford: Stanford Univ. Press.

Eisenberg, J. F., and McKay, G. M. 1974. Comparison of ungulate adaptations in the new world and the old world tropical forests with special reference to Ceylon and the rainforests of Central America. In *The behaviour of ungulates and its relation to management*, ed. V. Geist and F. Walther, pp. 585–602. Morges: Int'l Union for Cons. Nature and Nat. Resources.

Eldredge, N., and Gould, S. J. 1972. Punctuated equilibria: An alternative to phyletic gradualism. In *Models in Paleobiology*, ed. T. J. M. Schopf, pp. 82–115. San Francisco: Freeman, Cooper & Co.

Empel, W. 1962. Morphologie des Schädels von *Bison bonasus* (Linnaeus 1758). *Acta Theriol.* 6:53–111.

Empel, W., and Roskosz, T. 1963. Das Skelett der Gliedmassen des Wisents, *Bison bonasus* (Linnaeus, 1758). *Acta Theriol.* 7:259–300.

Espenshade, E. B., Jr. 1960. *Goode's World Atlas*. 12th ed. Chicago: Rand McNally & Co.

Estes, R. D. 1974. Social organization of the African Bovidae. In *The behaviour of ungulates and its relation to management*, ed. V. Geist and F. Walther, pp. 166–205. Morges: Int'l Union for Cons. Nature and Nat. Resources.

Figgins, J. D. 1933. The bison of the western area of the Mississippi Basin. *Proc. Colo. Mus. Nat. Hist.* 12:16–33.

Fink, J., and Kukla, G. J. 1977. Pleistocene climates in central Europe: At least 17 interglacials after the Olduvai event. *Quat Res.* 7:363–371.

Flerow, C. C. 1965. Sravnetyelnaya craneologeya sovremennih predstaveetelyae roda Bison. *Biul. Mock. Obshz. Ispt. Prirodi, otd. biol.* 70:40–54.

———. 1967. On the origin of the mammalian fauna of Canada. In *The Bering land bridge,* ed. D. M. Hopkins, pp. 271–280. Stanford: Stanford Univ. Press.

———. 1971. On the history of *Bison. Abh. hess. L. -Amt Bodenforsch.* 60:59-63.

Flerow, C. C., and Zablotski, M. A. 1961. O prichinah izmyenyeniya areal bizonob. *Biul. Mock. Obshz. Ispt. Prirodi, otd. biol.* 66:99–109.

Frenzel, B. 1968. The Pleistocene vegetation of northern Eurasia. *Science* 161:637–649.

Frick, C. 1930. Alaska's frozen fauna. *Nat. Hist.* 30:71–80.

———. 1937. Horned ruminants of North America. *Bull. Am. Mus. Nat. Hist.* 69:1–669.

Fries, M. 1962. Pollen profiles of late Pleistocene and Recent sediments at Weber Lake, northeastern Minnesota. *Ecology* 43:295–308.

Frison, G. C. 1973. *The Wardell buffalo trap 48SU301: Communal procurement in the upper Green River Basin, Wyoming.* Anthropol. Papers, Mus. Anthropol., Univ. Mich., no. 48.

———. 1975. Man's interaction with Holocene environments on the plains.

Quat. Res. 5:289–300.

Frison, G. C., Walker, D. N., Webb, S. D., and Zeimens, G. M. 1978. Procurement of *Camelops* on the northwestern plains. *Quat. Res.* 10:385–400.

Frison, G. C., Wilson, M., and Wilson, D. J. 1974. The Holocene stratigraphic archaeology of Wyoming: An introduction. In *Applied geology and archaeology: The Holocene history of Wyoming*, ed. M. Wilson, pp. 108–27. Geol. Survey Wyoming, Rept. Invest., no. 10.

———. 1976. Fossil bison and artifacts from an early Altithermal period arroyo trap in Wyoming. *Am. Antiq.* 41:28–57.

Frobenius, L., and Obermaier, H. 1965. *Hadschra Maktuba, Urzeitliche Felsbilder Kleinafrikas*. Graz: Akademische Druck—u. Verlagsanstalt.

Frye, J. C., and Willman, H. B. 1958. Permafrost features near the Wisconsin glacial margin in Illinois. *Am. J. Sci.* 256:518–524.

Fuller, W. A. 1950. Aerial census of northern bison in Wood Buffalo Park and vicinity. *J. Wildl. Mgmt.* 14:445–451.

———. 1959. The horns and teeth as indicators of age in bison. *J. Wildl. Mgmt.* 23:342—344.

———. 1960. Behaviour and social organization of the wild bison of Wood Buffalo National Park, Canada. *Arctic* 13:3–19.

Fuller, W. A., and Bayrock, L. A. 1965. Late Pleistocene mammals from central Alberta, Canada. In *Vertebrate Paleontology in Alberta*, ed. R. E. Folinsbee and D. M. Ross, pp. 53–63. Edmonton: Univ. Alberta.

Galbreath, E. C., and Stein, H. S. 1962. Bison occidentalis in South Dakota. *Proc. S. Dak. Acad. Sci.* 41:41–43.

Gazin, C. L. 1955. *A review of the upper Eocene Artiodactyla of North America*. Smithsonian Miscellaneous Collections 128, no. 8.

Geist, V. 1963. On the behaviour of the North American moose in British Columbia. *Behaviour* 20:377–416.

———. 1966. The evolution of horn-like organs. *Behaviour* 27:175–214.

———. 1971a. The relation of social evolution and dispersal in ungulates during the Pleistocene, with emphasis on the old world deer and the genus *Bison*. *Quat. Res.* 1:283–315.

———. 1971b. *Mountain sheep: A study in behavior and evolution*. Chicago: Univ. Chicago Press.

———. 1974. On the relationship of ecology and behaviour in the evolution of ungulates: Theoretical considerations. In *The behaviour of ungulates and its relation to management*, ed. V. Geist and F. Walther, pp. 235–246. Morges: Int'l Union for Cons. Nature and Nat. Resources.

———. 1978. On weapons, combat, and ecology. In *Aggression, Dominance, and Individual Spacing*, ed. L. Krames, P. Pliner, and T. Alloway, pp. 1–30. New York: Plenum.

Geist, V., and Karsten, P. 1977. The wood bison (*Bison bison athabascae* Rhoads) in relation to hypotheses on the origin of the American bison (*Bison bison* Linnaeus). *Z. Säugetierk.* 42:119–127.

Gentry. A. W. 1967. *Pelorovis oldowayensis* Reck, an extinct bovid from east Africa. *Bull. Brit. Mus. (Nat. Hist.), Geol.* 14:245–299.

Getty, R. 1975. *Sisson and Grossman's "The anatomy of the domestic animals."* 5th ed. Philadelphia:

W. B. Saunders.

Giterman, R. E., and Golubeva, L. V. 1967. Vegetation of eastern Siberia during the Anthropogene period. In *The Bering land bridge,* ed. D. M. Hopkins, pp. 232–244. Stanford: Stanford Univ. Press.

Gmelin, J. F. 1788. *Systema naturae per regna tria naturae, secundum classes ordines, genera, species, cum characteribus, differentiis, synonymis, locis.* 30th ed. Leipzig: Georg. Emanuel Beer.

Goodwin, G. G. 1936. Big game in the northeastern United States. *J. Mammal.* 17:48–50.

Gould, S. J. 1974. The origin and function of "Bizarre" structures: Antler size and skull size in the "Irish elk," *Megaloceros giganteus. Evolution* 28:191–220.

———. 1977. *Ontogeny and phylogeny.* Cambridge: Belknap Press, Harvard Univ.

Graham, A. 1972. Outline of the origin and historical recognition of floristic affinities between Asia and eastern North America. In *Floristics and paleofloristics of Asia and eastern North America,* ed. A. Graham, pp. 1–18. New York: Elsevier Pub. Co.

Grayson, D. K. 1973. On the methodology of faunal analysis. *Am. Antiq.* 39:432–439.

Green, M. 1962. Comments on the geologic age of *Bison latifrons. J. Paleont.* 36:557–559.

Green, M., and Martin, H. 1960. *Bison latifrons* in South Dakota. *J. Paleont.* 34:548–550.

Griffin, J. W., and Wray, D. E. 1945. Bison in Illinois archaeology. *Ill. Acad. Sci., Trans.* 38:21–26.

Grigson, C. 1974. The craniology and relationships of four species of *Bos*: 1. basic craniology: *Bos taurus* L. and its absolute size. *J. Arch. Sci.* 1:353–379.

———. 1976. The craniology and relationships of four species of *Bos*: 3. basic craniology: *Bos taurus* L. sagittal profiles and other nonmeasurable characters. *J. Arch. Sci.* 3:115–136.

Gromova, V. I. 1935. Pervobtnyi zubr (*Bison priscus* Bojanus) v SSSR. *Trudi Zool. Inst. Acad. Nauk SSSR* 2:77–204.

Grüger, E. 1972a. Late Quaternary vegetation development in south-central Illinois. *Quat. Res.* 2:217–231.

———. 1972b. Pollen and seed studies of Wisconsin vegetation in Illinois, U.S.A. *Geol. Soc. Am. Bull.* 83:2715–2734.

Grüger, J. 1973. Studies on the late Quaternary vegetation history of northeastern Kansas. *Geol. Soc. Am. Bull.* 84:239–250.

Gunderson, H. L. 1975. When the big bull bellows, turn your head. *Smithsonian* 6 (no. 7):42–47.

Gunnerson, D. A. 1974. *The Jicarilla Apaches: A study in survival.* DeKalb: Northern Illinois Univ. Press.

Guthrie, D. A. 1968. The tarsus of early Eocene artiodactyls. *J. Mammal.* 49:297–302.

Guthrie, R. D. 1966a. Bison horn cores—character choice and systematics. *J. Paleont.* 40:738–740.

———. 1966b. Pelage of fossil bison—a new osteological index. *J. Mammal,* 47:725–727.

———. 1967. Differential preservation and recovery of Pleistocene large mammal remains in Alaska. *J. Paleont.* 41:243–246.

———. 1968. Paleoecology of the large-mammal community in interior

Alaska during the late Pleistocene. *Am. Midl. Nat.* 79:346–363.

———. 1970. Bison evolution and zoogeography in North America during the Pleistocene. *Qtly. Rev. Biol.* 45:1–15.

———. 1978. Bison and man in North America. In *Abstracts, fifth biennial meeting, American Quaternary Association*, pp. 83–87.

Guthrie, R. D., and Matthews, J. V., Jr. 1971. The Cape Deceit fauna—early Pleistocene mammalian assemblage from the Alaskan arctic. *Quat. Res.* 1:474–510.

Hafsten, U. 1961. Pleistocene development of vegetation and climate in the southern high plains as evidenced by pollen analysis. In *Paleoecology of the Llano Estacado*, ed. F. Wendorf, pp. 59–91. Santa Fe: Mus. N. Mex. Press.

Halder, U. 1976. *Okologie und verhalten des Banteng* (Bos javanicus) *in Java: eine Feldstudie.* Hamburg: Paul Parey.

Hall, S. A. 1972. Holocene *Bison occidentalis* from Iowa. *J. Mammal.* 53:604–606.

Hamilton, T. D., and Porter, S. C. 1975. Itkillik glaciation in the Brooks Range, northern Alaska. *Quat. Res.* 5:471–497.

Hammond, J. H. 1955. *Progress in the physiology of farm animals.* London: Butterworth.

Hansen, R. O., and Begg, E. L. 1970. Age of Quaternary sediments and soils in the Sacramento area, California, by uranium and actinium series dating of vertebrate fossils. *Earth Planet. Sci. Letters* 8:411–419.

Harington, C. R. 1970. *A postglacial muskox* (Ovibos moschatus) *from Grandview, Manitoba, and comments on the zoogeography of Ovibos.* Nat'l. Mus. Nat. Sci., Publ. in Palaeont., no. 2. Ottawa.

Harington, C. R., and Clulow, F. V. 1973. Pleistocene mammals from Gold Run Creek, Yukon Territory. *Can. J. Earth Sci.* 10:697–759.

Harlan, R. 1825. *Fauna Americana: Being a description of the mammiferous animals inhabiting North America.* Philadelphia: Anthony Finley.

Harris, A. H., and Mundel, P. 1974. Size reduction in bighorn sheep (*Ovis canadensis*) at the close of the Pleistocene. *J. Mammal.* 55:678–680.

Hay, O. P. 1914. The extinct bisons of North America; with description of one new species, Bison regius. *Proc. U. S. Nat'l. Mus.* 46:161–200.

———. 1915. Contributions to the knowledge of the mammals of the Pleistocene of North America. *Proc. U.S. Nat'l. Mus.* 48:515–575.

———. 1924. *The Pleistocene of the middle region of North America.* Carnegie Inst. Wash. publ. 322a.

———. 1927. *The Pleistocene of the western region of North America and its vertebrated animals.* Carnegie Inst. Wash. publ. 322b.

Hay, O. P., and Cook, H. J., 1928. Preliminary descriptions of fossil mammals recently discovered in Oklahoma, Texas and New Mexico. *Proc. Colo. Mus. Nat. Hist.* 8:33.

———. 1930. Fossil vertebrates collected near, or in association with, human artifacts at localities near Colorado, Texas; Frederick, Oklahoma; and Folsom, New Mexico. *Proc. Colo. Mus. Nat. Hist.* 9:4–40.

Haynes, C. V., Jr. 1967. Carbon-14 dates and early man in the new world. In

Pleistocene extinctions: The search for a cause, ed. P. S. Martin and H. E. Wright, Jr., pp. 267–286. New Haven: Yale Univ. Press.

Hays. W. J. 1871. Notes on the range of some of the animals in America at the time of the arrival of the white men. *Am. Nat.* 5:387–392.

Hediger, H. 1964. *Wild animals in captivity: An outline of the biology of zoological gardens.* New York: Dover.

Heine, K. 1973. Studies of glacial morphology and tephro-chronology on the volcanoes of the central Mexican highland. In *Abstracts, Int'l. Union for Quat. Res., 9th Cong.,* pp. 144–145.

Heptner, W. G., Nasimovitsch, A. A., and Bannikov, A. G. 1961. *Die Säugetiere Sowjetunion.* Jena: Gustav Fischer.

Hernandez, F. 1651. *Rerum medicarum Novae Hispaniae thesaurus; seu, Plantarum animalium mineralium Mexicanorum historia.* Rome: Vitalis Mascardi.

Herrig, D. M. and Haugen, A. O. 1969. Bull bison behavior traits. *Proc. Iowa Acad. Sci.* 76:245–262.

Hershkovitz, P. 1957. The type locality of *Bison bison* Linnaeus. *Proc. Biol. Soc. Wash.* 70:31–32.

Hester, J. J. 1960. Late Pleistocene extinction and radiocarbon dating. *Am. Antiq.* 26:58–77.

———. 1972. Summary and conclusions. In *Blackwater locality no. 1.: A stratified, early man site in eastern New Mexico,* ed. J. J. Hester, pp. 164–180. Fort Burgwin Research Center, Publ. no. 8/Taos.

Heusser, C. J. 1966. Pleistocene climatic variations in the western United States. In *Pleistocene and post-Pleistocene climatic variation in the Pacific area: A symposium,* ed. D. I. Blumenstock, pp. 9–36. Honolulu: Bishop Mus. Press.

Hibbard, C. W. 1955a. *The Jinglebob interglacial (Sangamon?) fauna from Kansas and its climatic significance.* Contrib. Mus. Paleont. Univ. Mich., vol. 12, no. 10.

———. 1955b. *Pleistocene vertebrates from the Upper Becerra (Becerra Superior) formation, Valley of Tequixquiac, Mexico, with notes on other Pleistocene forms.* Contrib. Mus. Paleont. Univ. Mich., vol. 12, no. 5.

———. 1970. Pleistocene mammalian local faunas from the Great Plains and Central Lowland provinces of the United States. In *Pleistocene and recent environments of the central Great Plains,* ed. W. Dort, Jr. and J. K. Jones, Jr., pp. 395–433. Dept. Geol. Univ. Kan. Spec. Publ. 3. Lawrence: Univ. Press of Kansas.

Hibbard, C. W., and Villa R. B. 1950. El bisonte gigante de Mexico. *An. Inst. Biol.* 21:243–253.

Hillerud, J. M. 1970. Subfossil high plains bison. Ph.D. Dissertation, Univ. Nebraska. Lincoln.

———. 1978. Bison and man in North America: Abstract—bison as indicators of geologic age. In *Abstracts, fifth biennial meeting, American Quaternary Association,* pp. 89–93.

Hilzheimer, M. 1918. Dritter Beitrag zur Kenntnis der Bisonten. *Arch. Naturgesch.* 84 (div. A, no. 6):41–87.

Hind, H. Y. 1860. *Narrative of the Canadian Red River exploring expedition of 1857 and of the Assiniboine and Saskatchewan exploring expeditions of 1858.* London: Longman,

Green, Longman and Roberts.

Hirth, D. H. 1977. *Social behavior of white-tailed deer in relation to habitat.* Wildl. Soc., Wildl. Monog. no. 53.

Hodge, F. W., and Lewis, T. H. eds. 1907. *Spanish explorers in the southern United States, 1528–1543.* New York: Chas. Scribner's Sons.

Hoffmann, R. S. 1978. Comments on "Faunal exchanges between Siberia and North America" by C. A. Repenning. In *Abstracts, fifth biennial meeting, American Quaternary Association,* pp. 78–80.

Hoffmann, R. S., and Jones, J. K., Jr. 1970. Influence of late-glacial and post-glacial events on the distribution of recent mammals on the northern Great Plains. In *Pleistocene and recent environments of the central Great Plains,* ed. W. Dort, Jr. and J. K. Jones, Jr., pp. 355–394. Dept. Geol. Univ. Kan. Spec. Publ. 3. Lawrence: Univ. Press Kan.

Hoffmann, R. S., and Taber, R. D. 1967. Origin and history of Holarctic tundra ecosystems, with special reference to their vertebrate faunas. In *Arctic and alpine environments,* ed. H. E. Wright, Jr. and W. H. Osburn, pp. 143–170. Bloomington: Indiana Univ. Press.

Hopkins, D. M. 1967a. The Cenozoic history of Beringia—a synthesis. In *The Bering land bridge,* ed. D. M. Hopkins, pp. 451–484. Stanford: Stanford Univ. Press.

———. 1967b. Quaternary marine transgressions in Alaska. In *The Bering land bridge,* ed. D. M. Hopkins, pp. 47–90. Stanford: Stanford Univ. Press.

———. 1970. Paleoclimatic speculations suggested by new data on the location of the spruce refugium in Alaska during the last glaciation. In *Abstracts, first biennial meeting, American Quaternary Association,* p. 67.

Hopkins, M. L. 1951. *Bison (Gigantobison) latifrons* and *Bison (Simobison) alleni* in southeastern Idaho. *J. Mammal.* 32:192–197.

Hopkins, M. L., Bonnichsen, R., and Fortsch, D. 1969. The stratigraphic position and faunal associates of *Bison (Gigantobison) latifrons* in southeastern Idaho, a progress report. *Tebiwa* 12:1–8.

Hornaday, W. T. 1889. The extermination of the American bison, with a sketch of its discovery and life history. *Ann. Rept. (1887), Smithsonian Inst.,* pp. 367–548.

Horowitz, A. 1975. The Pleistocene paleoenvironments of Israel. In *Problems in prehistory: North Africa and the Levant,* ed. F. Wendorf and A. E. Marks, pp. 207–227. Dallas: SMU Press.

Houston, D. B. 1974. Aspects of social organization of moose. In *The behaviour of ungulates and its relation to management,* ed. V. Geist and F. Walther, pp. 690–696. Morges: Int'l Union for Cons. Nature and Nat. Resources.

Huang, W.-P., and Chang, Y.-P. 1966. The Quaternary mammalian fossil localities of Lantian district, Shensi. *Vert. PalAsiatica* 10:35–46.

Hughes, S. S. 1978. Bison diminution on the Great Plains. *Wyoming Contrib. Anthropol.* 1:18–47.

Jakle, J. A. 1968. The American bison and the human occupance of the Ohio Valley. *Proc. Am. Phil. Soc.* 112:299–305.

Janis, C. 1976. The evolutionary strategy of the Equidae and the origins of rumen and cecal digestion. *Evolution* 30:757–774.

Jennings. J. D. 1957. *Danger Cave*. Univ. of Utah Anthropol. Papers, no. 27.

Jennings, J. D., and Norbeck, E. 1955. Great Basin prehistory: A review. *Am. Antiq.* 21:1–11.

Johnson, A. W., and Packer, J. G. 1967. Distribution, ecology, and cytology of the Ogotoruk Creek flora and the history of Beringia. In *The Bering land bridge*, ed. D. M. Hopkins, pp. 245–265. Stanford: Stanford Univ. Press.

Johnson, C. 1974. Geologic investigations at the Lubbock Lake site. *The Mus. J.* 15:79–106.

Johnson, D. L. 1977. The late Quaternary climate of coastal California: Evidence for an ice age refugium. *Quat. Res.* 8:154–179.

Johnson, E. 1974. Zooarchaeology and the Lubbock Lake site. *The Mus. J.* 15:107–122.

———. 1976. Investigations into the zooarchaeology of the Lubbock Lake site. Ph.D. Dissertation, Univ. Calif. Berkeley.

Johnson, P. 1974. Los Colinas site report: Appendix. Typed MS, Ariz. State Museum, Tucson.

Kapp, R. O. 1970. Pollen analysis of pre-Wisconsin sediments from the Great Plains. In *Pleistocene and recent environments of the central Great Plains*, ed. W. Dort, Jr. and J. K. Jones, Jr., pp. 143–155. Dept. Geol. Univ. Kan. Spec. Publ. 3. Lawrence: Univ. Press Kan.

Kapp, R. O., and Gooding, A. M. 1964. Pleistocene vegetational studies in the Whitewater Basin, southeastern Indiana. *J. Geol.* 72:307–326.

Katz, P. R. 1973. Report: Radiocarbon dates from the Sutter site, northeastern Kansas. *Plains Anthropol.* 18:167–168.

Kelley, L., Spiker, E., and Rubin, M. 1978. U.S. Geological Survey, Reston, Virginia, radiocarbon dates XIV. *Radiocarbon* 20:283–312.

Keyser, J. D. 1977. The role of seasonal bison procurement in the prehistoric economic systems of the Indians of north central Montana. Ph.D. Dissertation, University of Oregon. Eugene.

Khan, E. 1970. *Biostratigraphy and palaeontology of a Sangamon deposit at Fort Qu'Appelle, Saskatchewan.* Nat'l Mus. Can., Publ. in Palaeont., no. 5.

Kind, N. V. 1967. Radiocarbon chronology in Siberia. In *The Bering land bridge*, ed. D. M. Hopkins, pp. 172–192. Stanford: Stanford Univ. Press.

King. J. E. 1973. Late Pleistocene palynology and biogeography of the western Missouri Ozarks. *Ecol. Monog.* 43:539–565.

Klein, J. T. 1732. A letter from Mr. Jac. Theod. Klein. *Phil. Trans.* 37:427–429.

(Knight, C.) 1849. Sketches in natural history: History of the mammalia. London: C. Cox.

Knight, R. 1968. *Ecological factors in changing economy and social organization among the Rupert House Cree.* Nat'l Mus. Can., Anthropol. Papers, no. 15.

Koch, W. 1935. The age order of epiphyseal union in the skeleton of the European bison (*Bos bonasus* L.). *Anat. Record* 61:371–376.

Kowalski, K. 1967. The evolution and fossil remains of the European bison. *Acta Theriol.* 12:335–338.

Krasinska, M., and Pucek, Z. 1967. The

state of studies on hybridisation of European bison and domestic cattle. *Acta Theriol.* 12:385–389.

Krasinski, Z. 1967. Free living European bisons. *Acta Theriol.* 12:391–405.

Kruuk, H. 1972. *The spotted hyena: A study of predation and social behavior.* Chicago: Univ. Chicago Press.

Kurten, B. 1950. Chinese Hipparion fauna. *Soc. Scient. Fennica, Comment. Biol.* 13:1–82.

———. 1968. *Pleistocene mammals of Europe.* Chicago: Aldine Pub. Co.

———. 1972. *The age of mammals.* New York: Columbia Univ. Press.

Lahren, L. A. 1971. Archaeological investigations in the upper Yellowstone Valley, Montana: Preliminary synthesis and discussion. In *Aboriginal man and environments on the plateau of northwest America*, ed. A. H. Stryd and R. A. Smith, pp. 168–182. Calgary: Students' Press.

LaMarche, V. C. 1973. Holocene climatic variations inferred from treeline fluctuations in the White Mountains, California. *Quat. Res.* 3:632–660.

Lande, R. 1976. Natural selection and random genetic drift in phenotypic evolution. *Evolution* 30:314–334.

———. 1979. Effective deme sizes during long-term evolution estimated from rates of chromosomal rearrangement. *Evolution* 33:234–251.

Leidy, J. 1852a. *Memoir on the extinct species of American ox.* Smithsonian Contrib. Knowledge, vol. 5, art. 3.

———. 1852b. (no title). [Report on bison from Big Bone Lick and vicinity.] *Proc. Acad. Nat. Sci. Phila.* 6:117.

———. 1854. (*Sus americanus/Harlanus americanus* referred to *Bison latifrons*.) *Proc. Acad. Nat. Sci. Phila.* 7:89–90.

Leonhardy, F. C. 1966. *Test excavations in the Magnum Reservoir area of southwestern Oklahoma.* Contrib. Mus. Great Plains, no. 2.

Leskinen, P. H. 1975. Occurrence of oaks in late Pleistocene vegetation in the Mojave Desert of Nevada. *Madroño* 23:234–235.

Lewis, G. E. 1970. New discoveries of Pleistocene bisons and peccaries in Colorado. *U.S. Geol. Survey Prof. Paper* 700-B:B137–B140.

Lewis, H. T., and Schweger, C. 1973. Paleo-Indian uses of fire during the late Pleistocene: The human factor in environmental change. *Abstracts, Int'l Union for Quat. Res., 9th Cong.*, p. 210.

Linnaeus, C. 1758. *Systema naturae per regna tria naturae, secundum classes, ordines, genera, species, cum characteribus, differentiis, synonymis, locis.* 10th ed. Stockholm: Laurentii Salvii.

Lintz, C. 1974. An analysis of the Custer focus and its relationship to the Plains Village horizon in Oklahoma. *Papers in Anthropol.* 15 (no. 2): 1-72.

Long, A., and Martin, P. S. 1974. Death of American ground sloths. *Science* 186:638–640.

Lott, D. F. 1972a. Bison would rather breed than fight. *Nat. Hist.* 81 (no. 7):40–45.

———. 1972b. The way of the bison: fighting to dominate. In *The marvels of animal behavior*, ed. T. B. Allen, pp. 321–333. Washington, D.C.: Nat'l Geog. Soc.

——. 1974. Sexual and aggressive behaviour of adult male American bison (*Bison bison*). In *The behaviour of ungulates and its relation to management*, ed. V. Geist and F. Walther, pp. 382–394. Morges: Int'l Union for Cons. Nature and Nat. Resources.

Lucas, F. A. 1898. (no title). [The fossil bison of North America, with description of a new species.] *Science* 8:678.

———. 1899a. The fossil bison of North America. *Proc. U. S. Nat'l Mus.* 21:755–771.

———. 1899b. The characters of *Bison occidentalis*, the fossil bison of Kansas and Alaska. *Kan. Univ. Qtly.*, ser. A 8:17–18.

Lundelius, E. L., Jr. 1967. Late-Pleistocene and Holocene faunal history of central Texas. In *Pleistocene extinctions: The search for a cause*, ed. P. S. Martin and H. E. Wright, Jr., pp. 287–319. New Haven: Yale Univ. Press.

———. 1972. Fossil vertebrates from the late Pleistocene Ingleside fauna, San Patricio County, Texas. *Bur. Econ. Geol., Univ. Texas, Rept. Invest.*, no. 77.

Lydekker, R. 1878. *Crania of ruminants from the Indian Tertiary*. Memoirs Geol. Survey India. Palaeontologica Indica. Ser. 10: Indian Tertiary and post-Tertiary vertebrata, vol. 1, part 3.

———. 1912. *The ox and its kindred.* London: Methuen and Co.

McAndrews, J. H. 1967. Paleoecology of the Seminary and Mirror Pool peat deposits. In *Land, life and water*, ed. W. J. Mayer-Oakes, pp. 253–272. Winnipeg: Univ. of Manit. Press.

McClung, C. E. 1904. The fossil bison of Kansas. *Trans., Kan. Acad. Sci.* 19:156–159.

McDonald, H. G., and Anderson, E. 1975. A late Pleistocene vertebrate fauna from southeastern Idaho. *Tebiwa* 18:19–37.

McDonald, J. N. 1974. The southwestern range of *Bison bison*: Two approaches to a biogeographic problem. Paper read at Southwestern Division, Ass. Am. Geog., March 30, Dallas, Texas.

———. 1976. Toward an understanding of *Bison antiquus*: An evolutionary, ecological, and geographical synthesis. Poster paper presented at fourth biennial meeting, Am. Quat. Ass., Oct. 9, Tucson, Ariz.

———. 1978a. The North American bison: A revised classification and interpretation of their evolution. Ph.D. Dissertation, Univ. Calif. Los Angeles.

———. 1978b. Man and bison in North America: Comments on a paper by R. D. Guthrie. *Abstracts, fifth biennial meeting, American Quaternary Association*, pp. 94–96.

McDonnell, L. J., Gartside, D. F., and Littlejohn, M. J. 1978. Analysis of a narrow hybrid zone between two species of *Pseudophryne* (Anura: Leptodactylidae) in south-eastern Australia. *Evolution* 32:602–612.

McHugh, T. 1958. Social behavior of the American buffalo (*Bison bison bison*). *Zoologica* 43 (part 1):1–40.

———. 1972. *The time of the buffalo.* New York: Knopf.

Mack, R. N., Rutter, N. W., Bryant, V. M., Jr., and Valastro, S. 1978. Reexamination of postglacial vegetation history of northern Idaho: Hager Pond, Bonner County. *Quat. Res.* 10:241–255.

McKay, G. M., and Eisenberg, J. F. 1974. Movement patterns and habitat utilization of ungulates in Ceylon. In *The behaviour of ungulates and its relation to management*, ed. V. Geist and F. Walther, pp. 708–721. Morges: Int'l Union for Cons. Nature and Nat. Resources.

McMillan, R. B. 1976. Rodgers Shelter: A record of cultural and environmental change. In *Prehistoric man and his environments: A case study in the Ozark highland*, ed. W. R. Wood, pp. 111–122. New York: Academic Press.

Mahan, B. R. 1977. Harassment of an elk calf by bison. *Can. Field-Nat.* 91:418–419.

———. 1978. Aspects of American bison (*Bison bison*) social behavior at Fort Niobrara National Wildlife Refuge, Valentine, Nebraska, with special reference to calves. M.S. Thesis, Univ. of Nebraska. Lincoln.

Markgren, G. 1974. The question of polygamy at an unbalanced sex ratio in moose. In *The behaviour of ungulates and its relation to management*, ed. V. Geist and F. Walther, pp. 756–758. Morges: Int'l Union for Cons. Nature and Nat. Resources.

Marsh, O. C. 1877. New vertebrate fossils. *Am. J. Sci.* 14:247–256.

Martin, H. T. 1924. A new bison from the Pleistocene of Kansas, with notice of a new locality for *Bison occidentalis. Kan. Univ. Sci. Bull.* 15:273–278.

Martin, L. D., Gilbert, B. M., and Adams, D. B. 1977. A cheetah-like cat in the North American Pleistocene. *Science* 195:981–982.

Martin, P. S. 1958. Pleistocene ecology and biogeography of North America. In *Zoogeography*, ed. C. L. Hubbs, pp. 375–420. Am. Ass. Adv. Sci. Publ. 51.

———. 1963. *The last 10,000 years: A fossil pollen record of the American Southwest.* Tucson: Univ. Ariz. Press.

———. 1967. Prehistoric overkill. In *Pleistocene extinctions: The search for a cause*, ed. P. S. Martin and H. E. Wright, Jr., pp. 75–120. New Haven: Yale Univ. Press.

Martin, P. S., and Mehringer, P. J., Jr. 1965. Pleistocene pollen analysis and biogeography of the Southwest. In *The Quaternary of the United States*, ed. H. E. Wright, Jr. and D. G. Frey, pp. 433–451. Princeton: Princeton Univ. Press.

Mathews, W. H. 1978. The geology of the ice-free corridor: Discussion—northwestern B. C. and adjacent Alberta. *Abstracts, fifth biennial meeting, American Quaternary Association*, pp. 16–18.

Matthew, W. D. 1921. Urus and bison. *Nat. Hist.* 21:598–606.

Matthews, J. V., Jr. 1976. Arctic-steppe—an extinct biome. *Abstracts, fourth biennial meeting, American Quaternary Association*, pp. 73–77.

Maxwell, J. A., and Davis, M. B. 1972. Pollen evidence of Pleistocene and Holocene vegetation on the Allegheny Plateau, Maryland. *Quat. Res.* 2:506–530.

Mayr, E. 1969. *Principles of systematic zoology*. New York: McGraw-Hill.

Meagher, M. 1971. Snow as a factor influencing bison distribution and numbers in Pelican Valley, Yellowstone National Park. *Proc., Snow and ice in relation to wildlife and recreation symposium, Feb. 11–12, 1971*, pp. 63–67. Ames: Iowa State Univ.

———. 1973. *The bison of Yellowstone National Park*. Nat'l Park Service, Scient. Monog. Ser., no. 1.

Mehringer, P. J., Jr. 1967. The environment of extinction of the late-Pleistocene megafauna in the arid southwestern United States. In *Pleistocene extinctions: The search for a cause,* ed. P. S. Martin and H. E. Wright, Jr., pp. 247–266. New Haven: Yale Univ. Press.

———. 1977. Great Basin late Quaternary environments and chronology. In *Models and Great Basin prehistory: A symposium*, ed. D. D. Fowler, pp. 113–167. Reno: Desert Res. Inst. Publ. in Social Sci., no. 12.

Mehringer, P. J., Jr., and Ferguson, C. W. 1969. Pluvial occurrence of bristlecone pine (*Pinus aristata*) in a Mohave Desert mountain range. *J. Ariz. Acad. Sci.* 5:284–292.

Mehringer, P. J., Jr., Schweger, C. E., Wood, W. R., and McMillan, R. B. 1968. Late-Pleistocene boreal forest in the western Ozark highlands? *Ecology* 49:567–568.

Mercer, J. H. 1972. The lower boundary of the Holocene. *Quat. Res.* 2:15–24.

Meyer, E. R. 1973. Late-Quaternary paleoecology of the Cuatro Cienegas Basin, Coahuila, Mexico. *Ecology* 54:982–995.

Middleton, W. G., and Moore, J. 1900. Skull of fossil bison. *Proc., Indiana Acad. Sci. for 1899*, pp. 178–181.

Miller, W. E. 1971. *Pleistocene vertebrates of the Los Angeles Basin and vicinity (exclusive of Rancho La Brea)*. Bull. Los Angeles County Mus. Nat. Hist., Science, no. 10.

———. 1976. Late Pleistocene vertebrates of the Silver Creek local fauna from north central Utah. *Great Basin Naturalist*, 36:387–424.

Moir, D. R. 1958. Occurrence and radiocarbon date of coniferous woods in Kidder County, North Dakota. *N. Dak. Geol. Survey, Misc. Ser.* 10:108–114.

Moodie, D. W., and Ray, A. J. 1976. Buffalo migrations in the Canadian Plains. *Plains Anthropol.* 21:45–52.

Mooser, O., and Dalquest, W. W. 1975. Pleistocene mammals from Aguascalientes, central Mexico. *J. Mammal.* 56:781–820.

Moran, J. M. 1973. The late-glacial retreat of "arctic" air as suggested by onset of *Picea* decline. *Prof. Geog.* 25:373–376.

Moran, J. M. 1976. *Glacial maximum tundra: A bioclimatic anomaly.* Occasional Publ. Dept. Geog. Univ. Illinois, Paper 10.

Morris, D. 1965. *The mammals: a guide to the living species.* New York: Harper & Row.

Movius, H. L., Jr. 1944. Early man and Pleistocene stratigraphy in south and east Asia. *Papers Peabody Mus. Archaeol. Ethnol.* 19, no. 3.

Mross, G. A., and Doolittle, R. F. 1967. Amino-acid sequence studies on artiodactyl fibrinopeptides. *Arch. Biochem. Biophys.* 122:674–684.

Müller-Beck, H. 1967. On migrations of hunters across the Bering land bridge in the upper Pleistocene. In *The Bering land bridge*, ed. D. M. Hopkins, pp. 373–408. Stanford: Stanford Univ. Press.

Newcomb, W. W., Jr. 1950. A re-examination of the causes of plains warfare. *Am. Anthropol.* 52:317–330.

Nichols, H. 1975. The time perspective in northern ecology: Palynology and the history of the Canadian boreal forest. *Proc., Circumpolar Conf. on Northern Ecology, Sept. 15–18, 1975*, pp. 157–165. Ottawa: Nat'l Res. Council Can.

Ogden, J. G., III. 1977. The late Quaternary paleoenvironmental record of northeastern North America. *Ann. N. Y. Acad. Sci.* 288:16–34.

Olsen, S. J. 1964. Mammal remains from archaeological sites: Part I—southeastern and southwestern United States. *Papers Peabody Mus. Archaeol. Ethnol.* 56, no.1.

Olson, E. C., and Miller, R. L. 1958. *Morphological integration.* Chicago: Univ. Chicago Press.

Otvos, E. G., Jr. 1978. Comment on "The Tunica Hills, Louisiana-Mississippi: late glacial locality for spruce and deciduous forest species" by Paul A. Delcourt and Hazel R. Delcourt. *Quat. Res.* 9:250–252.

Owen, R. 1847. Observations on certain fossil bones from the collection of the Academy of Natural Sciences of Philadelphia. *J. Acad. Nat. Sci. Phila.,* ser. 2 1:18–20.

Painter, F. 1978. Bison remains from the Currituck site. *The Chesopiean* 16 (nos. 1–3):28–31.

Palmieri, R. P. 1968. Yak hybridization: Its mechanism and significance. Typed MS in possession of author.

Patton, J. L. 1973. An analysis of natural hybridization between the pocket gophers *Thomomys bottae* and *Thomomys umbrinus* in Arizona. *J. Mammal.* 54:561–584.

Peale, R. 1803a. *An historical disquisition on the Mammoth.* London: E. Lawrence.

———. 1803b. Account of some remains of a species of gigantic oxen found in America and other parts of the world. *Phil. Mag.* 15:325–327.

Peterson, R. L. 1955. *North American moose.* Toronto: Univ. Toronto Press.

Petrov, O. M. 1967. Paleogeography of Chukotka during late Neogene and Quaternary time. In *The Bering land bridge,* ed. D. M. Hopkins, pp. 144–171. Stanford: Stanford Univ. Press.

Péwé, T. L. 1975. *Quaternary geology of Alaska. U. S. Geol. Survey Prof. Paper* no. 835.

Péwé, T. L., and Hopkins, D. M. 1967. Mammal remains of pre-Wisconsin age in Alaska. In *The Bering land bridge,* ed. D. M. Hopkins, pp. 266–270. Stanford: Stanford Univ. Press.

Pilgrim, G. E. 1947. The evolution of the buffaloes, oxen, sheep and goats. *J. Linnaean Soc. London (Zool.)* 41:272–286.

Powers, W. R., and Hamilton, T. D. 1978. Dry Creek: A late Pleistocene human occupation in central Alaska. In *Early man in America from a circum-pacific perspective,* ed. A. L. Bryan, pp. 72–77. Occasional Papers Dept. Anthropol. Univ. Alberta, no. 1.

Ranov, V. A., and Davis, R. S. 1979. Toward a new outline of the Soviet central Asian paleolithic. *Cur. Anthropol.* 20:249–270.

Rasmussen, D. L. 1974. Bison occidentalis from northeastern Montana. *Northwest Geol.* 3:59–61.

Raup, D. M., and Stanley, S. 1978. *Principles of paleontology.* 2d ed. San Francisco: Freeman.

Ray, C. E., Cooper, B. N., and Benninghoff, W. S. 1967. Fossil mammals and pollen in a late Pleistocene deposit at Saltville, Virginia. *J. Paleont.* 41:608–622.

Ray, J. 1693. Synopsis methodica Animalium Quadrupedum et Serpentini Generis, etc. London: Smith and Walford.

Reed, E. K. 1952. The myth of Montezuma's bison and the type locality of the species. *J. Mammal.* 33:390–392.

———. 1955. Bison beyond the Pecos. *Texas J. Sci.* 7:130–135.

Reeves, B. O. K. 1973. The concept of an Altithermal cultural hiatus in northern plains prehistory. *Am. Anthropol.* 75:1221–1253.

———. 1978a. Bison killing in the southwestern Alberta Rockies. In *Bison procurement and utilization: A symposium*, ed. L. B. Davis and M. Wilson, pp. 63–78. Memoir 14, *Plains Anthropol.*

———. 1978b. Head-Smashed-In: 5500 years of bison jumping in the Alberta plains. In *Bison procurement and utilization: A symposium*, ed. L. B. Davis and M. Wilson, pp. 151–174. Memoir 14, *Plains Anthropol.*

Reher, C. A. 1978. Buffalo population and other deterministic factors in a model of adaptive process on the shortgrass plains. In *Bison procurement and utilization: A symposium*, ed. L. B. Davis and M. Wilson, pp. 23–39.

Repenning, C. A. 1967. Palearctic-Nearctic mammalian dispersal in the late Cenozoic. In *The Bering land bridge*, cd. D. M. Hopkins, pp. 288–311. Stanford: Stanford Univ. Press.

———. 1978. Faunal exchanges between Siberia and North America. *Abstracts, fifth biennial meeting, American Quaternary Association*, pp. 40–55.

Rhoads, S. N. 1897. Notes on living and extinct species of North American Bovidae. *Proc. Acad. Nat. Sci. Phila.* 49:483–502.

Richardson, J. 1852–1854. *The zoology of the voyage of H. M. S.* Herald. London: Lowell Reeve.

Richmond, G. M. 1970. Comparison of the Quaternary stratigraphy of the Alps and Rocky Mountains. *Quat. Res.* 1:3–28.

Ritchie, J. C. 1964. Contributions to the Holocene paleoecology of westcentral Canada. I. the Riding Mountain area. *Can. J. Bot.* 42:181–196.

———. 1978. The paleoecology of the ice-free corridor. *Abstracts, fifth biennial meeting, American Quaternary Association*, pp. 24–30.

Ritchie, J. C., and deVries, B. 1964. Contributions to the Holocene paleoecology of west central Canada: A late glacial deposit from the Missouri Coteau. *Can. J. Bot.* 42:677–692.

Ritchie, J. C., and Hare, F. K. 1971. Late-Quaternary vegetation and climate near the arctic tree line of northwestern North America. *Quat. Res.* 1:331–342.

Ritchie, J. C., and Lichti-Federovich, S. 1968. Holocene pollen assemblages from the Tiger Hills, Manitoba. *Can. J. Earth Sci.* 5:873–880.

Robertson, J. S., Jr. 1974. Fossil *Bison* of Florida. In *The Pleistocene mammals of Florida*, ed. S. D. Webb, pp. 214–246. Gainesville: Univ. Florida Press.

Roe, F. G. 1951. *The North American buffalo: A critical study of the species in its wild state.* Toronto: Univ. Toronto Press.

Romer, A. S. 1966. *Vertebrate paleontology*. Chicago: Univ. Chicago Press.

Roosma, A. 1958. A climatic record from Searles Lake, California. *Science* 128:716.

Rostlund, E. 1960. The geographic range of the historic bison in the southeast. *Ann. Ass. Am. Geog.* 50:395–407.

Rutter, N. W. 1978. Geology of the ice-free corridor. *Abstracts, fifth biennial meeting, American Quaternary Association*, pp. 2–12.

Sahni, M. R., and Khan, E. 1968. Probison dehmi n. g. n. sp. A recent find of an Upper Sivalik bovid. *Mitt. Bayer.*

Staatssamml. Paläont. hist. Geol. 8:247–251.

Sainsbury, C. L. 1967. Quaternary geology of western Seward Peninsula, Alaska. In *The Bering land bridge*, ed. D. M. Hopkins, pp. 121–143. Stanford: Stanford Univ. Press.

Salthe, S. N. 1975. Problems of macroevolution (molecular evolution, phenotypic definition, and canalization) as seen from a hierarchical view point. *Am. Zool.* 15:294–314.

Sartore, G., Stormont, C., Morris, B. G., and Grunder, A. A. 1969. Multiple electrophoretic forms of carbonic anhydrase in red cells of domestic cattle (*Bos taurus*) and American buffalo (*Bison bison*). *Genetics* 61:823–831.

Sauer, C. O. 1956. The agency of man on the earth. In *Man's role in changing the face of the earth*, ed. W. L. Thomas, Jr., pp. 49–69. Chicago: Univ. Chicago Press.

Saunders, J. J. 1977. Lehner Ranch revisited. *The Mus. J.* 18:48–64.

Savage, D. E., and Curtis, G. H. 1970. The Villafranchian stage-age and its radiometric dating. In *Radiometric dating and paleontologic zonation*, ed. O. L. Bandy, pp. 207–231. *Geol. Soc. Am. Spec. Paper* no. 124.

Schaffer, W. M. 1968. Intraspecific combat and the evolution of the Caprini. *Evolution* 22:817–825.

Schaller, G. B. 1967. *The deer and the tiger.* Chicago: Univ. Chicago Press.

———. 1972. *Serengeti: A kingdom of predators.* New York: Knopf.

———. 1977. *Mountain monarchs: Wild sheep and goats of the Himalaya.* Chicago: Univ. Chicago Press.

Schloeth, R. 1958. Cycle annual et comportement social du taureau de Camargue. *Mammalia* 22:121–139.

———. 1961. Das Socialleben des Camargue-Rindes. *Z. Tierpsych.* 18:574–627.

Schroedl, G. F. 1973. *The archaeological occurrence of bison in the southern plateau.* Laboratory of Anthropol., Wash. State Univ., Rept. of Invest., no. 51.

Schueler, F. W., and Rising, J. D. 1976. Phenetic evidence of natural hybridization. *Syst. Zool.* 25:283–289.

Schultz, C. B., and Frankforter, W. D. 1946. The geologic history of the bison in the Great Plains (a preliminary report). *Bull. Univ. Neb. State Mus.* 3:1–10.

Schultz, C. B., and Hillerud, J. M. 1977. The antiquity of *Bison latifrons* (Harlan) in the Great Plains of North America. *Trans. Neb. Acad. Sci.* 4:103–116.

Schultz, C. B., and Stout, T. M. 1948. Pleistocene mammals and terraces in the Great Plains. *Bull. Geol. Soc. Am.* 59:553–588.

Schultz, C. B., Tanner, L. G., and Martin, L. D. 1972. Phyletic trends in certain lineages of Quaternary mammals. *Bull. Univ. Neb. State Mus.* 9:183–195.

Schultz, G. E., and Cheatum, E. P. 1970. *Bison occidentalis* and associated invertebrates from the late Wisconsin of Randall County, Texas. *J. Paleont.* 44:836–850.

Schultz, G. E., and Lansdown, C. H. 1972. A skull of *Bison latifrons* from Lipscomb County, Texas. *Texas J. Sci.* 23:393–401.

Scott, G. R., and Lindvall, R. M. 1970. Geology of new occurrences of Pleistocene bisons and peccaries in Colorado. *U. S. Geol. Survey Prof.*

Paper 700-B:B141–149.

Seton, E. T. 1885–1886. The wood buffalo. *Proc. Can. Inst. Toronto*, ser. 3 3:114–117.

Shackleton, D. M. 1968. Comparative aspects of social organization of American bison. M.S. Thesis, Univ. Western Ontario, London.

Shackleton, D. M., and Hills, L. V. 1977. Post-glacial ungulates (*Cervus* and *Bison*) from Three Hills, Alberta. *Can. J. Earth Sci.* 14:963–986.

Shackleton, D. M., Hills. L. V., and Hutton, D. A. 1975. Aspects of variation in cranial characters of plains bison (*Bison bison bison* Linnaeus) from Elk Island National Park, Alberta. *J. Mammal.* 56:871–887.

Shaw, D. H., and Patel, J. R. 1962. Demonstration of antigenic difference between the American bison (*Bison bison*) and domestic cattle (*Bos taurus*). *Nature* 196:498–499.

Shay, C. T. 1967. Vegetation history of the southern Lake Agassiz during the past 12,000 years. In *Land, life and water*, ed. W. J. Mayer-Oakes, pp. 231–252. Winnipeg: Univ. Manit. Press.

Sher, A. V. 1974. Pleistocene mammals and stratigraphy of the far northeast USSR and North America. Translated by Dorothy Vitaliano. *Int'l Geol. Review* 16, nos. 7–10.

Sherman, H. B. 1954. The occurrence of bison in Florida. *Q'tly J. Florida Acad. Sci.* 17:228–232.

Shoemaker, H. W. 1915. *A Pennsylvania bison hunt*. Middleburg: Middleburg Post Press.

Shult,M. J. 1972. American bison behavior patterns at Wind Cave National Park. Ph.D. Dissertation, Iowa State Univ., Ames.

Simoons, F. J. 1968. *A ceremonial ox of India: The mithan in nature, culture, and history.* Madison: Univ. Wisc. Press.

Simpson, G. G. 1953. *The major features of evolution.* New York: Columbia Univ. Press.

Simpson, G. G., and Roe, A. 1939. *Quantitative zoology*. New York: McGraw-Hill.

Sinclair, A. R. E. 1977. *The African buffalo: A study of resource limitation of populations.* Chicago: Univ. Chicago Press.

Sirkin, L. 1977. Late Pleistocene vegetation and environments in the middle Atlantic region. *Ann. N. Y. Acad. Sci.* 288:206–217.

Sisson, S., and Grossman, D. 1953. *The anatomy of the domestic animals.* 4th ed. Philadelphia: W. B. Saunders.

Skinner, M. F., and Kaisen, O. C. 1947. The fossil *Bison* of Alaska and preliminary revision of the genus. *Bull. Am. Mus. Nat. Hist.* 89:123–256.

Slaughter, B. H. 1966. The Moore Pit local fauna; Pleistocene of Texas. *J. Paleont.* 40:78–91.

Slaughter, B. H., and Hoover, B. R. 1963. Sulphur River formation and the Pleistocene mammals of the Ben Franklin local fauna. *J. Grad. Res. Center* (SMU) 31:132–148.

Slaughter, B. H., and McClure, W. L. 1965. The Sims Bayou local fauna: Pleistocene of Houston, Texas. *Texas J. Sci.* 17:404–417.

Smiley, F. E. 1978. Changes in the cursorial ability of Wyoming Holocene bison. *Wyoming Contrib. Anthropol.* 1:105–126.

Smith, C. H. 1827. Supplement to the order Ruminantia. In *The animal kingdom arranged in conformity with its organization*, by G. Cuvier, vol. 4. London: Geo. B. Whittaker.

Soper, J. D. 1941. History, range and home life of the northern bison. *Ecol. Monog.* 11:347–412.

Sorenson, C. J. 1977. Reconstructed Holocene bioclimates. *Ann. Ass. Am. Geog.* 67:214–222.

Speth, J. D., and Parry, W. J. 1978. Late prehistoric bison procurement in southeastern New Mexico: The 1977 season at the Garnsey site. *Mus. Anthropol., Univ. Mich., Tech. Rept.* no. 8.

Stalker, A. M. 1978. The geology of the ice free corridor: The southern half. *Abstracts, fifth biennial meeting, American Quaternary Association,* pp. 19–22.

Stanley, S. M. 1974. Relative growth of the titanothere horn: A new approach to an old problem. *Evolution* 28:447–457.

Stearns, C. E. 1976. Estimates of the position of sea level between 140,000 and 75,000 years ago. *Quat. Res.* 6:445–449.

Steen-McIntyre, V., Fryxell, R., and Malde, H. E. 1973. Unexpectedly old age of deposits at Hueyatlaco archaeological site, Valsequillo, Mexico, implied by new stratigraphic and petrographic findings. *Abstracts, Geol. Soc. Am.* 5 (for 1973):820–821.

Stevens, A. 1978. Sexual dimorphism in some post-cranial elements of *Bison latifrons*. M.S. Thesis, Calif. State Univ., Los Angeles.

Stewart, O. C. 1951. Burning and natural vegetation in the United States. *Geog. Review* 41:317–320.

———. 1956. Fire as the first great force employed by man. In *Man's role in changing the face of the earth,* ed. W. L. Thomas, Jr., pp. 115–133. Chicago: Univ. Chicago Press.

Storer, J. E. 1971. *Bison alaskensis* from the Pleistocene of Alberta. Typed MS in possession of author.

Stormont, C., Miller, W. J., and Suzuki, Y. 1961. Blood groups and the taxonomic status of American buffalo and domestic cattle. *Evolution* 15:196–208.

Szabo, B. J., Malde, H. E., and Irwin-Williams, C. 1969. Dilemma posed by uranium-series dates on archaeologically significant bones from Valsequillo, Puebla, Mexico. *Earth Planet. Sci. Letters* 6:237–244.

Teilhard de Chardin, P., and Piveteau, E. J. 1930. Les mammifères fossiles de Nihowan (Chine). *Ann. Paleont.* 19:80–86.

Thomas, O. 1911. The mammals of the tenth edition of Linnaeus; an attempt to fix the types of the genera and the exact bases and localities of the species. *Proc. Zool. Soc. London 1911,* pp. 120–158.

Thorson, R. M., and Hamilton, T. D. 1977. Geology of the Dry Creek site; a stratified early man site in interior Alaska. *Quat. Res.* 7:149–176.

Tulloch, D. G. 1978. The water buffalo, *Bubalus bubalis*, in Australia: Grouping and home range. *Aust. Wildl. Res.* 5:327–354.

Valdez, R., Nadler, C. F., and Bunch, T. D. 1978. Evolution of wild sheep in Iran. *Evolution* 32:56–72.

Vanderhoof, V. L. 1942. *A skull of Bison latifrons from the Pleistocene of northern California*. Univ. Calif. Publ., Bull. Dept. Geol. Sci. vol. 27, no. 1.

VanDevender, T. R., and King, J. E. 1975. Fossil Blanding's turtles, *Emydoidea*

Blandingi (Holbrook), and the late Pleistocene vegetation of western Missouri. *Herpetologica* 31:208–212.

VanDevender, T. R., and Spaulding, W. G. 1979. Development of vegetation and climate in the southwestern United States. *Science* 204:701–710.

Van Horn, R. 1979. *The Holocene Ridgeland formation and associated Decker soil (new names) near Great Salt Lake, Utah.* U.S. Geol. Survey Bull. 1457-C.

Van Valen, L. 1971. Toward the origin of artiodactyls. *Evolution* 25:523–529.

Vialli, G. S. 1954. I bisonti fossili delle alluvioni Quaternarie Pavesi. *Atti Istit. Geol. Univ. Pavia* 5:1–27.

Waddington, J. C. B., and Wright, H. E., Jr. 1974. Late Quaternary vegetational changes on the east side of Yellowstone Park, Wyoming. *Quat. Res.* 4:175–184.

Walther, F. 1974. Some reflections on expressive behaviour in combats and courtship of certain horned ungulates. In *The behaviour of ungulates and its relation to management*, ed. V. Geist and F. Walther, pp. 56–106. Morges: Int'l Union Cons. Nature and Nat. Resources.

Waring, R. H., and Franklin, J. F. 1979. Evergreen coniferous forests of the Pacific Northwest. *Science* 204:1380–1386.

Watts, W. A. 1967. Late-glacial plant microfossils from Minnesota. In *Quaternary Paleoecology*, ed. E. J. Cushing and H. E. Wright, Jr., pp. 59–88. New Haven: Yale Univ. Press.

———. 1970. The full-glacial vegetation of northwestern Georgia. *Ecology* 51:17–33.

———. 1973. The vegetation record of a mid-Wisconsin interstadial in northwest Georgia. *Quat. Res.* 3:257–268.

———. 1975. A late Quaternary record of vegetation from Lake Annie, south-central Florida. *Geology* 3:344–346.

Watts, W. A. and Bright, R. C. 1968. Pollen, seed, and mollusk analysis of a sediment core from Pickerel Lake, northeastern South Dakota. *Bull. Geol. Soc. Am.* 79:855–876.

Watts, W. A., and Wright, H. E., Jr. 1966. Late-Wisconsin pollen and seed analysis from the Nebraska Sandhills. *Ecology* 47:202–210.

Wayne, W. J. 1967. Periglacial features and climatic gradient in Illinois, Indiana, and western Ohio, east-central United States. In *Quaternary Paleoecology*, ed. E. J. Cushing and H. E. Wright, Jr., pp. 393–414. New Haven: Yale Univ. Press.

Webb, S. D. 1977. A history of savanna vertebrates in the new world. Part I: North America. *Ann. Rev. Ecol. Syst.* 8:355–380.

Wells, P. V. 1965. Scarp woodlands, transported soils, and concept of grassland climate in the Great Plains region. *Science* 148:246–249.

———. 1966. Late Pleistocene vegetation and degree of pluvial climatic change in the Chihuahuan Desert. *Science* 153:970–975.

Wells, P. V., and Berger, R. 1967. Late Pleistocene history of coniferous woodland in the Mohave Desert. *Science* 155:1640–1647.

Wells, P. V., and Jorgensen, C. D. 1964. Pleistocene wood rat middens and climatic change in Mohave Desert: A record of juniper woodlands. *Science* 143:1171–1174.

Wendland, W. M. 1978. Holocene man in North America: The ecological setting and climatic background. *Plains*

Anthropol. 23:273–287.

Wendorf, F. 1970. The Lubbock subpluvial. In *Pleistocene and recent environments of the central Great Plains*, ed. W. Dort, Jr. and J. K. Jones, Jr., pp. 23–35. Dept. Geol. Univ. Kan. Spec. Publ. 3. Lawrence: Univ. Press of Kan.

Western, D. 1979. Size, life history and ecology in mammals. *Afr. J. Ecol.* 17:185–204.

Wheat, J. B. 1979. *The Jurgens site.* Memoir 15, *Plains Anthropol.*

Whitehead, D. R. 1973. Late-Wisconsin vegetational changes in unglaciated eastern North America. *Quat. Res.* 3:621–631.

Wilber, C. G., and Gorski, T. W. 1955. The lipids in *Bison bison. J. Mammal.* 36:305–308.

Wilmsen, E. N. 1974. *Lindenmeier: A Pleistocene hunting society.* New York: Harper & Row.

Wilson, D. E., and Hirst, S. M. 1977. *Ecology and factors limiting roan and sable antelope populations in South Africa.* Wildl. Soc., Wildl. Monog. no. 54.

Wilson, E. O. 1975. *Sociobiology: The new synthesis.* Cambridge: Belknap Press, Harvard Univ.

Wilson, M. 1969. Problems in the speciation of American fossil bison. In *Post-Pleistocene man and his environment on the northern plains*, ed. R. G. Forbis, L. B. Davis, O. A. Christensen, and G. Fedirchuk, pp. 178–199. Calgary: Univ. Calgary Archaeol. Ass.

———. 1972. Review of "Biostratigraphy and palaeontology of a Sangamon deposit at Fort Qu'appelle, Saskatchewan." *Univ. Wyoming Contrib. Geol.* 11:87–92.

———. 1974a. History of the bison in Wyoming, with particular reference to early Holocene forms. In *Applied geology and archaeology: The Holocene history of Wyoming*, ed. M. Wilson, pp. 91–99. Geol. Survey Wyoming, Rept. Invest. no. 10.

———. 1974b. The Casper local fauna and its fossil bison. In *The Casper site: A Hell Gap bison kill on the high plains*, ed. G. C. Frison, pp. 125–171. New York: Academic Press.

———. 1974c. Fossil bison and artifacts from the Mona Lisa site, Calgary, Alberta—part 1: stratigraphy and artifacts. *Plains Anthropol.* 19:34–45.

———. 1975. Holocene fossil bison from Wyoming and adjacent areas. M.A. Thesis, Univ. Wyoming, Laramie.

———. 1978. Holocene geology and archaeology of the Bow River floodplain at Calgary, Alberta. *Abstracts, fifth biennial meeting, American Quaternary Association*, p. 181.

Winship, G. P. 1896. The Coronado expedition; 1540–1542. *Fourteenth Ann. Rept., Bur. Ethnol.*, part 1, pp. 329–613.

Wint, G. B. 1960. A record of hybrid bobwhite X scaled quail. *Proc. Okla. Acad. Sci.* 40:151–152.

Woillard, G. M. 1978. Grande Pile peat bog: A continuous pollen record for the last 140,000 years. *Quat. Res.* 9:1–21.

Wolfe, J. A. 1972. An interpretation of Alaskan Tertiary floras. In *Floristics and paleofloristics of Asia and eastern North America*, ed. A. Graham, pp. 201–233. New York: Elsevier.

———. 1978. A paleobotanical interpretation of Tertiary climates in the northern hemisphere. *Am. Sci.* 66:694–703.

Woodruff, D. S. 1973. Natural hybridization and hybrid zones. *Syst. Zool.* 22:213–218.

Woodward, S. L. 1976. Feral burros of the Chemehuevi Mountains, California: The biogeography of a feral exotic. Ph.D. Thesis, Univ. Calif. Los Angeles.

———. 1979. The social system of feral asses (*Equus asinus*). *Z. Tierpsych.* 49:304–316.

Wormington, H. M. 1957. *Ancient man in North America.* 4th ed. Denver Mus. Nat. Hist., Popular Ser., no. 4.

Wright, H. E., Jr. 1968. The roles of pine and spruce in the forest history of Minnesota and adjacent areas. *Ecology* 49:937–955.

———. 1970. Vegetational history of the central plains. In *Pleistocene and recent environments of the central Great Plains*, ed. W. Dort, Jr. and J. K. Jones, Jr., pp. 157–172. Dept. Geol. Univ. Kan. Spec. Publ. 3. Lawrence: Univ. Press Kan.

———. 1971. Late Quaternary vegetational history of North America. In *The late Cenozoic glacial ages,* ed. K. K. Turekian, pp. 425–464. New Haven: Yale Univ. Press.

Yablokov, A. V. 1974. *Variability of mammals.* New Delhi: Amerind Pub. Co.

Young, S. 1976. Is steppe tundra alive and well in Alaska? *Abstracts, fourth biennial meeting, American Quaternary Association*, pp. 84–88.

Yurtsev, B. A. 1972. Phytogeography of northeastern Asia and the problem of Transberingian floristic interrelations. In *Floristics and paleofloristics of Asia and eastern North America*, ed. A. Graham, pp. 19–54. New York: Elsevier.

INDEX